ENTANGLED ENCOUNTERS AT THE NATIONAL ZOO

ENVIRONMENT AND SOCIETY

KIMBERLY K. SMITH, EDITOR

Entangled Encounters at the National Zoo

Stories from the Animal Archive

Daniel Vandersommers

UNIVERSITY PRESS OF KANSAS

© 2023 by the University Press of Kansas
All rights reserved

Published by the University Press of Kansas (Lawrence, Kansas 66045), which was organized by the Kansas Board of Regents and is operated and funded by Emporia State University, Fort Hays State University, Kansas State University, Pittsburg State University, the University of Kansas, and Wichita State University.

Library of Congress Cataloging-in-Publication Data

Names: Vandersommers, Daniel, 1984– author.
Title: Entangled encounters at the National Zoo : stories from the animal archive / Daniel Vandersommers.
Description: Lawrence, Kansas : University Press of Kansas, 2023 | Series: Environment and society | Includes bibliographical references and index.
Identifiers: LCCN 2023000413 (print) | LCCN 2023000414 (ebook)
ISBN 9780700635689 (cloth)
ISBN 9780700635696 (paperback)
ISBN 9780700635702 (ebook)
Subjects: LCSH: National Zoological Park (U.S.)—History—19th century. | National Zoological Park (U.S.)—History—20th century. | Zoo animals—Washington (D.C.)—History—19th century. | Zoo animals—Washington (D.C.)—History—20th century.
Classification: LCC QL76.5.U62 W378 2024 (print) | LCC QL76.5.U62 (ebook) | DDC 590.73753—dc23/eng/20230413
LC record available at https://lccn.loc.gov/2023000413.
LC ebook record available at https://lccn.loc.gov/2023000414.

British Library Cataloguing-in-Publication Data is available.

Printed in the United States of America

10 9 8 7 6 5 4 3 2 1

The paper used in this publication is acid free and meets the minimum requirements of the American National Standard for Permanence of Paper for Printed Library Materials Z39.48–1992.

for Erica
and all the ways we are together

Contents

Acknowledgments

This book began as a dissertation project in the Department of History at The Ohio State University. Leslie Alexander, Greg Anderson, Ken Andrien, Steve Conn, Jared Gardner, Stephen Kern, Margaret Newell, and Walter Rucker all shaped this book in important ways. Their influences lie deep in this book's methodology. I would like to especially acknowledge and thank Randy Roth, Alan Gallay, John Brooke, and Chris Otter for their teaching, support, guidance, encouragement, and friendship while I was a graduate student. They taught me how to think between the boundaries of disciplines and how to see both the micro and the macro simultaneously. And they taught me that each word matters. With both scrutiny and care, they read drafts of many chapters, and their thoughts, feedback, critiques, and praise played an important role in shaping this book.

I want to thank Tracy McDonald, who chose me to work with her as a postdoctoral fellow in animal history during the 2016–2017 academic year. Our conversations, which stretched over several years, our shared desire of uniting history and animal studies, and our productive collaborations greatly influenced *Entangled Encounters*. The zoo studies workshop that she and I hosted in Hamilton, Ontario, in 2016, and our work editing the resulting volume, *Zoo Studies: A New Humanities* (McGill-Queens University Press, 2019), pushed my thinking in new and entangled directions. Thank you, Tracy, for your support. And thanks to all who were involved in this Hamilton workshop and volume: Ralph Acampora, Ron Broglio, David Clark, Sheri Crawford, Kathryn Denning, Stefan Dolgert, Jesse Donahue, Angela Fernandez, Guro Flinterud, James Gillett, Takashi Ito, John Kinder, Alice Kuzniar, Randy Malamud, Susan Nance, Judith Nicholson, Jonathan Osborn, Violette Pouillard, Nigel Rothfels, Matthew Senior, Marianna Szczygielska, Erica Tom, and Zeb Tortorici. Your voices and scholarship have shaped me significantly, and thus the stories that follow. I also want to acknowledge the Social Sciences and

Humanities Research Council of Canada (SSHRC) for funding this postdoc and supporting the workshop. Thanks also to the Department of History, the Office of the Dean of Humanities, and the Wilson Institute for Canadian History at McMaster University. It is amazing how support for a single scholarly collaboration can prove generative for very many years to follow.

I want to acknowledge the Consortium for History of Science, Technology, and Medicine (CHSTM), as well as the National Endowment for the Humanities, for generous support as a postdoctoral fellow during the 2017–2018 academic year. This NEH fellowship allowed me to finish the archival research represented in the following pages. Even more, the discussions and reading groups at the CHSTM were (and continue to be) deeply influential. I want to thank Kevin Baker, Nicole Belolan, Theodora Dryer, Nabeel Hamid, Larry Kessler, Taylor Moore, Sheila O'Shaughnessy, Aaron Slater, and Michelle Smiley, as well as Frederick Davis, Mark Hersey, and Jeremy Vetter's Earth and Environmental Sciences working group. Our conversations at the CHSTM were intellectually exciting, productive, and important to this book. What a special place in Philadelphia, where so many diverse scholars can come together and share ideas! Thank you, especially, to Babak Ashrafi, for his support and encouragement.

Many institutions supported this book. First, I cannot imagine its existence without the scholarly homes of the American Society for Environmental History (ASEH) and the Society for Literature, Science, and the Arts (SLSA). These two wonderful communities inspired so many of the ideas I present here (unless they are bad ideas, then they are, of course, my own). I would also like to thank Albrecht Schulte-Hostedde, Tanis Mercer, ReNewZoo, and Canada's Accredited Zoos and Aquariums (CAZA) for inviting me, on multiple occasions, to share my work with Canadian zoo communities. I would like to thank the Helen Riaboff Whiteley Center at the University of Washington's Friday Harbor Laboratories for an inspiring place to write and edit, before the countenances of harbor seals and bald eagles. I would like to acknowledge the Department of History at the University of Dayton for its financial support in acquiring the images of this text. And thanks to the many institutions, and dedicated staffs, who assisted with research: the Chicago Public Library, Chicago Public Library Special Collections (especially Johanna Russ and Liz Vogel), Cincinnati Zoo Archives (especially Deb Zureick),

Harvard University Archives, Historical Society of Pennsylvania, Library Company of Philadelphia, Lincoln Park District Archives, Manuscript Division of the Library of Congress, Massachusetts Historical Society, Museum of Comparative Zoology at Harvard University, Museum of Science in Boston, National Anthropological Archives, National Air and Space Archives, Newberry Library, Northwestern University Archives, Philadelphia Zoo Archives (especially Rachael DeCaro), Temple University Archives, University of Chicago Library, and the Wildlife Conservation Society Library and Archives at the Bronx Zoo (especially Maddie Thompson). Most important, I would like to thank the Smithsonian Institution Archives (SIA), where nearly all the stories here were first discovered. I want to acknowledge the entire staff of the SIA, but especially Tad Bennicoff and Ellen Alers, who were so kind and helpful during early stages of this research.

I am deeply grateful to those who read segments, versions, and drafts of the chapters that follow—Mark Barrow, John Brooke, Joan Cashin, Miguel Gomez, Dolly Jørgensen, Larry Kessler, Susan Nance, Chris Otter, Karen Raber, Harriet Ritvo, Randy Roth, Nigel Rothfels, Bobbi Sutherland, Erica Tom, Antoine Traisnel, and Daniel Watkins. Your thoughtful feedback and constructive criticisms were so important to shaping this book. Thank you. I also want to thank the blind peer reviewers of the University Press of Kansas as well as my editor, David Congdon, for thoughtful feedback that helped transform my manuscript into this book. Many others provided conversation, correspondence, or advice along the way. While it would be impossible to remember all the influences that shaped these pages in both large and small ways, I would like to thank, in addition to those named above, Abel Alves, Katey Anderson, Janet Bednarek, Amílcar Challú, Kate Christen, Hatley Clifford, Istvan Csicsery-Ronay, Allyson DeMaagd, Albert Dzur, Andy Flack, Earl Hess, Michael Kimaid, Xinghua Li, Bernard Lightman, Katherine McLeod, Marya McQuirter, Michele Navakas, Catherine Nichols, Joelle Papillon, Rocky Penate, Lisa Richards, Eric Rominger, Mark Rosenthal, James Secord, Stephanie Honchell Smith, Paula Tarankow, Joe Wachtel, Evan Ward, and Leigh Ann Wheeler. Also, I appreciate the feedback of the members of the Department of History Research Colloquium at the University of Dayton, the Modern US History Workshop at Binghamton University, and the History of Technology graduate seminar at The Ohio State University, where I shared various chapters. Thank you to Penelope

Cray for masterful copyediting. And I would like to thank the students of my wilderness and animal history courses at the University of Mississippi; McMaster University; The Indiana Academy for Science, Mathematics, and Humanities; and the University of Dayton. So many of the stories below were first told or forged or thought through in my classrooms with wonderful students.

Portions of this book were previously published in slightly different forms. A part of chapter 1 appears as "The Sectionalism of the National Zoo: Animals, Language, Politics, and Laughter," in *Animal Histories of the Civil War Era*, ed. Earl Hess (Baton Rouge: Louisiana State University Press, 2022). A portion of chapter 3 was published as "Failed Zoo Experiments: Primatology, Aeronautics, and the Animality of 'Modern' Science, 1891–1903," in *Zoo Studies: A New Humanities*, eds. Tracy McDonald and Daniel Vandersommers (Montreal: McGill-Queens University Press, 2019): 65–92. An earlier version of chapter 4 was published as "Animal Activism and the Zoo-Networked Nation," *Humanimalia* 6, no. 2 (March 2015): 111–165. And a significant portion of chapter 5 was published as "Narrating Animal History from the Crags: A Turn-of-the-Century Tale about Mountain Sheep, Resistance, and a Nation," *Journal of American Studies* 51, no. 3 (August 2017): 751–777.

I would like to thank Carl Briggs, Lisa Lawrence, and the horses, goats, and cockatoos of Hillcrest Farm. I would like to thank my parents, Aaron Vandersommers and Dana Giron, who first sparked my interest in history as a child. And I would like to acknowledge the loved ones who believed in me and this project, who opened their homes to an itinerant researcher in an insanely competitive job market, and who shared laughs and beers and political rantings and board games and rounds of disc golf and wonderful stories and intellectual exchange. I will always think about your encouragement and support when I see this book on my shelf—Meghan and Dan Watkins, Donna and Walter Tom, Holly and Mark Brunetti, Marilyn and Ed Battiste, and especially Debbie and Steve McCarter, and Jeff Battiste.

Miles, this book has unfolded across the entirety of your life, written alongside our wonderful memories, across so many places.

INTRODUCTION

At the Entrance Gate

Ten years into the twentieth century, as the urban bustle of Washington, DC, spread in every direction, naturalist Robert Wilson Shufeldt hunched over the work desk at his residence just northwest of the White House. One of the most productive ornithologists in the world, and a former honorary curator of the Smithsonian Institution, you can imagine Shufeldt nestled in his study just after his morning walk. Some mornings, he would meander along the pathways of the Soldier's Home, next to Trinity College and Glenwood Cemetery. Other mornings, he would explore the woods along Rock Creek. Or he would take jaunts down Georgetown's avenues, or sit quietly in Dupont Circle. Always, he carried back the inspiration of his local fellows—red-headed woodpeckers, cardinal grosbeaks, purple finches, robins, catbirds, and flickers.[1] As morning sun cracked through his study window, local songs propelled Shufeldt through his prolific writings on the extermination of wild turkeys, on avian paleontology, and on the classification of the eggs of shore-inhabiting birds like sandpipers and plovers.[2] The birds that Shufeldt encountered on his morning walks ushered him into days, into a lifetime, of learning about the birds of the world. Creatures at hand led him to creatures beyond. And by 1924, Robert Shufeldt had produced an astounding fifteen hundred publications, making contributions not only to ornithology in general but also to the wide-ranging yet ever-specializing fields of paleopathology, myology, oology, photography, and especially osteology.[3]

Occasionally, on pleasant mornings, Shufeldt would guide his morning strolls to the National Zoological Park, a ten-minute walk from his front door. Some mornings, he would go to the zoo. Sometimes, through his neighborhood's soundscape, the zoo would come to him. One forenoon in 1913, as Shufeldt was revising a fresh manuscript about the death of a toucan or the bones of an Australian goose, a steady breeze, with "delightful distinctness," carried the howlings of wolves and the barkings of seals through the open fenestra into his study. Zoos interested Shufeldt "most keenly," for they had been racing across the urban landscape since the 1890s, and he had written about them before. This time, though, these voices urged him to write for the largest audience yet, *The Popular Science Monthly*.[4]

Founded in 1872, *Popular Science* served as a public forum for leading scientific thinkers—Charles Darwin, Thomas Edison, Louis Pasteur, and William James. For the November 1913 edition, Shufeldt contributed a seven-page article titled "The National Zoological Garden," which depicted the zoo as a multifarious place. On the one hand, Shufeldt wrote in support of zoos everywhere, claiming that they were "extremely valuable" "to any civilized community of people." On the other hand, Shufeldt claimed that the zoo needed larger enclosures that did not "torture" the animals. He wrote from many perspectives. He wrote as a conservationist when he lamented the failure of the zoo's aquarium. He believed that the collapse of the national fisheries made it "vitally important" to study fish "in nature and in aquaria." Shufeldt wrote as a passionate zoo booster. Despite hosting some substandard exhibits, the National Zoo's "flying-cage" and deer, moose, and caribou paddocks were pioneering enclosures modeled on naturalistic settings. Shufeldt wrote as a zoo lobbyist. If allocated more funding, the National Zoo could be "an educational center of the greatest magnitude." He wrote as a supporter of science and medicine who argued that the zoo offered the perfect setting for anatomical laboratories. And Shufeldt wrote as a patron of the arts who smiled approvingly upon the "enthusiastic boy, vigorously at work with pencil, color or brush, in front of one of the cages, doing his best to faithfully portray its inmate."[5]

The zoo, for Shufeldt, was a place of opposites. He acknowledged that the salaries of National Zoo employees "would make a car conductor blush." He acknowledged that a third of the animals were "housed in hat-boxes." Nonetheless, Shufeldt also acknowledged that "so much [good] had been accomplished" in the zoo. For "the sculptor . . . the scientific taxidermist . . . the artist and biologist," the National Zoo represented a golden opportunity.[6] Shufeldt saw the National Zoo as a place filled with different, even contradictory, stories. How appropriate for a man who may seem to be, from a twenty-first-century vantage, composed of hypocrisies himself, for while Shufeldt extended empathy toward plants and animals, he poured hate into a white nationalist agenda, typified by a vile three-hundred-page, 1907 monograph titled *The Negro: A Menace to American Civilization*.[7] And he poured misogyny into essays like *On Female Impotency*, which he published to blackmail his wife, Florence Audubon, and which forced the Smithsonian to revoke his prestigious position.[8] Of course, individuals are always many—capable of both curious openness and violent closure toward the Other, capable of finding solace in black-bellied tree ducks yet danger in Blackness. Nonetheless, contrasting strands in the present often pull from the laced conjunctures of the past. As anthropologist Yuka Suzuki, sociologist Dorceta Taylor, environmental historian Miles Powell, and cultural historian Alexandra Minna Stern have established, conservation, animal protection, and white nationalism shared intimate historical ties, bound to the eugenic desire of "preserving the best"—environment, species, or humans—for the nation.[9] African diasporic scholar Bénédicte Boisseron demonstrates in *Afro-Dog: Blackness and the Animal Question* that repressive social systems based in whiteness always try to control the "movability" of those marked by race and animality, lives owned and displayed "like captive animals in a zoo."[10] The zoo, then—for Robert Wilson Shufeldt—may have been the perfect place to rehearse a privileged stroll upon the diversely Other while simultaneously calling attention to the opposing legacies he experienced on its surface.

The zoo always produced opposing experiences, as well as sublimated ones. This book is a biography of a single historical zoo, a site study of the National Zoological Park of Washington, DC, from 1887 to 1920. African and animal historian Sandra Swart has stated recently that "human-animal histories" have a "strong sense of place," and the stories that follow will try to continue this tradition.[11] Each chapter centers animals, and each looks at, from a different angle, an influential popular science that I will refer to broadly as "popular zoology." The book's mission is twofold. First, through narrative, this book shows how zoo animals always ran away from the zoo. I mean this literally. Animals escaped frequently, the topic of chapter 2. Even more, though, I mean this figuratively. Living, breathing, historical, zoo animals ran away from their cultural constructions, and these constructions ran away from the living bodies they were supposed to represent. Though Western culture (in general) and the zoo (in particular) worked to make escapes invisible, real animals and their imagined counterparts always parted company. Who animals were and what they were thought, hoped, and made to be never lined up, even if zoogoers themselves did not notice the incongruencies. This book shows that the resulting gaps produced by escapes—by runaway animals—contain concealed, distorted, and erased histories worthy of uncovering. Second, this book shows how the popular zoology fostered by the National Zoo shaped every aspect of American science, culture, and conservation during the Gilded Age and Progressive Era.

This gives you everything you need to know to pass through the entrance gate. The rest of the opening chapter will contextualize the two central threads that run through this work—the first will be of most interest to scholars working across animal studies and the second, to scholars of history. The first section lays out the theoretical assumptions that shape this biography, offering three metaphors (*knots*, *layers*, and *gaps*) that help us construct posthuman lenses through which to read the historical zoo.[12] I begin with theory because I believe with historiographic theorist Ewa Domańska and animal historian Hilda Kean that simply "writing animals into history" is insufficient.[13] Instead, "interpretive frameworks need to be reworked,"[14] and these metaphors might help us destabilize the Western (white, male, elite, European) "anthropocentrism" derivative of Renaissance humanism, its "assumed and idealized" universal *human*,[15] and its hierarchies established to "judge, classify, and eliminate, both persons and things" in the grand ordering of Nature.[16] Contemporary notions of time, space, causality, subjectivity, and personhood shaped by this Renaissance project are still etched deeply into the academic discourses that shape the sciences, social sciences, and most historical thought.[17] The second section frames American popular zoology within the broader history of science, and animal history within the broader historical profession. The introduction concludes with a summary of the chapters that follow.

Zoo Biography: Theory and Methods

I seek to resurrect the National Zoo from history. Or, more precisely, I seek to bring to life a few traces of this zoo. Or, rather, a few traces of the entangled lives buried within and beneath this institution. To do this, I have chosen to hold "the archive" close to the pages that follow and, thus, close to you. I have tried to assemble (curate) the documents left by a historical zoo into stories that best capture the traces of animals (their voices, their predicaments, their histories) without "processing" these away through argumentation ("re-presentation") too far removed from the primary sources themselves.[18] With animal studies and Renaissance scholar Erica Fudge and animal historian Zeb Tortorici, I challenge the assumption that "animals are 'inarticulate'; they do not leave documents."[19] Zoo archives are bursting with life. Feeling and following animals as they bolt from folios and bound from boxes, tucked away in dusty collections, has given me great joy for more than a decade. Feeling and following these animals has also produced great sadness in me. I wish I could accompany you to the zoo (archive), but since I cannot, I bring it to you with the hope that you may have a few meaningful experiences of your own. This approach, of keeping the archive near the text, goes against much of my training. Historians are taught to *use* archives to form historical arguments that "further" the conversations (historiography) of academic historians. While this dialogue is important, it all-too-often remains enclosed within history departments, rather than reaching outwardly in all directions, as the late-nineteenth-century zoo certainly did. Furthermore, the zoo is always-already a site of extraction, and I want to prevent, to the greatest extent possible, the continued colonization of its animals by putting them to work for furthered disciplinary ends. By lessening the gap between the (zoo) archive and these pages, I hope to provide more space for animals (and readers) to roam through what ecofeminist and animal studies scholar Donna J. Haraway might call a "polyphony of stories." The potential experiences, the possibilities, of this book will twist around each other as they lead in new directions.[20]

Knots, Tentacles, Stories

Haraway in *Staying with the Trouble: Making Kin in the Chthulucene* holds up string games, string figures, knots as "*thinking* as well as *making* practices."[21] Knots can teach us a lot. They can teach us, or rather *unteach* us, about both our universalizing and atomizing tendencies, our joint desires to either make beyond or take apart, desires encoded within virtually every product of Western meaning-making (including the zoo). How would it change things if we envisioned time, culture, and history as knotted rather than flat, intact, and progressive?[22] There is something more real about tracing the turns of lives, of meanings, of opposites, around each other rather than forcing these onto a timeline stretched taut or into an argument distilled neatly. I strive for knotted ecological thinking in the pages that follow. The National Zoo and its animals meant different things to different people. Indeed, the zoo represented a universe of meanings,[23] and knots can help us see how different lived experiences and different histories—how multiple realities, so to speak—can become "stuck together" through and within a manifold institution that burst into the urban world at the end of the nineteenth century.

If I must place this zoo biography into a taxonomic category, I would classify it within the genus of "new institutional history" and the family of "historical institutionalism," a framework embraced by some sociologists, political scientists, economists, and new political historians of the twenty-first century. Though the methodologies embraced under this academic umbrella are quite diverse, they all strive to show how institutions can lead to multiple outcomes. In the words of political economist Sven Steinmo, historical institutionalists "are something like the environmental biologist who believes that in order to understand the specific fate of a particular organism or behavior, she must explicitly examine that organism or behavior in the ecology or context in which it lives."[24] To think with Haraway, the National Zoo, between 1887 and 1920, became a tentacular institution, reaching in many directions at once.[25] Institutions, like individuals, are rarely intact. They are entangled. This site study will begin with the records of the National Zoo and will then follow their connections. The chapters below will move along tentacles toward different ecologies. In housing a vast heterogeneity, the National Zoo became a wild cultural entity, one that escaped within the captive society, the society of captivity, that created it. The zoo grasped at every arena of American culture—politics, marketplaces, activisms, landscapes, and laboratories. Ironically, this unintended wildness permanently cemented the zoo into American life.

Like most zoological collections, the animals here have been curated—framed through scholarly discourses undergirding animal studies, environmental history, and science studies. Unlike most zoological collections, however, these curations are intended to be open systems—to call attention to moats and fences, to turn ordered bodies inside out, to lead you behind the scenes, to encourage escapes, and to inspire knotted contemplations that cannot quite fit on zoo signage. Multispecies scholars Deborah Bird Rose, Thom van Dooren, and Matthew Chrulew in *Extinction Studies: Stories of Time, Death, and Generations* assert that "unlike many other modes of giving an account, stories can allow multiple meanings to travel alongside one another."[26] For this reason, I use narrative as a method because I believe, with essayist Caroline Randall Williams, that stories can "hold multiple truths together."[27] "Storytelling," according to philosopher Hannah Arendt, "reveals meaning without committing the error of defining it."[28] Storytelling can open new ways of seeing, new ways of feeling, and thus new possibilities for the future. This book is meant to be read first and foremost as a collection of stories with animals, as a collection of "zoo stories," to borrow from ecocriticist Randy Malamud's *Reading Zoos: Representations of Animals and Captivity*. Yet where for Malamud "zoo stories" names a genre, a "diverse range of literary and popular cultural artifacts," produced *about* zoos, here I assemble zoo stories that emerge *within* zoos, especially within the National Zoo in the decades around 1900.[29]

Layers, Lions, Enfolding

This book will underscore the significances of zoos to American cultural history, environmental history, and the history of science and medicine, thus joining an expanding body of critical zoo histories and zoo studies that has emerged since environmental historian Harriet Ritvo's 1987 classic study *The Animal Estate: The English and Other Creatures in the Victorian Age* and Bob Mullan and Garry Marvin's 1987 foundational anthropology *Zoo Culture: The Book about Watching People Watch Animals*.[30] The zoo stories here will take us along lines of flight[31] to less-bounded, more wild, more interdisciplinary scholarly terrains where questions are more important than answers, where thinking and feeling coalesce, where knowledges merge with ways of knowing, and where both are always animate, always with spirit, always alive. These lively grounds, cultivated over the last two decades by the interdisciplinary humanities, are the perfect ones upon which to examine the zoo as an historical institution. Animal historian Andy Flack has depicted zoos as the embodiment of "layers of human-animal relationship."[32] Envisioning human-animal relationships as a tidy stack of strata, one on top of the other, opens our eyes to "the animal," any animal, as more than and many. By allowing for the viewing of human-animal relationships to be a stratigraphic or sedimentological exercise, by locating layers of meanings that settle imperfectly upon bodies, we can conceptualize humans, animals, and their relations as aggregates produced over human-cultural time. Employing geologic metaphor to see historic depth in our relationships with animals is not meant, of course, to suggest that we can discover a "real" bedrock "animal" who lurks somewhere beneath the layers. Nor is it meant to suggest that a "real" animal, or our "real" relationship with animals, is merely the sum of its layers. Both conclusions would reproduce the Western fetish for Authenticity so encoded into notions of Nature and Other, those symbolic landscapes where Euro-Americans believed, in the words of environmental historian Gregg Mitman, "the trappings of civilization might be shed, the purity of God's hand felt, and the real self found."[33] Institutions (zoos) are knotted ones, and individuals are layered. Geologic metaphor, then, serves as conceptual resistance to Singularity, to help resist the temptation to see human and animal individuals as single, autonomous, neoliberal Subjects, just as employing knot metaphor helps resist the temptation to see the zoo as an intact institution that can occupy a single space or context along the horizontal plane of a timeline.

Seeing human-animal relationships as we might see the lithic layers exposed on the side of a cliff—where colors, types, and textures settle, over time, in discernible parallel order, occasionally disrupted by slides, collapses, folds, and faults—can help us *see* animals how zoogoers *experienced* them, namely through meanings assembled across the entirety of human-animal history yet collapsed into single moments. Human-animal relations should appear to us as layered with disruptions. The meanings given to animals, in this view, never disappear as time supposedly progresses. Layers are always-already there, supporting new symbolic systems that develop upon older surfaces. Take the lion, for instance. The lion could appear throughout medieval bestiaries, across British imperial media, in Nietzsche's writing, upon the Iranian national flag (before the Revolution), and within Brazilian income tax regimes only because these particular usages stacked up on deeper symbolic layers, first established by ancient North African and Near Eastern civilizations, that attributed regal significances to big cats. And between this strata and recent surfaces lay many other layers still, for lions range throughout oral traditions across Africa, within the world's monotheisms, and across Hinduism, classical religions, Buddhism, mysticism, and alchemy. Dig even deeper, and we find Paleolithic dreamscapes with lions sketched on cave walls, black manganese and red ocher inscriptions that outline the fear or awe that members of the genus *Homo* held toward this feline predator.

No wonder animals, as queer theorist Mel Chen acknowledges, "bear the burden of symbolic weight."[34] Zoo animals meant so many different things to so many different people. When touring the zoological park, zoogoers not only walked through different layers of history but also wandered across these layers in uniquely ordered routes. To historicize the zoo as a knotted, tentacular institution is to also accept that the zoo (by enfolding the experiences of its visitors into many timelines of animal-human history) resists historicization—which, frankly, may explain the institution's longevity and persistence into our present day.

Gaps, Curation, and Recovering Repressed Lives

Stories that are held together can do for *thinking* history what quantum superposition does for *thinking* matter, what trans theory does for *thinking* gender, what ecology does for *thinking* Nature—namely enabling viewers to see multiple positionalities and possibilities. All human actors (keepers and visitors alike) can always be found residing at different places at once. The zoo meant many things to the human characters of this book, as did the animals. Certainly, the same was true for the animals themselves. Surely, they had a multitude of experiences shaped by an immense range of species-specific thoughts, feelings, choices, and perspectives. And I hope that fragments or traces or shimmers of their experiences will be felt obliquely, carried through the human voices that always narrate, but never fully enclose, the Other. Nonetheless, despite equal uniqueness, the human and animal characters here were, of course, never equal at all, for the zoo always sought to contain the latter for the pleasure of the former—animals at the pleasure of zoogoers.[35] Humans always possessed the greater latitude of agency in the space of the zoo. The National Zoo was always a site of power, where animals were captured and displayed for humans. Yet, as these stories will demonstrate, power—no matter how forceful, no matter how normalized, no matter how concealing—always remains cleft, incomplete, open.[36] Thinking with philosopher Bruno Latour, animals exert agencies *beyond* human control that still "provoke" or "induce" "humans or (other) nonhumans to do something," to "respond."[37] And these actions almost always remain hidden within (the power of) the zoo.

This simple contention lays at the heart of this book. In the zoo, power was truly omnipresent yet always never quite all-encompassing. The zoological park was a unique institution not because it was curated for its visitors (*what institution is not curated?*) but rather because its curation was *not* meant to remain entirely hidden from its consumers. While holding their maps, following signage, looking upon ordered displays, zoogoers knew quite well that they were immersed in a curation. Indeed, zoogoers craved this curation. Its power *was* (the structuring principle of) the zoo, literally an essential characteristic. Yet the power of the curation never just emanated from its presence, from the raw ability of an elite, or a city, or a nation *to display* a world of animals. Rather, this power magnified exponentially as zoogoers repressed the curation they consciously consumed. For curation to work, to genuinely *feel* authentic, zoogoers had to forget, or pretend as if, they did not know what they knew—that the Nature manufactured around them was not Nature at all.[38] This doubled process of *knowing/not-knowing*, this chosen forgetfulness, this act of play *as if*, served as the psychological landscape (the economy of desire) that undergirded zoogoing and its pleasures. On the one hand, put into words, this dynamic of the zoo may feel excessively abstract or philosophical. On the other hand, this dynamic is also something we know all too well (indeed, *so* well that it is necessary to pause and shine a light on it), for it is simple suspension of disbelief that undergirds so many orchestrated ideological and cultural experiences—from a fun night at the theater to a game of *Fortnite* to the rehearsal of a patriotic origin story. *I know. But I cannot acknowledge it. I know that I choose not to know what I know.* This is pleasure. And, to a certain extent, this *is* the zoo.

Calling attention to this goofy property of curation is important at the outset of this book because I would like to suggest that within this quirky and internal game of repression, of conscious not-knowing, lie entire realms of forgotten and repressed histories. Zoogoers sought the curative theater of the zoo only to *repress* what they sought, and through this process, they predictably experienced and witnessed a great deal that was also forgotten or erased, lost from memory and thus History, though marked on primary sources. And I would like to suggest that locating the subtle fissures on the surface of power—the places where things just do not appear as they should or do not quite line up—may provide an approach for animal history. Acknowledging from the outset that power is cracked is theoretically and methodologically essential for this biography of the zoo, where multiple meanings travel alongside one another, for it helps author and readers alike to resist the desire to "frame" zoo animals merely with or against the *power of curation*. For example, there is a lot to be said about the historical significance of *how* animals were curated, how they were constructed by and presented in the zoo. There is also a lot to be said about the individual, historical animals themselves, the ones partially erased by their curated avatars; these specific living beings were always and completely divorced from how zoos framed them. In this way, the zoo stories here will speak both with and against curation. However, with these stories, I also hope to open up spaces between these two ways of seeing, within the discourse of power so central to the zoo. The zoo stories here will twist in many directions, yet as they turn, they will point toward, through, and beneath subtle cracks on the surface of containment, where less-curated ecologies reside. These spaces, once located, open to expansive zones, and the laws that govern the spaces beneath power may operate by different rules than the ones that reign on its surface. For example, zoogoers always split (they always *know/not-know*), and animals subsequently split too. Also, individuals (animal or otherwise) can inhabit multiple places (timelines, meanings) at once. Things are never what they seem to be at the zoo.[39] As you watch the processes of capture and display, as you look closely at captive animals, notice how capture always remains incomplete. Seek the remainder. Locate the excess. Search for the wildness that resides in the blind spot of the zoo. Zoo animals are always more than, or other than, what zoos say they are, and zoo stories told through close scrutiny may be able to lift the curtain. Look closely at the knots that tether humans and animals and you may see, along their bights, spaces where "the human" and "the animal" dissipate into more lively and less-scripted encounters.[40]

This book will raise philosophical questions from beginning to end regarding what Derrida has famously called "the question of the animal."[41] In so doing, the knots, layers, and gaps of this book will provide textured case studies of historicized and archivally sourced human-animal relationships that I hope are useful for theorizing these encounters. Paying close attention to the complexities of both how layered animals were framed in zoos and how they always exceeded those frames will begin an exploration of dialectical scrutiny that has important ethical and ontological implications emerging from the gap between the symbolic ("the animal") and its lack (the *wild* behaviors, characteristics, lives, and lifeworlds of an animal-individual that resists or falls out of symbolization). Bioethicist and environmental philosopher Ralph Acampora in his important "Zoos and Eyes: Contesting Captivity and Seeking Successor Practices" traces an obvious paradox produced by curating over this gap—a contradiction that shapes the very experience of zoological display—namely that the zoo "presents as wild . . . what is captive." For Acampora, this discrepancy, structurally encoded within nearly all zoo exhibits, causes animals "to disappear precisely by overexposing them."[42] The disjuncture, the alterity, created by *re*-presenting animals not only produces semiotic dissonance and phenomenological confusion—an inability to truly *see* the inhabitant of an enclosure—but can culminate in historical violence against those very alive and breathing distortions, as well as against humans outside the zoo, as we saw with Robert Wilson Shufeldt and his conservation racism.[43] In the eye-opening *Capture: American Pursuits and the Making of a New Animal Condition*, critical theorist Antoine Traisnel examines the same gap between "the wild" and "the captive," showing that across the nineteenth century "the regime of capture"—and its accompanying illusion of preserving, even saving, animals—"folds the explicit violence of conquest into biopolitical protocols of management and regulation."[44] Maybe more than any other nineteenth-century institution, zoos and their animals embodied the aesthetic of capture and management in the name of conservation, and the stories that follow track their historical impact on American science, culture, and environmental practice. The zoo, you will discover, is a place of opposites, of ironies, of contradictions arising from curated and systemic distortion, repression, and forgetfulness. And zoo stories can twist strangely around the resulting tensions, producing diverse, opposing, tentacular legacies. In other ways, for animals themselves (despite their infinite differences), the curative powers of the zoo created a more simple and straightforward condition—that of containment. The stories below will move between and around this given, calling attention to gaps, to spaces between, to runaway animals.

Zoo Biography: Popular Zoology and Animal History

At the beginning of the nineteenth century, "biology" emerged as a "general science of living things" that "united the study of botany and zoology over against the study of nonliving nature."[45] On paper, both "natural history" and "biology" meant the same thing—the scientific examination of plants and animals, of all living things. After mid-century, though, "biology" came to connote studies undertaken by professional scientists, often in laboratories, whereas "natural history" came to mean studies undertaken by educated amateurs, often in outdoor environments. After the

1830s, the usage of the term "natural history" declined steadily, and after the 1860s, "biology" began to fragment into specializations (morphology, cytology, hematology, systematics, and so on).[46] In this context at the end of the century, within the semantic-epistemologic register that housed a waning "natural history," popular zoology emerged in the American public sphere as a wide-ranging, heterogeneous, democratic zoological discourse. This term was used occasionally, and flexibly, during the period to describe a particular work of zoology (the study of animal life) that was intended for a large audience—like, for example, J. Dorman Steele and J. W. P. Jenks's 1887 textbook *Popular Zoology*.[47] However, the rapidly changing and far-reaching scientific milieu represented by this term was much larger than its specific usage. This book will show that the popular zoology experienced within the National Zoo between 1887 and 1920 transformed how zoogoers thought about animals. Charles Darwin had certainly influenced every realm of the life sciences with his 1859 publication *On the Origin of Species* by showing how all life emerged from a common ancestor. But Darwin's classic alone cannot explain why by the end of the nineteenth century increasing numbers of Americans had rethought the animal kingdom and their proximity to it. To see how ideas truly change, we must look beyond texts and scientists and into the realm where these meet popular culture, a realm composed of stories and storytelling, a realm that does not follow the "rules" of Science but was (and is) at the center of "knowledge-production."

Between the 1880s and World War I, as intellectuals debated Darwinism and scientists institutionalized the laboratory, zoological parks appeared suddenly at the heart of nearly every major American city, captivating tens of millions of visitors. Darwin's theory of evolution pushed scientists, philosophers, and theologians to *theorize* about humans and animals. However, the public zoo movement allowed the multitudes to *experience*, to feel, the similarities and differences between the human world and the animal world. Zoos brought animals physically closer to urban Americans. They allowed visitors to hear and see the creatures that were—previously in the nineteenth century—merely figments of the zoological imagination. And the timing could not have been better for an anxious and industrializing people who had long believed that they had depleted their Wilderness and who had long felt divorced (as programmed by a statecraft derived by Hobbes, Locke, and Rousseau) from (a state of) Nature. For nineteenth-century Americans, wild animals were simply no longer in view. The zoo—through its pursuit of science, education, and recreation—offered to change this.

Upon entering the zoo, these Americans saw (or believed that they saw) the world's exotic species for the first time. Their bright colors, mammoth bodies, sharp fangs, thick coats of fur, spots, scales, and stripes produced feelings of newness, of otherness, of authenticity, of wildness. And through these imagined encounters, Americans experienced a new, global animal kingdom. These diverse experiences of animality and globality in the aggregate produced a new, more complex, more liminal "animal" for twentieth-century scientific, cultural, and conservation thinking. Zoo animals led zoogoers in surprising directions—to the halls of Congress and policymaking, to the exhibits of museums, to global trade networks, to the big business of print journalism, to the birth of the airplane, to conservation politics, to the formation of primatology, to tuberculosis outbreaks and preventative medicine, to the rise of animal rights discourses, and to the genesis of ecological thought. Zoos, in turn, ushered animals, real and symbolic, into American institutions. All of these topics, and more, appear in this book. Weaving these seemingly disparate subjects together are the threads of popular zoology brought to life by zoo animals. Before the emergence of zoos, exotic animals existed as the imagined subjects of natural history, art, and literature, influenced greatly by the intellectual currents of the Enlightenment. Yet in zoos imagined animals became tangible, though imprisoned, and their countenances became powerful, though constrained. (As the previous section stated, zoo animals and humans are always double, so they need to be viewed through a sort of double vision.) Individual animals, therefore, despite their captivity, ran wildly and unpredictably through American culture.

Popular zoology provided a malleable and accessible cultural-scientific genre that helped Americans make sense of their new experiences with, and knowledges about, animals arising from zoos. The historiography of the life sciences is vast, diverse, and expanding rapidly in the twenty-first century, and this book offers this literature a collection of zoo stories that will exhibit popular zoology, a diverse field of public-scientific inquiry that blurred the lines between culture and science. In this book, I embrace an expansive definition of science inspired by two trends across the history and sociology of science and science studies. First, I consider popular zoology as a part of the broader world of "popular science" that emerged throughout the transatlantic Western world in the late eighteenth century.[48] Over the last three decades, the importance of public places and popular culture to the formation of science has become well-established. The historical scholarship of popular science has examined public places where science was practiced and exhibited, has analyzed public mediums (like the namesake mass market magazine that featured Shufeldt's zoo writing) through which scientific knowledge was transmitted to mass audiences, and has excavated the reasons why these audiences voraciously consumed certain types of science, like mesmerism and phrenology in the nineteenth century.[49] Despite its persistence across public discourses today, this literature has greatly problematized the diffusionist myth or deficit model of understanding popular science. In this model, scientists hypothetically practice "pure" science and then cultural media (books, lectures, and museums in the nineteenth century, with televisions added in the twentieth, and the internet in the twenty-first) disseminate science to the public, where it becomes "popular," diluted, and less "scientific." This hierarchical, trickle-down manner of conceptualizing science is historically inaccurate, philosophically simplistic, and generally presumptive and elitist. Popular cultures, public spheres, and scientific spaces—fields, laboratories, and "truth spots" (cultural locations that lend credibility to supposedly scientific claims)—have always been in constant and mutual exchange, despite common misnomers otherwise.[50]

The popular zoology practiced in the zoo was a popular science in all regards—in its place, its mediums, and its audiences. And zoo stories will reveal a persistent cross-pollination between realms too often labeled "Culture" and "Science." Popular zoology surrounding the National Zoo will direct our attention to the entanglement of these domains, and the chapters here will frame a "popular zoology," to think with historian of science James Secord, as *both* a "science in context" (one enmeshed in the cultures, institutions, and ideas of its time) and a "science in transit" (one that is constantly mediating with and changing within its surroundings).[51] Thinking with historian of science Andreas Daum, popular zoology, like all science, was a product of "transformations of public knowledge across time, space, and cultures."[52] In this book, I will show how the zoo shaped science from the ground up, how science appeared on the ground in the zoo, and how the zoo sent science moving to other settings, from Congress to the laboratory. To do this, I will resist placing "science," "popular science," and "popular zoology" into conceptual straightjackets and instead follow zoological thinking wherever it may lead.

Second, I treat "popular zoology" as a science composed of stories. Microhistory resurrected the mode of "the story" for both cultural history and the history of science in the 1970s and 1980s, in the wake of anthropologist Clifford Geertz's call for "thick description" and alongside the rise of the social history movement.[53] I believe that looking closely at the particular, the minute, and the local is the best way to see the entire world.[54] And I believe that stories, their vivid characters and twisting plots, not only assist in "zooming in" on the particular but also hold an imaginative and theoretical potential that argument-driven analysis often struggles to find. Anthropologist Anna Tsing, in her 2015 *The Mushroom at the End of the World*, states that "to listen to and tell a rush of stories is a method. And why not make the strong claim and call it a science, an addition to knowledge?"[55] This book makes this strong claim. Science is a story composed of stories, which are embedded in and surrounded by others. Historian of science Robert Kohler has recently and rightly declared that "like living, residing, and observing, storytelling is another everyday activity that becomes a practice in science."[56] Therefore, the zoo stories here will shed light on the intimate and interior practices of science that so frequently get processed away through professional overture, scientific scaffolding, and veneers of objectivity. Zoo stories, though, will not simply recover lost narratives *within* science, but they will also lead outwardly to show that the domain called "Science" was always intimately tied to domains that were perceived as anything but scientific. For example, as sociologist of education and animal studies scholar Donna Varga has demonstrated for the period of the late-nineteenth century, children's stories and toys were always a "reproduction of adult beliefs, values, and actions and as such they provide a means for exploring the history of ideas about animal-human relationships." A litany of panorama, picture, and pop-up books produced about zoos beginning in the 1870s and stuffed-cloth zoo animal toys manufactured in the 1880s paved the way for the mass consumption of material and literary "animaltainment." This market of children's trinkets was crucial to teaching future adults not only how to "participate in animal captivity"[57] but also how to transform wild animals (notions of wildness) into tamed civic pets (notions of control), a historical process animal historian Susan Nance has documented thoroughly in regards to Jumbo the elephant.[58] Zoo stories bound scientists to children, and animals to both. The point is that popular zoology, like the zoo itself, wove all the institutions of American society into a zoological latticework.

Each chapter below employs a different approach to animal history.[59] As observed by historian of genetics and zoology Daniel Gamito-Marques, in science "each actor has an agency and intrinsic motivations . . . knowledge is frequently co-constructed, and . . . is not only able to travel and act at a distance, but also circle back and have an impact on the places where encounter occurred."[60] This book follows popular zoology across such circulating routes through and around the zoo, and as an intervention in animal history, it conceptualizes zoo animals themselves as actors possessing a broad range of agencies within the co-construction of zoological thought.[61] The stories here frame animals as protagonists, yet they do so in distinctive and sometimes contrasting ways, together showing that animals and animality always intersect our supposedly "human" domain and supposedly "human" history. While this book frames zoo animals as active agents, they should not be misunderstood as being free-acting subjects making choices in a vacuum of freedom. The stories here conceptualize animals (and humans, for that matter) as acting-in-confinement; they can "act" thoughtfully, coercively, deceptively, unintentionally, surprisingly, aggressively. Seeing animals as agents here should certainly not be read as a liberatory gesture. There is an awful lot of violence to account for in the zoo.

As Nance puts it, "animal history is really about what it has meant to be human throughout our interaction with other species."[62] Similarly, environmental historian Drew Swanson emphasizes that "humans and animals inhabit the same world, living, dying, and creating history separately, on occasion, but much more often, in conjunction with one another."[63] And medieval historian Michael McCormick states simply that "the human and animal past is one."[64] Animal history takes these truths to be self-evident. These truths of sharedness, of intersection, of entanglement are almost always ignored, forgotten, or repressed, yet they remind us that it is often the things that we take for granted—that we see every day but do not see at all—that matters the most to us. How would human history look different if "the human," and its subjectivity, was not staged upon a grand disavowal? Animal history in 2023 has made inroads into virtually every corner of the historical profession, asking pressing questions of environmental history;[65] cultural history;[66] economic/

commodity history;[67] urban history;[68] histories of marginalization, examining the intertwined categories of race, gender, ethnicity, sexuality, ability, class, and species;[69] histories of science, technology, medicine;[70] and of global, deep, "big," and evolutionary histories.[71] Historians must continue to embrace all nonhumans, from stones to spruces to salamanders to systems, if we are to push through and past our anthropocentric traditions and face the challenges of our Anthropocene, global dilemmas caused by (some) humans yet bigger than "the human."[72] We must find new ways of seeing, new ways of speaking, new ways of listening. We must find new ways of being. We must create narratives that open sympoietic, collectively produced ways of knowing rather than obsess over "enlightened," outmoded, and autopoietic, self-producing knowledges.[73] Animal history nudges us toward new symbiotic ideals. Etienne Benson, in "Animal Writes: Historiography, Disciplinarity, and the Animal Trace," challenges the animal historian to "forge a (real, genuine, authentic, ethical) relationship with the embodied traces of past animal life."[74] Let's follow the traces of past lives and seek new ways of relating.

As we saw through Robert Wilson Shufeldt, the National Zoo is a place where opposites entangle, where surprising histories conjoin. The chapters below will present seven different stories, with seven different historical arguments, embedded in seven different historiographies, concerning the National Zoo between 1887 and 1920. Each of these will look at the zoo from a unique perspective. None will offer a single interpretation of the zoo. None will be ironed onto a timeline that moves toward Progress.[75] The historical zoo, like the contemporary zoo, is a messy place. As is so often the case in both history and ethics, good intentions may lead to bad outcomes, bad intentions may (eventually) lead to good outcomes, right and wrong blur when you look closely, and ethical judgment always depends upon the scale at which, and the vantage from which, you assess. The zoo will be a challenging place to make sense of, and this book, by looking closely, will provide a knotted, layered, and gapped case study from which we can begin to evaluate an institution that today sells hundreds of millions of tickets every year. Zoo stories will lead to layered lessons and knots and open spaces within the zoo animals that we thought were enclosed.

Sometimes, though, lessons will feel quite easy to discern, like when zoo superintendent Frank Baker described a moose that was shipped to the zoo in July 1891. The young moose, which had not yet reached adolescence, was placed inside a wooden crate, which was loaded into a steel train car in the scorching heat of summer. Then, over the course of five or six days, the moose hurtled across the country, jostled upon Gilded Age rails, from Yellowstone to the nation's capital. Upon arrival at the zoo, Baker stated simply that the moose "could not stand." He died shortly thereafter, and the superintendent blamed the "baggage smashing express system" that all zoos relied upon for animals.[76] At the end of the day, despite all of the meanings sketched by this biography, the city zoo was always an institution filled with sad, sick, and dying animals—a place created by and symbolic of the systematic destruction of places and peoples, animals and ecosystems, that characterized so much of American history. You will encounter here all types of stories—tragedies, mysteries, biographies, adventures, comedies, stories of improvement, and ones of decline, stories of both sadism and love. The saddest part, the most confusing part, is that all of these are always entangled.[77] This biography will chronicle some of these knotted relations and will mourn the loss of the animals caught in them.

Zoo Guide: Chapter Summary

Entangled Encounters at the National Zoo is a thematically organized biography, examining the institution and its entanglements between the 1880s and 1920s. Only the first chapter adheres strictly to the chronological imperative, documenting the establishment of the National Zoo. Chapter 1 concerns animal commodities and trading networks, museums and Congress, politics and symbolisms. The next six chapters concern animal escapes, sciences in the public sphere, animal activism, conservation and capture, tuberculosis medicine, and the historical zoo movement, respectively. Chapter 1 begins in a taxidermic zoo. It tells the story about how William Hornaday, chief taxidermist of the Smithsonian's United States National Museum, established a Department of Living Animals with the hopes that it would evolve into a national zoo. The fact that a famous hunter and taxidermist would build an institution that sought to preserve living animals by preventing their destruction may seem paradoxical. However, a logical connection existed between the processes of public taxidermy and national zoo–building. As Hornaday's department grew in popularity, its eventual evolution into a zoological park appeared to be inevitable. Nonetheless, before the department could fulfill its destiny, the Smithsonian needed to gain congressional support. Whereas other zoos of the day simply needed funding and public sanction, a national zoo constructed and administrated by a federal institution required approval by the federal government. The idea of a national zoo thrilled the masses from the beginning. Yet even seemingly uncontroversial and nonpartisan issues, broadly supported by constituents, can become contentious and complicated in the hands of policymakers. Chapter 1 also examines the political controversy of the National Zoo. This controversy concerned some southern Democrats' refusal, in a nation still reconciling after the Civil War, to support a zoo that embodied, for them, federal power. This chapter finally addresses the American park movement and the material processes of zoo-building that took place along Washington, DC's Rock Creek.

Starting with chapter 2, the biography becomes thematically organized. This chapter tells the stories of literal runaway animals, one of the two central metaphors of the book. Escapes demonstrated that building a secure zoo was a hard thing to do. Zoo animals frequently got away. In fact, the entire history of the public zoo movement is filled with runaway animals, especially prior to World War I. This chapter examines the dramatic moments of animal escapes, using these episodes as windows into Americans' beliefs, agendas, anxieties, fears, and desires. It examines zoo escapes nationwide and tells the stories of two episodes specific to the National Zoo, the grizzly escape of 1892 and the wolf escape and rabid dog debacle of 1902. Both absconding animals caused media frenzies. The latter even reached the doorstep of the federal court system. The tales of runaway animals demonstrate how *popular* zoology could become when etched into sensational narratives.

It should come as no surprise that a zoo sponsored by the Smithsonian found a place at the epicenter of science. Conveniently located in the nation's capital and in the midst of a Smithsonian community, the National Zoo quickly forged inroads into myriad branches of the life sciences. Chapter 3 explores some of these relations, demonstrating the centrality of a zoo to the ever-specializing world of science. The chapter opens with an examination of anthropologist Frank Baker, who molded the zoo as an institution of and for popular science for sixteen years. It then explores the zoo-primate experiments of Richard Garner and his attempt to popularize primatology. The chapter's third section paints a general portrait of popular zoology in the space of the zoological park. In its fourth section, the chapter moves from the general back to the particular, illuminating a series of important experiments concerning flying machines and the birth of human aviation. This chapter picks up where Anita Guerrini's *The Courtiers' Anatomists* left off, shifting zoo-science attention from the seventeenth century, menageries, and dissection tables to the late nineteenth century, public zoos, and science in public spaces.[78]

Chapter 4 tells activist stories. By focusing attention on the public space of the zoo, this chapter demonstrates what concern for animals first looked like on the ground and argues that this concern was far more widespread in the decades surrounding 1900 than previously understood. Zoos surely functioned as sites of violence toward animals, but they simultaneously and ironically functioned as incubators of compassion and concern. Zoos networked Americans into larger discourses about animal welfare. Most surprising, this chapter locates early iterations of the phrase "animal rights" used by anti-zoo, anti-Cartesian women activists living near the National Zoo.

One reason the National Zoo arose in the first place was to save North American animals from extinction, as chapter 1 discusses. Ironically, though, many of these campaigns endangered the very species they sought to save. Even though the National Zoo (and other zoos) hoped to protect animals, the process of catching them inevitably caused high casualties. Chapter 5 tells a story of mixed legacies by zooming in on one of the species facing extinction—bighorn sheep. By following bighorns, this chapter shows what early zoo-driven conservation looked like on the ground through a surprising interspecific caper of pursuit, adventure, escape, and capture.

Chapter 6 examines the rise of a new type of medicine born in and around the National Zoo. On the one hand, zoo medicine directed an agriculturally based veterinary tradition to address the needs of a new urban institution. On the other hand, zoo medicine adapted human medical practices to fit the needs of zoo animals. This chapter tells a story about tuberculosis and zoos and makes several arguments regarding the history of tuberculosis, the history of zoo medicine, the history of public health in America, and the connectivity of zoo knowledge networks. This chapter also establishes the emerging importance of "air" to shifting pathological understandings of tuberculosis. It finally raises crucial questions about ethics and the place of animals in science and medicine. The conclusion will consider the place of the National Zoo within the American zoo movement at the beginning of the twentieth century. It will also pose a few questions of you and of zoos and will encourage us to move collectively together toward more ecological futures.

CHAPTER ONE

The Origins of a National Zoo

Taxidermic Beginnings

Across the late nineteenth and early twentieth centuries, William Temple Hornaday gained fame as a hunter, naturalist, and conservationist. He is popularly attributed the honor (surely an exaggeration, but the lore is noteworthy nonetheless) of single-handedly discovering the American crocodile and saving the bison and the Alaskan fur seal from extinction. There are mountains and Boy Scout medals named after him. He published widely, and *The Extermination of the American Bison* was a hit. Yet his most profound legacy was his driving influence in the American zoo movement. He founded the National Zoological Park. He served as director of the New York Zoological Park (now the Bronx Zoo). He organized the first successful zoo reintroduction program. And he carried an international reputation as one of the world's foremost experts on zoo-building.[1] However, as we saw with Robert Shufeldt, legacies often serve to mask the violence within them, and Hornaday is no exception. He should be remembered also for his vehement hatred for immigrants, Africans, African Americans, and Native Americans. And he should be remembered for displaying Ota Benga, a Congolese man, in the monkey house in New York.[2] A history of the National Zoo must begin with its entangled beginnings, where this white taxidermist built a zoo to save the animals that he himself took part in decimating. Let us begin with this fitting contradiction for a zoo biography.

Hornaday was born in 1854 on his father's farm in Hendricks County, Indiana. When he was three years old, his family moved to the prairies of Iowa, where young William cultivated a passion for the passenger pigeons, turkey vultures, and prairie chickens that soared overhead. At some point during this agrarian childhood, William visited a half-brother who stayed behind in Indiana. It was then that he, for the first time, encountered taxidermied birds on display, their bodies perching, roosting, suspended in flight. At fifteen, William was orphaned, and in 1870 he attended Oskaloosa College, later to enroll in the Iowa State Agricultural College. Here, Hornaday discovered natural history through Audubon's classic *The Birds of America*. Soon after, when he amazed the college president with an exquisitely mounted pelican of his own, Hornaday seized a taxidermist position in the campus museum and began training as a naturalist. In 1873, he left Iowa State to serve as an assistant to prominent naturalist and geologist Henry Augustus Ward in Rochester, New York, where he made a name for himself during high-profile collecting trips to the Caribbean, South America, South Asia, and Indonesia. Shortly thereafter, the assistant secretary of the Smithsonian Institution George Brown Goode recognized Hornaday's ambition and recruited him to help the Smithsonian meet a swelling demand for taxidermy.[3]

By the last quarter of the nineteenth century, the Smithsonian, under the leadership of its second secretary Spencer Baird, was expanding its priorities beyond research to include the development of the National Museum and the expansion of its natural history and ethnography collections. As chief taxidermist of the mammal collection, Hornaday commanded, arguably, the nation's most prestigious taxidermy job. Goode, an ichthyologist and taxonomist who studied at Harvard's Museum of Comparative Zoology under Louis Agassiz, was pushing the Museum's collection from two hundred thousand specimens to more than three million. And he placed Hornaday at the helm of this great expansion.[4]

Upon hire, Hornaday shifted his own focus from the exotic animals of Asia and Africa, his primary objective in Rochester, to the fauna of North America. His years in Washington were devoted especially to the American bison. In 1886, the Smithsonian funded an expedition to Montana to collect individual specimens before the rapidly disappearing herds of the American West reached the inevitable extinction that naturalists had been predicting. On the trip, Hornaday experienced the gravity of these forecasts.[5] The herds he witnessed roaming the plains during his childhood had been reduced to single individuals thinly scattered throughout the West, and it took him months to locate and shoot twenty-five representative specimens. In a letter to Secretary Baird, mailed from Miles City, Montana, Hornaday penned:

> We killed very nearly all we saw and I am confident there are not over thirty-head remaining in Montana, all told. By this time next year the cowboys will have destroyed about all of this remnant. We got in our Exploration just in the nick of time—the last day in the evening, so to speak. . . . I think we can have a series of mounted specimens that will be envied by all other museums.[6]

Along with these specimens, which he would transform into a famous bison exhibit for the halls of the museum, Hornaday brought a new purpose back to the nation's capital.[7] The irony here, though, is unavoidable. As historian Nigel Rothfels has shown with antelope, big-game hunting and zoos share common origins.[8] In the words of animal historian Susan Nance, "as the logic went, it was necessary to kill endangered animals in order to save them."[9] The same lesson is shown here. On the rolling fields of Montana, as he hunted bison, Hornaday cultivated a conservationist ethic that would become etched into the zoo that he would establish. To make his turn to conservation official, Hornaday informed the American public of the genocide taking place through a book-length report titled *The Extermination of the American Bison*.[10]

The desire to protect bison led directly to the genesis of the National Zoo. In an 1887 letter to Goode, Hornaday stated:

> In view of the fact that this government has done nothing to preserve alive any specimens of the American Bison, the most striking and conspicuous species on this continent, I have the honor to propose that the Smithsonian . . . take immediate steps to procure either by gift or purchase, as may be necessary, the nucleus of a herd of live buffaloes . . . which may, in a small measure, atone for the national disgrace that attaches to the heartless and senseless extermination of the species.[11]

Hornaday echoed this sentiment in *Extermination*, suggesting that the Smithsonian acquire several bison to keep the "breed absolutely pure."[12] Soon after, Hornaday purchased six individuals, which he placed in a corral on the lawn behind the Smithsonian Castle.

As bison muddied their makeshift pasture along the National Mall,

Figure 1.1. Hornaday and baby bison Sandy, acquired from a collecting trip to Montana, walking on the grounds near the Smithsonian Castle. Courtesy of the Smithsonian Institution Archives.

other living animals crowded the inside of the Smithsonian. The museum had long hoped to develop a "Department of Vivaria, to afford materials for the prosecution of scientific studies," and in 1887 Hornaday took the initiative to finally establish the Department of Living Animals, addressing the "imperative" need to keep living animals on hand while he worked on taxidermic displays for the museum. By using living animals as references, as models, Hornaday could make the dead look alive, giving his specimens realistic poses and vivid expressions. The ideal of the artistic taxidermist—like Michelangelo in sculpture and painting—was to resurrect past lives in and through form.[13] Taxidermy, for Hornaday and

Figure 1.2. Hornaday collected and mounted these bison for the United States National Museum in 1887. The Department of Living Animals allowed Hornaday to use living specimens as models. Despite realist aspirations, however, he still arranged them into an "ideal" American-human family unit—father, mother, child—instead of a herd. Courtesy of the Smithsonian Institution Archives.

the class of taxidermists to which he belonged, was as much about the living as about the dead.[14] However, "the experiment of keeping live animals in the taxidermic laboratory," he admitted, "had proven expensive, and in many ways unsatisfactory." Living, breathing, defecating animals should not be haphazardly scattered around a workplace, so eventually Hornaday "deemed it necessary to form a small collection of living animals for purposes of study, and, if practicable . . . for exhibition to the public."[15]

To collect animals for this purpose, in the fall of 1887, Hornaday accompanied J. Frank Ellis of the United States Fish Commission to the Pacific coast with the hopes of collecting "as many specimens of mammals as could be secured by gift" or purchased cheaply. Passing through St. Paul, Minnesota; Fargo and Mandan, Dakota; Helena, Montana; Tacoma, Washington Territory; Portland, Oregon; Salt Lake City, Utah; and Cheyenne, Wyoming, Hornaday acquired the first sixteen animals for the department. The National Zoo, then, emerged humbly from two red foxes, one cinnamon bear, one white-tailed deer, one Columbian black-tailed deer, one cross fox, one mule deer, two badgers, five prairie dogs, one golden eagle, and one spotted lynx (which, most likely,

Figure 1.3. Hornaday (*center*) working in his taxidermic laboratory located in a shed on the South Yard of the Smithsonian Castle. Courtesy of the Smithsonian Institution Archives.

was a bobcat). Hornaday brought these individuals back to Washington, DC, and instead of caging them in his laboratory, he housed them in a temporary structure (25 x 106 feet) that he had built frugally from the scrap wood of the recently demolished New Orleans Annex building. This makeshift department, located on the south side of the eastern wing of the Smithsonian, was filled with temporary cages, heated, and opened to the public on New Year's Eve, 1887.[16]

As soon as the facility opened, individuals from around the United States presented Hornaday with "rare and valuable" animal gifts, and the population of the Department of Living Animals increased rapidly, tripling in size by the end of the month. This collection demanded "constant attention," and on February 1, 1888, Hornaday hired N. R. Wood as "keeper," to feed, care for, and protect the animals. In February and March, the number of animal occupants doubled again, populating the wooden building to capacity. At the department's western end, a pit was dug and graveled for the black bears. And in the center of this pit stood a single cage to hold a cinnamon bear. Another large cage was subdivided into individual compartments to hold the large birds of prey: eagles, owls, and vultures.[17]

Animal donations bombarded Hornaday. Temporary cages were hastily assembled, "and every effort was made to make the best showing possible under the circumstances." The department was soon bursting at its seams, and "with the rapid increase in the size and value of the collection came an increase in the amount of labor and attention it absolutely required." Especially important to keeping the department operational was W. C. Weeden, assistant engineer of the Smithsonian, who as a favor to Hornaday took on the extra responsibilities of caring for the animals. According to Hornaday, without Weeden, he "would have suffered very serious embarrassment." Yet, the collection became a magnetic local attraction in the District of Columbia, drawing thousands of visitors. And on May 12, 1888, William Temple Hornaday was officially anointed curator of the Department of Living Animals.[18]

Most of the world's zoos in the 1880s and 1890s purchased their collection's celebrity animals from exotic animal dealers, like Germany's famous Carl Hagenbeck, who had forged a well-established and profitable industry that bought and sold rhinoceroses, lions, elephants, and giraffes.[19] However, as historian Elizabeth Hanson notes, "wherever new zoos were built, an enthusiastic public deluged them" with donations.[20] And this was especially the case in Washington. Hornaday's department, and later the National Zoo itself, arose from quite humble origins indeed. Between October 8, 1887, and June 30, 1888, the department purchased thirty-two mammals, five birds, and one reptile, while receiving as donations, forty-two mammals, fifty-six birds, and twenty-five reptiles. In addition, eleven birds were born in the department. After nine months, it possessed 172 specimens, representing sixty-eight species. Of these, 123 were gifts, three-fourths of which were simply dumped on Hornaday, "offered voluntarily, and without either suggestion or personal solicitation."[21] In this way, the department, like many zoos, functioned as a public repository, a dumping ground, for unwanted pets.

The first animals exhibited on the Smithsonian grounds were not the exotic species typical of many zoos' founding generations. While the zoos of the world possess unique institutional histories, local zoological societies, comprised of an urban intellectual elite, often played an important role in establishing collections that sought to transform crowded metropolitan centers into the refined "capitals of modernity" of the nineteenth century.[22] As zoo historians have documented, nation-states and city elites founded zoological parks and gardens as part of the grand planning programs that accompanied the democratization, urban-reform, and civic-consciousness movements fashioned in the West following the age of revolutions. While nations blanketed the rhetoric of "modernity," an updated "civilizing" discourse, over all the institutions of the nineteenth century, the "modern zoo" was also built upon the ancient template of the menagerie, as old as empire itself.[23] Yet this "modern zoo" constructed itself as something new, an institution in pursuit of three ends: science, education, and recreation. Hornaday's department, however, emerged directly from practical needs. The founding generation of animals at the National Zoo was less defined by grandiosity than were other zoos' first animals. Of the individuals donated to Hornaday, "the most noteworthy mammals" were:

> A fine pair of American Bison, from Mr. E. C. Blackford, New York; a very large male Jaguar, from Mr. J. W. Riddle, Eagle Pass, Texas; two black bears from Mr. J. J. E. Lindberg, El Paso, Texas; a male Virginian Deer, from Capt. R. L. Hoxie, Montgomery, Alabama, and a female of the same species from Dr. P. Glennan, Washington, D.C.; two black bear cubs from Mr. J. S. Miller, Commissioner of Internal Revenue, Washington; a coyote from Mr. P. D. Nowell, North Platte, Nebraska; a gray wolf from Mr. C. A. Dole, Glendive, Montana; a grivet monkey from Mr. L. Moxley, Washington, and an exceedingly interesting Spider Monkey (*Ateles vellerosus*) from Mr. C. H. Townsend, U. S. Fish Commission Steamer *Albatross*; a gray fox from Mr. Geo. E. Brown, Alexandria, Va.[24]

With the exceptions of the monkeys, all of these hailed from North America and would have been popular species for taxidermic displays around the United States. Even the animals that Hornaday purchased fit the same description—a puma from Fort Keogh, a grizzly bear cub from Billings, a black bear from South Carolina.[25]

Despite the rapid growth of the department, it seems that Hornaday, Wood, and Weeden kept most of the animals in relatively good health, at least in as far as menageries are concerned, where breathing and thriving were equated. Two spotted lynxes, respectively, died from pneumonia and uremic poisoning in the winter due to a poor heating system; six quails and two partridges were killed by the rats; and a gray wolf, pronghorn, and deer also died quickly for unknown reasons.[26] Other than these unfortunate deaths, some opossums and woodchucks managed to escape the premises, setting an early precedent for runaways.[27] All told, this casualty rate is not half bad for a fledgling animal collection.

The department required a great deal of labor; Wood and Weeden could not do it all. Feeding alone was a monumental task. Many mouths needed three to five meals per day, and each species required a specialized diet. Furthermore, they created an informative "label" for each animal (except for the reptiles) that instructed visitors on general zoological information. Hornaday even tried to create maps that would show the "geographical distribution" of each creature, but despite his commitment to popular zoology, the "size of the collection overwhelmed him with more imperative duties" than signage. In March, April, May, and June, Hornaday purchased only four animals; "the gifts that came in from day to day" alone "required diligent exertions." Of course, these duties squeezed the institution's finances. Hornaday saved the "fine clover hay" used by the Department of Public Grounds as feed for the ruminants, conserving probably one hundred dollars. Birds of prey were given the infested building's captured rodents. And bison grazed on the Smithsonian lawn. Nonetheless, despite Hornaday's abstemious attempts to save, feeding animals was inevitably expensive.[28]

Though maintenance and feeding were costly, transporting animals was surprisingly quite reasonable. Typically railroad companies in the late-nineteenth century would charge enormous fees for shipping living animals. However, due to the lobbying of S. C. Brown, registrar of the National Museum, the United States Express Company (then including the B. & O. Company) agreed to ship animals to the Smithsonian at an "ordinary merchandise rate." Despite the obdurate Adams Express and Southern Express Company, who refused the Smithsonian any discount, Hornaday announced that the "generous concession" of the US Express "will be of great value to us in building up the collection," and he urged all of his magnanimous donors to ship animals using the B&O lines![29] (Shipping for the animals, though, no matter the bargain, surely and always constituted a middle passage of sorts—defined by a litany of traumas we can only begin to imagine.)

Managing the department required practical solutions to everyday problems. All the while, the department began to make its first contributions to science. The bears especially piqued Hornaday's curiosity. Since the black, cinnamon, and grizzly bears shared an exhibit, the department provided Hornaday an opportunity to explore the controversies of ursine taxonomy. He also studied how different bear species interacted with one another.[30] In established zoos, black and cinnamon bears were never displayed together, so Hornaday planned to take advantage of this "opportunity." While the results of his research were never published, Hornaday made up for it with his conservationist activism. For the 1889 Cincinnati Exposition, for example, he prepared an exhibit to call attention to the disappearing species of North America. In the exhibit, Hornaday included masterfully mounted specimens of the American bison, West Indian seal, California elephant seal, walrus, moose, elk, mountain goat, mountain sheep, antelope, and beaver. These bodies were surrounded by a series of paintings and photographs documenting the bison's demise, a display of bison hides, a collection of rifles (what he called, the "modern weapons of destruction"), and seventy skins of Rocky Mountain goats. Covering eight hundred square feet, the display commemorated the deceased by displaying their remnants and pieces, relics and monuments for who would soon be no more.[31] It also showcased the ironic, but well-established, role of sportsmen in the rise of American conservation.[32]

The department grew and grew and soon lacked space for visitors. In fact, the "daily throng" of people, sometimes approaching three thousand individuals, had "been so great as to make it a matter of difficulty to pass through the building," which would have been steeped in stenches unimaginable. These herds required Hornaday to hire a watchman "to compel the small-boy element," as well as the "ragamuffin element," "to depart after reasonable time." By the summer of 1888, the department was already functioning as a proto-zoo while also serving the National Museum as a taxidermic workshop.[33] Hornaday knew, though, that the Smithsonian could not manage a zoo on the Washington Mall forever. Space and resources were simply not available for the long-term comfort of either the animals or the human visitors.

Zoos typically embodied the values of nation-states in an imperial world—the ability to conquer, control, and display.[34] Yet they represented a bourgeois hegemony too, the ability of the economic elite to amass and advance knowledge, to distance themselves from the uneducated and unrefined lower orders. These significations usually required the long necks, bright colors, spots, scales, and stripes of exotic and expensive animals from far-off places. The department, though, emerged to satisfy the *practical* needs of the Smithsonian's chief taxidermist. Many of its first animal

Figure 1.4. Bison herd, acquired by Hornaday for the Department of Living Animals, muddying their paddock along the National Mall. Courtesy of the Smithsonian Institution Archives.

inhabitants were quite commonplace and unspectacular, including rats, foxes, rabbits, deer, owls, doves, pigeons, and garter snakes. Some were considered weeds or pests. Hornaday even found it necessary to purchase two squirrels, two raccoons, two opossum, and a box tortoise. The jaguar, bison, bears, monkey, prairie dogs, eagles, and macaw would have attracted attention in the Philadelphia Zoo or the zoological gardens of Europe, but none fetched high prices in the international animal trade. Though the department emerged from a taxidermic need, it soon demanded too much from the Smithsonian. As Hornaday recognized, "The most important result accomplished by our collection of Living Animals has been in emphasizing the great need of a national zoological garden, established in the city of Washington."[35]

From Department to Zoo

It was June 1888, only nine months after the department's establishment, that Hornaday first commented on overcrowded conditions.[36] Nonetheless, between then and June 1889, he squeezed 171 additional specimens into the collection, of which 126 were donated. On the last day of the second fiscal year, the department boasted a teeming population of 341 living animals, many of whom were shipped by US Army officers stationed in Texas. Even Buffalo Bill Cody donated three elk. When Dr. V. T. McGillycuddy of Dakota offered the Smithsonian four American bison, Hornaday felt obligated to make room for them in the Smithsonian's herd. In most cases, though, donations were simply rejected.[37]

By its second year, Hornaday had cloaked the department with a national purpose. Though he credited taxidermy, he also began to emphasize the need to prevent the mass extinction of American wildlife.[38] Ultimately, though, it was the animals, not the ideals, that drew visitors to the cages. From the beginning, these animals entertained and occasionally educated. When onlookers expressed confusion about how the baby barred owls could digest raw meat with the "hair and bone intact," keeper Weeden explained that after four hours the bones would be "disgorged in the shape of small pellets." In fact, he emphasized that without consuming bone along with hair or feathers (roughage), owls would die. (Osteophagy, an approach for gaining necessary phosphates that first evolved in Late Triassic archosaurs, still makes owls and other raptors challenging to feed in captivity.[39]) Visitors to the department were captivated similarly by famous carrier pigeons named Liberty and Minstrel, a jaguar, the prairie dogs, a trio of tame black bears named after Confederate generals, and Jim the cinnamon bear. A cage that housed a monkey and opossum together especially attracted attention, for the monkey would "hug the 'possum until it would cry out, and then drag it around the floor until tired of the fun." Since the department lacked the room to confine individual species to their own cages, bizarre interspecific violence amused visitors in ways that a formal zoo rarely could. These interactions sometimes exceeded entertaining abuse—for example, when the monkey eventually kidnapped baby opossums from the mother's pouch only to bury them alive beneath the dirt floor.[40]

A black-tailed deer, remembered by Hornaday as a "cannibal deer," refused to be contained with any other animals, killing and eating the chickens and rabbits that shared their exhibit. When Hornaday attempted to move the deer into another enclosure, the deer attacked him, pinning him against the ground. Weeden luckily came to the rescue and hit the deer with his club. "Had it not been for me," he remembered, "the professor would, in all probability, now be chanting with the angels." Even a bighorn sheep had to be removed from a stall and relocated to the buffalo yard for repeatedly kicking the indoor structure, damaging the wall and creating a pounding havoc in the crowded building. Washington's newspapers never hesitated to print "zoo stories" like these, simple narratives featuring animal characters and plots of drama, suspense, and humor. Even though such stories drew attention to the department's shortcomings, "crowds continued to flock . . . to look in on an embryo zoo."[41]

Chicago's newspaper the *Evening Lamp* ran a two-page, six-column feature about the Department of Living Animals and its potential to become America's first national zoo. According to the author, the department began with "Mrs. Cleveland's fawn and Mr. Hornaday's buffaloes." First Lady Frances Cleveland received a spotted fawn as a gift from a friend in Upstate New York, and since the "White House grounds afforded no good place for the little thing," the First Lady donated the animal to Hornaday. President Grover Cleveland subsequently added a golden-headed eagle he received from a friend in Tennessee. The author dedicated a significant amount of space to recounting the numerous creatures received by the great American presidents (from lions, horses, moose, and panthers to racoons, prairie chickens, and grizzlies), asserting that "if all the animals that have been sent to the White House from first to last had been turned over to a zoological garden, the capital would have the finest collection of animals in the world."[42] Apparently, keeping animals came with the responsibilities inherited by the nation's chief executive. Surely, the animals of a national zoo could evoke, for its visitors, a patriotic pride for country, as well as a public admiration for the first family.

As Hornaday's department grew in popularity, its evolution into a zoological park appeared to be the next step in a logical progression. Before the department could fulfill this destiny, though, the Smithsonian would have to gain congressional support. Whereas other zoos simply required funding and public sanction, a national zoo built and administered by a federal institution required federal approval. Hornaday, Smithsonian leaders, and the public lobbied hard for a zoo, and they hoped that the transition from department to zoo would move seamlessly. Yet even seemingly uncontroversial issues, broadly supported by constituents, can become contentious in the hands of policy-makers. The next section will tell a story about the political lives that Hornaday's animals acquired within the House of Representatives between 1888 and 1891, as congressional delegates turned the zoo bill and later zoo appropriations into battlegrounds for sectional ideologies and differing ideas about federalism in a far-from-Reconstructed, Gilded Age America.

The Sectionalism of a National Zoo: Animals, Politics, Laughter

William Forney took to the floor of the House of Representatives on September 12, 1888, as a well-respected Democrat from Alabama. He stood before his peers in the Capitol chamber as a veteran politician. He also stood before them as a former brigadier general of the Confederate States of America, wounded at Dranesville, Williamsburg, Salem Church, and Gettysburg. Though he was captured twice, he was paroled and then exchanged. He returned to combat both times and eventually surrendered alongside Robert E. Lee at Appomattox. Over the course of his congressional career, Forney had been witness to controversial legislation centered on postwar nation-building projects like the Desert Land, Chinese Exclusion, Edmund-Tucker, Dawes, and Interstate Commerce Acts as well as the formation of the District of Alaska. On this day, though, he confronted legislation of a new kind. Bestowed upon the House by the Senate was an amendment that sought to establish a federally subsidized zoo.[43]

Forney did not rant against the proposition. Instead, he stated simply that the government had no business creating a zoo, especially when it was deeply in debt. Though the appropriation was for $200,000 (the equivalent of $5.43 million in 2022), Forney believed that the cost of a zoo would be unceasing: "Next year we will buy buffaloes, tigers, lions, monkeys, and other animals. After that, reptiles, snakes, and things of that kind." Laughter filled the chamber (which we know because "laughter" was regularly recorded in congressional transcripts). Forney exaggerated, yet he was not completely wrong. Maintaining a zoo was truly expensive, and many Southern Democrats found the "zoo bill" to be a laughing matter. Yet very seriously, the notion of a national zoo sat at the center of long-standing disagreements about the purpose of government.[44] The formal establishment of the National Zoological Park, as the only zoo in American history whose governance fell under the purview of the federal legislature, offers a unique window into a semiology that fused animals to debates that divided a nation through and after the American Civil War. Animals were used by congressmen in the House of Representatives to conjure sectional ideologies that could not be wielded as aggressively in an era of reconciliation.

On April 23, 1888, as the Department of Living Animals teemed with visitors, Senator James Beck, Democrat from Kentucky, proposed Senate Bill 2752 to establish a zoo in the District of Columbia. Science, education, recreation—and the passionate support of Justin Morrill (Republican from Vermont)—pushed the bill through the Senate. For zoo champions like Beck and Morrill, the bill seemed to be bipartisan. As historian Helen Horowitz has emphasized, "There was no discernable difference between the collection of living animals in a national zoological park and the stuffed dead ones in the National Museum. Goals were the same, the advancement of science; the authority was the same, the Smithsonian; thus the means of support should be the same, Federal sponsorship."[45] Through debate in the Senate, the zoo inherited additional purposes that resonated with the American public. It would, for example, protect American species by both preserving them in enclosures and spearheading breeding projects. And it would bolster the nationalist ambitions of the United States. What could be controversial about a national zoo?[46]

The opposition in the House answered simply that the government had no *right* to create an institution of this kind. The way in which Representatives made this claim is illuminating. Missouri Democrat Richard Bland pointed out, as many others did, that other cities established parks and zoos to beautify their landscapes. Why should Washington not have the same responsibility? He argued that it was not right to tax the People for an institution that would just benefit local real estate markets. To strengthen this position, Bland addressed the science mission that zoo boosters extolled, since this purpose (more than education and recreation) was a goal infused with *national* interest. He asserted sarcastically that it is not in the interest of science to bring monkeys, bears, and elephants to Washington. Then, he joked, "hundreds of the people of my State would prefer, if this circus is to be inaugurated, that it be put on wheels. Let us appoint the clown." Bland concluded, fittingly for a politician who backed silver miners, "I am opposed to the amendment and to all appropriations of public money that go simply to the benefit of the aristocratic classes."[47]

Others echoed these arguments, often with a sarcasm appropriate for a joke. Like Forney and Bland, Tennessee Democrat Benton McMillin (who, though prevented by his father from serving in the army, was still jailed by Union troops for refusing to take an oath of allegiance) believed that a zoo was purely a local matter. He considered it a crime for the national government to "sell the homes of the people in order to make them the habitations of monkeys"—a claim that led to laughter. McMillin further asseverated that the federal government had done plenty to beautify the capital, offering Judiciary Square, Lafayette Square, Franklin Square, and the Capitol and White House grounds as evidence. "If the people of Washington," he declared, "want a display of rabbits, snakes, and all kinds of animals, let them have it; but let them pay for it themselves." Congress had no right to force Americans who will never even visit Washington, DC, to "pay the fiddler for others' dancing." He, like Bland, invoked a proletarian critique, referencing both Cicero and Edmund Gibbon (like a good Southern statesman), as he asserted that "the display of wild beasts" had led to the "degeneracy" of the Roman Empire.[48]

Mississippi Democrat (and former Confederate lieutenant colonel) Thomas Stockdale was anti-zoo from the beginning. Not only did he, with Bland and McMillin, believe the zoo to be a scheme for the District's wealthy class, but he also declared the eminent domain that would be necessary to designate land for a zoo as blatantly unconstitutional. Only despots "condemn private property for the pleasure of citizens." Stockdale described the zoo as a stupendous folly and as a never-ending expense. He asked rhetorically, "Who will determine the price of a buffalo or a monkey or a rhinoceros? Why, the man who has it to sell; and these fellows that the Government sends out, how are they going to know the price of a rhinoceros or an elephant or a monkey?" Again, laughter filled the chamber. "Why, what does one know about the price of a rhinoceros?" Then modest laughter turned into a "great laughter." Amid the din, Marcus Smith, Democrat from the Arizona Territory, interjected, "Or sea lions." Stockdale, turning to his "friend," acknowledged the additional species tossed into the fray, "Yes, or of sea lions, or, as my friend from New Jersey said the other day, of coyotes?" He concluded his diatribe with a guilt trip: "Congress ought to be ashamed." Commotion consumed the House, and the Speaker's "hammer fell" to bring the menagerie to order before other Democrats started shouting out more unnamed species—"Or polar bears," "Or alligators," "Or, maybe, coatimundi!"[49]

The idea of a national zoo ignited an ire in those opposed, but not because politicians really cared about zoos or the animals themselves. The zoo as a symbol, though, reached into beliefs about government. This allowed it to be read, in part, through a sectional, North-South lens.

Usually, these sectional differences inscribed into the zoo lay dormant; most representatives did not wake up in the morning stewing about how zoo animals brought up old angers regarding federalism, slavery, or the politics of the Reconstruction period. Occasionally, though, sectional meanings projected *into* the idea of a national zoo would crack through the surface, for example, when James Blount, Georgia Democrat and former Confederate lieutenant colonel, stated that "the great agricultural sections of this country are agitated by straitened financial conditions. The cloud of despondency hangs over them." How could these poor Americans, Blount asked, support a zoo that was hundreds of miles away from them?[50]

Nothing explicit in this statement referenced the Civil War, and nothing explicit cast the regionalism (the *city-country* rhetoric) of it in terms of North-South sectionalism. The subtext—the Civil War—was, however, palpable to every statesman in the room. This burst onto the page when Massachusetts Republican (and former Union corporal) Elijah Morse, in response to Blount, gave it voice. "Prior to the war," he bellowed, "the doctrine of State rights prevailed. In 1861 with the election of Abraham Lincoln a change came. We began immediately to emphasize the fact that we were a nation by macadamizing the streets, by laying out parks, and by the erection of statuary . . . until this city has come to be the finest Capital city in the world." Morse placed the zoo strategically into this national postwar narrative, one that echoed the language of modernity that so frequently enveloped zoological parks. In his rehearsal of what would have been heard by southern statesmen as northern revisionism, especially through his invoking of Lincoln, Morse acknowledged what he saw as the real issues surrounding the zoo. Georgia Democrat (and former Confederate first lieutenant) Judson Clements responded to Morse's "reading" with humor, refusing to take the sectionalist bait: "I do not propose to be drawn by the gentleman from Massachusetts into the question as to whether this is a nation or a union of States, or both, nor will I now discuss the former tendency of Massachusetts toward secession."[51]

Missouri Democrat Alexander Dockery recoiled at Morse's rhetoric, though for different reasons. "This Government," Dockery reasoned, "has been maintained (passing meanwhile through a great civil war) without the aid of any zoological park." He continued that he knew "of no functions connected with a monkey, tiger, or lion which are in anywise essential to the public defense or necessary to discharge the functions of government." Again, the chamber resounded with laughter. Yet in this case, the laughter heightened (instead of punctuating) tension, prompting Dockery to launch a rant of his own concerning the question of whether the city of Washington itself should have financial responsibility for the zoo being bestowed upon it, since other cities like New York, Philadelphia, and Cincinnati capably supported their own zoos. After quoting from the Sermon on the Mount, he explained that though he was staunchly against the zoo, he was also (if there must be one) staunchly against forcing the capital city to pay for it. Since the District of Columbia was under taxation without representation, Dockery (who had been a Southern sympathizer in a border state) believed that "we may have right here under the very shadow of the Capitol an exemplification of the type of government which some gentlemen in this country are so anxious to impose upon the people of the South." He then spiraled into an anecdote about "colored troops," an attack on the Republican Party, and a derision of Morse's conjured Lincoln, who he referred to as "a Republican gentleman in the White House who was acting as President." Laughter consumed any further discussion, and the Speaker dropped the hammer.[52]

Sometimes anger over the National Zoo spiraled even further into sectional territory, directing attention to the discourses of animality upon which race was constructed.[53] On April 18, 1890, Democrat and former Confederate lieutenant colonel James McCreary, representing the border state of Kentucky, stated: "I am one of those not in favor of that Zoological Park. I do not think we came here to legislate for the purpose of furnishing homes for monkeys, zebras, and lions, but rather to legislate for *white* people and the people at large." Maryland Republican Louis McComas, whose support during the war from a different border state lay with the Union, interjected, "And for the colored people." Though McCreary respoke—"I said for the people at large. I say we came here to legislate for the people: white people, colored people, and all the people"—the gallery filled with hisses and an anonymous Representative shouted, interestingly, "And saddle-colored" (an adjective used to depict a "generic brown" color typified by leather riding gear). Then McCreary drove home his anti-zoo rhetoric with "But we did not come here to legislate for monkeys. Let it [the zoo] be strangled." The chamber again filled with hisses, prompting the Speaker to again bring the melee to order—"Unless demonstration ceases in the gallery, the Chair will order the gallery cleared." North Carolinian Henry Cheatham, a former slave and the only Black representative in the House, surely reacted to this language that bound race and animality to both each other and to a city-improving institution often viewed as Progressive. But at least on record, Cheatham never spoke a word about the zoo.[54]

It would be reductionist to claim that the Civil War determined purely a Representative supported or opposed the idea of a national zoo. It was never that simple. On February 28, 1889, the House voted in favor of the zoo, 131 to 98. Eighteen of the yea votes were indeed Southern votes, and one-third of these (six votes) were South Carolinian, since Democrats Samuel Dibble and John Hemphill broke ranks with their party and region to outspokenly support the zoo. As a public educator and a chairman of the Committee on the District of Columbia, respectively, they rallied their fellow South Carolinians to the zoo cause. These eighteen votes, though, were outliers when compared with the other forty Southern nay votes. More telling, only two of thirty-two Deep South votes (other than the South Carolina block) voted in support of the zoo. Of course, party also correlated to voting outcomes; only seventeen of 112 Republican votes were cast against the zoo. It was clear, though, that no matter the reason Representatives voted against the zoo, sectionalist rhetoric did undergird or overlay anti-zoo rhetoric.[55]

All of this explains why on June 1, 1890, the *Dallas Morning News* published a column-long parody about the "monkey park bill." In an imagined dialogue among representatives, the satire depicts a $10,000 "buffalo wallow," a "big hole filled with water." It describes the spending of "$50,000 for housing and feeding the monkeys," "$40,000 for a tank of seals and sea lions," and a whopping "$15,000 for an aviary for sea gulls." The story continues in hyperbolic fashion, imagining objections like "Sea lions, sea lions! My goodness, this is downright robbery!" It also depicts zoo supporter and Speaker of the House Tom Reed of Maine as speaking in a "broad yankee tone." After the "monkey bill passed," the caricature concludes with the House moving to the next item of business, a bill for constructing a fictional government building in the fictional town of Whapwallopy, Pennsylvania. Clearly, this sketch drew both its political incision and comedy from Southern and Democratic objections to the federal government. Nonetheless, the citizens of Dallas could not have been too opposed to zoos in general. Just a few months later, the same newspaper announced in excitement that Dallas's own zoo in Marsalis Park (which, established in 1888, was one of the first of the public-park animal displays to evolve into a zoo) had acquired new black bears.[56]

Laughter serves many purposes. A deep history of laughter would surely find its origins in primate social behavior, if not mammalian behavior generally, since studies have suggested that rats and dolphins laugh alongside chimpanzees, bonobos, orangutans, and humans. The latest evolutionary studies suggest that laughter gained utility (reproductive advantage) when primates began to live in cohesive groups of twenty or more individuals. So when laughter interrupted congressional proceedings, these outbursts were everything but noise. They were saturated with meaning, which is why the *Congressional Record* noted laughter in the first place. Laughter signaled unease. Laughter marked contention. Laughter gave voice to the unspeakable. Laughter served as borders between ideologies, as signposts that signaled "stop" or "don't go any further." As a cultural theorist might say, laughter demarcates, or dances around, the Real. And it certainly serves as windows into humans' semi-conscious, yet embodied, reactions to ideological discord. When the House debated the zoo, members laughed for two reasons. Representatives laughed when debate turned to the war, and they laughed when animals were named.[57]

What lessons can we glean from this zoo opposition? First, though they were instrumental to every aspect of American society throughout the nineteenth century, and though they were taken seriously within their prescribed roles, animals themselves were not considered subjects worthy of discussion. They may reside in a taxidermist's shed, they may dwell behind museum glass, but they surely did not belong in a legislature, in statecraft, or in the public imaginary. Animals did not deserve this type of subjectivity. This particular fact should come as no surprise, though, since the perception of animals as automata—lacking emotion, personality, volition, and spirit—had been established, intellectually, in the West since Descartes. It makes sense, then, that the act of *naming* the animals of the proposed zoo, that acknowledging the unacknowledgeable—"monkey," "seal," "rhinoceros,"—always incited laughter. Philosopher Tom Tyler, in *CIFERAE: A Bestiary in Five Fingers*, interrogates the rhetorical animals of Western philosophy and demonstrates that they are almost always used as ciphers, "chosen placeholders, unwittingly serving some higher pedagogic purpose."[58] Laughter was the vehicle through which "the animal" entered politics. Laughter continually erupted into the House. Not because anti-zoo Representatives were using animals rhetorically. And not because legislators (actually, in reality) found the named animals comical. Rather, laughter filled the House because the named animals represented objects-turned-subjects, background-turned-foreground, tools-turned-persons. The act of naming felt new, felt blasphemous even, so laughter, to Representatives, felt like the appropriate response. Furthermore, these animals were named unconsciously—below consciousness—to delegitimize their *named* purpose in a national zoo. Detractors, for example, named "monkeys," foregrounded them in their rhetoric, for the purpose of undercutting their foregroundedness in a zoo that would be located just a few miles from the White House. Representatives felt (even, if surely, they did not recognize) the dissonance, the gap, between the monkey signifiers, both of which ran away from the living monkeys themselves. Even more, they felt the gap between the animal sensibilities of old, established since the Enlightenment, and the new ones emerging within the ethos of nineteenth-century Romanticism and conservation. Their physical reactions to change, embedded in the historical shift from timeless menagerie to zoological park, were simply knotted into and expressed through the ejaculations of laughter.[59]

The second lesson is more historical, drawn from the conscious purposes of anti-zoo representatives. After the Civil War, Southern representatives enlisted zoo animals into their language for a clear reason: they symbolized the ever-expanding State. Not only did monkeys, seals, and rhinoceroses embody the "expansion of federal power" that historian Drew Gilpin Faust claims "characterized the transformed postwar nation," but the language of zoo animals also allowed representatives to disguise sectionalism when, in an age of reconciliation, straightforward North-South talk upon the House floor would not have been tolerated.[60] The nature of federalism was the quintessential American question of the nineteenth century, a question that led to civil war. Zoo animals gave representatives a voice to talk about what could not be spoken about—the resentments of a white nation divided by slavery. As historian David Blight has demonstrated in *Race and Reunion*, memories of the war had always "clashed or intermingled in public memory."[61] Whereas Blight emphasizes that much of postwar remembering was actually about forgetting, historian Caroline Janney, in *Remembering the Civil War*, has argued the opposite. In the decades after the war, remembering was indeed about remembering—"forgetting was what they all feared most."[62] The metonymic animals of a new national zoo, once paraded from Hornaday's department into Congress, allowed representatives to both remember *and* forget a national trauma that was still relevant to interpreting the purpose of government.

The final lesson is methodological. Concerning the House of Representatives and the controversy of the National Zoo, looking at traditional primary sources (direct records of governance) through the lens of animal studies can help locate new meanings surrounding zoo animals, as well as new languages assembled by legislating humans. For the representatives in the House, zoo animals were merely discursive, symbolic animals who had run away from their living counterparts. Opposition had nothing to do with *real* rhinoceroses, *real* sea lions, but had everything to do with how the zoo, and thus its animals, opened sectional wounds. If these congressmen were to look at *real*, living and breathing, zoo animals, they would not have seen symbols of the State, imposing its will onto the nation. Instead, they might have seen animals, as agents in the North American animal trade surely did, as objects that offered potential business to those living on the margins across American regions. As we have seen with the Department of Living Animals, many animals came from the American South or West. Cultural-environmental histories of the animal trades of the American South still need to be written. Nonetheless, the story of the National Zoo will become a story of copperhead snakes from Hillsboro, Virginia; pocket gophers from Stovall, Texas; bobcats from St. Augustine; Sand Hill cranes from Orlando; chipmunks from Sewanee, Tennessee; and pet kinkajous from New Orleans.[63] If House representatives were looking at *real* zoo animals, they would have acknowledged that not only was a zoological park underway in Dallas, but there was also one underway in Atlanta, formed when a bankrupt circus donated its animals to the city. By the first decade of the twentieth century, southern cities—like Chattanooga, Augusta, Norfolk, and San Antonio—will be swept up in the American zoo movement. While the National Zoo could be viewed in terms of *national* reconciliation, the broader zoo movement transcended regions and nation. In 1901, at least fifty-six living zoological collections existed in the United States,[64] and formal and established zoos could be found in places as distant as South Perth, Sofia, Rio de Janeiro, Havana, and Istanbul.[65] Nonetheless, within the House of Representatives, zoo animals were not animals, and a zoo was not a zoo. They both composed a language through which the unacknowledgeable could be given a voice, and a lingering sectional trauma could be embodied and then enacted for the purpose of expressing beliefs about the duty of the State.

The Zoo's Popular Mandate: Practical Ideologies and Popular Zoology

Historian Andy Flack in *The Wild Within: Histories of a Landmark British Zoo* depicts zoos as institutions of "the loved, the loathed, and—increasingly—the (nearly) lost."[66] Whereas in the House of Representatives, zoo animals led delegates to debates about government, in the Senate, zoo animals led to the fear of permanent loss. The call to prevent extinction itself was nonpartisan. The Democratic *Charleston News*, for example, exclaimed:

> Several of the leading Northern newspapers are advocating [for] a National Zoological Garden at Washington. . . . The scheme is well worthy of consideration. . . . The buffalo has disappeared from the plains. The megatherium [a Pleistocene ground sloth] is with us no longer and . . . many of the rarest specimens of the animal kingdom have been lost for the reason that no well-organized attempt has ever been made to preserve them.

Not only did the *Charleston News* pledge allegiance to the cause of preventing extermination, but so did other Southern, staunchly Democratic papers, including the *Petersburg Index-Appeal* and the *St. Louis Globe-Democrat.* As the Democratic *Boston Globe* put it, "there should be no partisan difference of opinion" on rescuing moose or manatees from extinction. Without intervention, as the Independent *San Francisco Call* exclaimed, they will disappear "as completely as the pterodactyl."[67]

Newspapers that endorsed the zoo bill unanimously extolled the nostalgic cause of saving animals, but they cited other causes as well.[68] Second to the prevention of extinction, journalists praised the zoo as an opportunity to provide, in the words of the *New York Tribune*, "instruction and pleasure to a great number of visitors to the capital," a sentiment that pervaded the zoo movement that stretched from Melbourne and Moscow to Tokyo and Philadelphia.[69] This defense of zoos echoed the boosterist rhetoric for public parks, botanical gardens, aquariums, outdoor promenades, and museums. Within these institutions, Nature was fashioned, ironically, "by modernity into a consumable remedy for modernity's own excesses."[70] The zoo served as an escape. Toward this end, zoos offered the urbanite a "middle landscape" in which to get away from the industrializing quagmires of the city.[71] Yet this landscape, with living animals, felt more alive, closer to Nature, than the halls of a taxidermy museum.

Third, boosters also described the zoo as a new staging ground for American science. The *Journal of Comparative Medicine and Surgery* especially described the national zoo as "greatly beneficial to education in popular zoology."[72] A national zoo would contribute to science in multiple ways. It would, first, serve as a governmental publishing house for science. According to the *Washington Star*, "the published results of scientific work done in the London Zoo made over 75 volumes, so valuable that no zoologist can afford to be without them."[73] A national zoo would hopefully produce a similarly impressive literature. Zoos, after all, were "apparatuses of inspectability," and a national zoo would enable scientists to inspect exotic animals without having to travel to faraway savannahs, jungles, and deserts.[74] In this sense, the zoo made the world smaller and more visible while simultaneously making it (seemingly) more diverse. Furthermore, a national zoo would function as a "storehouse" where scientists could accumulate living specimens. The dusty corridors of university biology departments and the crowded hallways of museums surely had little room for extra elephants, bison, and giraffes—either dead or alive. Last and most important, a national zoo would ignite an American popular zoology, a popular science that would reach far beyond the zoo. According to one editorial, the zoological gardens in San Francisco and Chicago possessed "few animals." The zoos in St. Louis and Cincinnati needed more funds. The zoo in Central Park was "cramped in space." And the nation's first zoo in Philadelphia was struggling with finances. Therefore, a "national zoo" could truly usher "popular zoology . . . on this side of the Atlantic" out of its state of "infancy."[75] The National Zoo, as we will see, was certainly successful in accomplishing this mission.

Fourth, journalists from coast to coast lauded a national zoo as a symbol of American pride. The *Boston Globe* described the United States as "the only great nation in the world which does not possess such an institution." Similarly, according to the *Minneapolis Journal*, "All the great nations of Europe have their 'Zoos,' but so far the United States has made no move in that direction." And the *American Naturalist* lamented that the "United States, with greater territory, greater riches, and with undoubtedly greater opportunities, is behind third and fourth-rate kingdoms in the matter of zoological gardens." Even the "effete monarchies" of Europe, in the gendered words of the *Pittsburgh Dispatch,* were "superbly equipped" with zoos.[76] At the frustration of Southern Democrats in the House, the press seemed to agree with all of the official justifications for a national zoo outlined in Beck's bill to the Senate—preventing extinction, enriching the lives of citizens through education and entertainment, spreading nationalism, and improving science, especially popularizing zoology.

Newspapers, however, imagined many additional purposes for a national zoo not spelled out in Beck's bill. The *Charleston News* predicted that a national zoo would become "the most interesting museum of natural history in the world," one with living displays. The *Petersburg Index-Appeal* suggested that maybe the zoo could also accommodate a large botanical garden, another attraction required in a capital city. The *Washington Star* believed a national zoo could practice "the free distribution of its duplicate specimens to educational institutions," fulfilling a philanthropic and pedagogical function by donating living animals to American schools. The *Washington Post* saw the zoo as an opportunity to protect the pristine Rock Creek from local "real estate vandals" who wanted to build new housing subdivisions for the city's poor.[77] And the *Public Opinion* portrayed the zoo as an asset for Washington's tourist industry, a "leading attraction" that would draw both American citizens and foreign visitors to the capital. This was especially enticing since the 1892 World's Fair would be held in Chicago, encouraging, in the words of the Smithsonian secretary, "hundreds of thousands of intelligent people from abroad" to travel to Washington en route to Illinois.[78] To most Americans, Beck's zoo bill seemed to be uncontroversial. The People wanted a zoo, and on March 2, 1889, after congressional setbacks, the first zoo bill passed.[79] A zoo was officially underway.

Despite outliers that considered the legislation a breach of power, a majority of congressmen viewed the zoo favorably. Nonetheless, the following summer, issues of funding reignited a debate that slowed the process of zoo-building, embroiling District residents, federal legislators, and the disgruntled anti-zoo bloc into another contentious quarrel over finances when the Organic Act of 1878, which set the precedent for financing public works (roads, sidewalks, etc.) in the District of Columbia through equal proportions of federal dollars and District taxes, was raised in regard to the zoo.[80] Those who believed that the zoo should be the full responsibility of the federal government (the District commissioners especially vocal on the issue) argued that "it was not the intention of Congress to place any portion of this burden upon the taxpayers of the District," and they assembled a convincing defense through close readings of the bill.[81] Simply put, in the words of the *Washington Post*, "It would be quite as fair to charge the people of the District with half the expenditures authorized by Congress for the Botanical Garden, the Army and Medical Museum, the Congressional Library, or the new Capitol terraces as with any portion of the cost of the 'National Zoo.'"[82] The controversy, according to this argument, was spawned by political theater, as the anti-zoo bloc sought to trip up the zoo on technicalities and legal-hermeneutical gymnastics. Several subcommittees had worked on many drafts of the zoo bill (which never explicitly mentioned District taxes) for quite some time, and as the session of Congress was coming to a close, the bill was conveniently attached to the District appropriation bill (governed by the Organic Act) in order to pass the legislation so that the next session of Congress would not have to muddle through old issues. Congress expressed no desire in making half of the $200,000 zoo bill the responsibility of the District, a significant expense for a city smaller than Cincinnati. Its placement in the District appropriation bill, under the "water department" heading specifically, was only for expediency's sake. Clearly, neither District taxes nor water department funds would pay for a national zoo! Or, so went the sober consensus of the Senate.[83]

Comptroller of the Currency William Trenholm, who was a Confederate Army veteran from South Carolina and the first Democrat and Southerner appointed to this position, took advantage of the loophole. In August 1889, Trenholm ruled that the District of Columbia would have "to pay one-half the expenses incurred in the establishment of the Zoological Park," and once this issue was ordained, the precedent was set for future expenditures.[84] Though no evidence indicates that any senators or representatives intended on this understanding, only formulated ex post facto of the passing of the bill, the penurious House members acted *as if* this decision was the original intent, predetermining their stances to future zoo appropriations, always to the dismay of the Senate and the zoo. When the Senate, in 1890, drafted and approved a bill that would appropriate $92,000 (which was, in fact, far too little) of federal monies for buildings, cages, fences, infrastructural expenses, and salaries, the House approved the bill only after adding an amendment clearly stating

Figure 1.5. A political cartoon depicting the unfairness of taxing DC citizens for the National Zoo from the *Sunday Herald*, May 4, 1890. Courtesy of the Smithsonian Institution Archives.

that $46,000 would be financed by District taxes.[85] Sectionalism, in this way, affected the early history and structure of the zoo. Its influence even trickled down into the daily policies of zoo management; for example, in 1891 zoo leaders chose to receive animal donations in higher volume, and at an accelerated pace, for the sole purpose of placing pressure on the House to pass an appropriations bill. Neglect would be bad politics, even for anti-zoo Southern Democrats.[86] All voters, regardless of party, loved animals, or at least the idea of animals.

Throughout the zoo's first years, sectionalism was most often experienced (after congressional politics was translated into financial structure) as lack. After the zoo was established, it lacked the means to purchase animals, having to rely entirely on donations instead, especially from institutions like Yellowstone, as historian Diane Smith has shown.[87] The frustration that the lack of purchasing power produced for zoo employees saturated zoo correspondence, as the zoo had to repeatedly reject the offers for animal sales, which came in torrents after the official opening. *Sadly, the zoo can't purchase your bears, but you can donate them. Sadly, the zoo can't purchase your manatee, but you can donate him. Sadly, the zoo can't purchase your rattlesnakes, but you can donate them.* This response was given again, and again, throughout the first months of the zoo's existence—for mocking birds, for monkeys, for armadillos, for quail doves, for flamingoes. Interwoven throughout these letters, increasing in frequency as time passed, were telling admissions like "This embarrasses the Park very much." The paucity of funds not only prevented the zoo from buying animals, but it hindered the institution from buying almost anything. Even rubber boots for the employees proved to be too expensive.[88]

Purchasing animals. Paying salaries. Rubber boots. The everyday concerns within the zoo were mundane. Similarly, many of the lofty ideals projected upon the National Zoo were seen, by zoo proponents, as straightforwardly necessary, as *practical.* Rarely are ideals labeled "practical." And seldom are "practical" ends seen as ideals. Yet "practical" in the late nineteenth century exuded a democratic spirit. To make an intellectual pursuit "practical" was to make it useful to the masses. Swept up in Benthamian, utilitarian, cultural currents, zoo leaders and naturalists sought to maximize the utility of zoological pedagogy by widening its scope—increasing the number of zoogoers reached and the number of species examined by zoological education. Making zoology "practical" suggests the democratization of a life science, replacing a focus on the "advancement of knowledge" (an Enlightenment teleology) with a focus on the "dissemination of knowledge" (a Progressive teleology). Hornaday's experience establishing a national zoo shaped directly his own developing commitment to "practical zoology," culminating in his 1910 publication of "The Right Way to Teach Zoölogy" in the widely read *Outlook* magazine.[89] Throughout its genesis, the National Zoo and its discourses were tethered to practicality, both in its promotion by Hornaday and the Smithsonian and in its ideological life within the halls of Congress.

Like the museum, the National Zoo was built for the public.[90] As historian Steven Conn has demonstrated, "museums functioned as the most widely accessible public fora to underscore a positivist, progressive and hierarchical view of the world, and they gave that view material form and scientific legitimacy." Through an "object-based epistemology" (the assumption that objects "could tell stories"), the museum could transport the museumgoer into wider worlds.[91] Similarly, according to its boosters, the zoo, with its living "objects," embodied the same logic—which makes sense for an institution literally carved out of a museum.[92] Supporters of the National Zoo (like supporters of new libraries, theaters, universities, arcades, and hospitals) believed that such an institution would deepen American culture and democracy.[93]

Though cloaked in practicality, the discourses concerning the foundation of the zoo rarely referenced the specific everyday needs required by this new institution. Rarely did legislators discuss real, breathing animals, mingling crowds, transportation, barbed wire, moats, animal feed, or concessions. Congressmen chirped about issues of funding, and thus the role of the federal government in establishing institutions for the "benefit of the people." National deliberation imagined a national zoo as one institution among many, a by-product of the great age of institution-building. However, the *idea* of the National Zoo as an institution rarely altered the lived experiences of the zoogoers, zookeepers, laborers, scientists, tour guides, businessmen, and, most importantly, the animals that composed this institution's population. The rhetoric surrounding the establishment of the zoo may have given zoogoers prefigured notions about the zoo as a monolithic entity, an institution, that merged easily with other national dialogues. Though once the zoo was established and zoogoers entered through its gates, it far exceeded ideology. The rhetoric, arrogance, ignorance, and destinies forged by zoogoers, within and through Culture, could never fully make sense of their actual zoo experiences. Experiences always test the integrity of their cultural containers. And occasionally historically contingent, discursive-ontological cultural capsules rupture under the pressures of experiences within them, especially wild ones.[94]

Changing the Land for a Zoo

At four o'clock in the afternoon, on March 12, 1889, John Willock Noble (secretary of the interior), Samuel Langley (secretary of the Smithsonian), and William Benning Webb (president of the board of commissioners of the District of Columbia), comprising the "commission for the establishment of a zoological park," traveled together in an open carriage, accompanied by John Wesley Powell, the superintendent of the US

Geological Survey, to the beautiful Rock Creek valley. These gentlemen, on this "exceedingly fine" day, had the task of choosing the one hundred acres best suited for a zoo. The picturesque expanse of Rock Creek made the decision a difficult one, and since Noble had never toured the area and his hectic schedule prevented him from seeing the entire park, the foursome decided it was best to take several excursions.[95]

The congressional bill prescribed the boundaries within which the zoo could be located—Massachusetts Avenue, the southernmost border, and Military Road, running west from Brightwood, the northernmost. The one-hundred-acre zoological park would be placed between these two avenues and must include a portion of the creek. The placement of the zoo was no straightforward matter. Langley hoped the commission would find a location as close to the city as possible. "The location," Langley argued, "should be such that the park could be easily reached by horse cars or afoot." The problem was that the properties nearest the city were expensive and would be difficult for the commissioners to acquire.[96]

The commission began surveying Rock Creek in mid-March 1889, and they spent the next two months deliberating with private property owners about issues of pricing. While some owners reached an agreement with the commission quickly, assuming they would earn a better price for their land through negotiation than by waiting for condemnation hearings to proceed, other owners either asked for outrageous amounts or simply refused to respond to the commissioners at all. On May 17, 1889, the commissioners met with appraisers to consider the value of every piece of property, except for ones deemed too expensive that lay close to Massachusetts Avenue. By the end of May, the issue was settled: the zoo would lie above Woodley Park, along Rock Creek. Since the land above Klingle Road was quite expensive, Hornaday himself traveled through the area and met with property owners, and he discovered that most were "in the most liberal spirit" and fully enthusiastic about the National Zoo.[97]

As Hornaday and the commissioners worked out the details for purchasing the land, surveyors were busy creating an official map.[98] On the morning of November 22, official plans were given to President Grover Cleveland. On average, the land cost $1,000 per acre, and the total expected purchase amounted to $177,128.76, leaving $22,871.24 of the original $200,000 allocated by Congress left for infrastructural expenditures. The land would be purchased from several owners, but Miss A. E. J.

Evans owned the majority of the property, ninety-four of 166 acres. Most of these landowners were, in the words of a Geo. Starkweather, "animated by a spirit of patriotic fair-mindedness," negotiating fair prices for their land and showing that Southern Democrat concerns about eminent domain were mostly misplaced.[99] Only thirty-three acres of privately owned land was acquired by condemnation, most of which was owned by the aging Pacificus Ord, who previously served as a member of California's Supreme Court. Also, the Colored Cemetery was seized, an unsurprising condemnation since "race" and "species," as we saw in the House, had long been mutually exploited. When all the purchases were complete, the landowners who chose to wait for the condemnation process (which dragged on for an entire year) instead of negotiating did receive more for their property. The final expenditure on land amounted to $190,824.59, and after legal fees of $8,789.57 were added to this amount, only $385.84 of the original $200,000 remained.[100]

The Rock Creek valley possessed many natural advantages that made it an ideal location for a zoo. First, as the *Evening Star* advertised, the valley boasted a stream of pure running water, whereas most zoos' water was "either collected in pools that became stagnant or a running supply was provided for at much expense." Rock Creek would flow throughout the entire park and could be widened artificially as needed for the zoo homes of aquatic animals like beavers, otters, and muskrats.[101] Second, a zoo located in the valley would become the largest in the world, especially compared to the European zoos cramped onto small tracts of land in the center of overpopulated cities. While zoos compacted into urban spaces allowed zoogoers to easily view an entire collection in a short amount of time, more acreage would allow for larger enclosures that could more closely model natural environments. For example, a zoo in the Rock Creek valley could accommodate herds of bison within large fields instead of forcing several individuals into tiny pastures. This extra space could also support breeding projects that would not only protect against extinction but would also provide the zoo with a steady supply of animal capital to ship to other zoos in exchange for different species. In this way, land would give the National Zoo a competitive edge in the international animal trade. Third, the Rock Creek valley would provide the zoo with a landscape that could be read as Nature, relieving the zoo from having to build that curation entirely from scratch. Hornaday believed that the valley offered more natural resources than any zoo ever established. And he was probably right. When interviewed by an *Evening Star* reporter, Hornaday proclaimed, "We will without any doubt have the best drained, best watered zoological park in the world. We will have more . . . deciduous and evergreen trees than any other park of this character. A million dollars expended on the Potomac flats could not do for a park there, what nature has done along the banks of Rock Creek."[102] The park site possessed rolling hills and dales; lakes and ponds; forests of oak, beech, locust, mulberry, hickory, persimmon, dogwood, and pine; deep ravines with jagged cliffs and high bluffs; a stream with banks both smooth and rocky; and granite and limestone boulders covered with mosses, lichens, and ferns—the perfect landscape for the assortment that a zoo required.[103]

The beauty of the Rock Creek valley so captivated America's congressmen that attempts to preserve it as an urban park began twenty years earlier. Following the Civil War, in 1867, Republican Senator B. Gratz Brown of Missouri presented Senate Bill 549, which sought to establish a public park in the valley. He pronounced before his peers in the Senate:

> I suppose all members of the Senate are familiar enough with the . . . beautiful and romantic valleys of Rock Creek. . . . It has rugged hills; it has picturesque scenery; it has a native undergrowth blushing with beauty; it has the tangled vine and clustering wild flower, and the quiet mosses gray with age, and indeed a thousand imprints of native adornment that no hand of art could ever equal in its most imitative mood.

Along with this Romanticism, Brown recited a report written by Brigadier General Nathaniel Michler, who similarly attested to the wonder of Rock Creek. A park of this sprawling size would, without doubt, rival some of the great parks of Europe—Dublin's Phoenix Park, Paris's Garden of Versailles, London's Hyde Park, Vienna's Prater, and Munich's Englischer Garten. Competing with more pressing Reconstruction concerns, this national park bill never passed, even when it was discussed again in 1886 and 1887. The zoo bill, then, was attractive not only because of the many benefits the zoo would offer but also because its location lay within this sublime valley of the District's public imaginary, which is why Hornaday sought and briefly hired the titan park architect Frederick Law Olmsted to fashion the first building and landscape plans.[104] Shortly after

Figure 1.6. Zoo boosters believed that the forests and rolling valleys of Rock Creek would make the National Zoo the most picturesque in the world. Courtesy of the Smithsonian Institution Archives.

zoo construction was underway, Congress finally protected the rest of the Rock Creek valley as a public park. Washington's Rock Creek Park would join New York's Central Park, Philadelphia's Fairmount Park, Boston's Franklin Park, San Francisco's Golden Gate Park, and Baltimore's Druid Hill Park as one of the finest parklands in urban America.[105]

While Nature could create the Rock Creek's wonders, only humans could transform them into a zoo. The location of the valley on the outskirts of the city required Washington to extend its urban infrastructure to the zoo rather than simply accommodate the zoo within the city. The National Zoo required, in the most literal terms, both the growth of the human city and the creation of an animal city. Of the $92,000 passed, in 1890, to support construction, $15,000 was earmarked for building animal shelters. One building, costing $10,000, would be a "half-permanent" house that could accommodate subtropical animals like lions, tigers, ostriches, kangaroos, and camels. The bill also funded a $5,000 aviary that would oddly and temporarily keep primates and reptiles until each could eventually be relocated into their own building. Even though representatives worried about these expenses, the amounts budgeted were quite low, worlds less than the buildings of the Cincinnati and Philadelphia Zoo.[106]

The bill designated $9,000 for the "custody of the animals"—a barn for the bison and elk ($1,500), fences and shelters for eight ruminant yards ($3,000), four cages and dens for the bears ($3,000), and fifteen iron cages for animals like wolves and foxes ($1,500). Two thousand dollars was earmarked for repairing the Holt mansion, equipped with offices and rooms for guests. And another $2,000 would establish artificial ponds for aquatic quadrupeds without access to the creek.[107]

The zoo required a water infrastructure for animals and humans alike. The 1890 bill provided $7,000 for water supply, sewerage, and drainage. Laying pipes, installing a supply tank, and placing a pump on the creek accounted for $4,500. The rest was devoted to the laying of sewer pipes and drains around the two buildings. These initial efforts to establish water infrastructure provided only the bare necessities; maintaining and expanding the water system would inevitably require more funds in the future. Zoogoers also needed a groundwork to stroll upon—roads, walkways, paths, and bridges—for which the bill apportioned $15,000 (an expense equal to the cost of animal residences). On opening day, visitors would gallivant upon 4,880 linear feet of macadamized footpaths and 5,000 feet of boardwalks.[108]

The transition from a Department of Living Animals to a zoological park, as historian of science David Edgerton has shown regarding other institutions of science and technology, did not represent the invention of a new type of institution as much as it marked the metamorphosis of an old type of institution.[109] The zoo represented an extension of both Hornaday's taxidermic practices and the Smithsonian's collecting and cataloguing practices. No longer could Hornaday's laboratory hold the materials, animals, and people needed to sustain the zoo that arose as a byproduct of museum displays. And no longer could his laboratory keep up with its own intellectual and scholarly goals. The National Zoo gave taxidermy room to expand, making its commitments accessible to the masses of an age that craved specimens that were even more real and more wild. The greatest cost of the 1890 zoo bill was not dedicated to the "building of the new" (paths, exhibits, buildings, water systems, etc.) but funded the expenses of old. In fact, $37,000 (40% of the bill's allocation) was dedicated to "current expenses" and the "maintenance of collections, food supplies, salaries of all necessary employees and the acquisition and transportation of specimens." In many respects, these everyday costs of Hornaday's department (as much as the lofty goals and rhapsody of zoo rhetoric) justified, in a very practical manner, the establishment of the National Zoo, a *new* branch of the Smithsonian that could manage *old* expenses and, subsequently, older ambitions, older ideals, and even older histories of display that stretch back to the genesis of the Smithsonian, if not back to Renaissance cultures forged upon cabinets of curiosity.[110]

Also, just as any human metropolis, the emerging animal metropolis required a working class. The 1890 bill approved the salaries for one superintendent, two keepers, two assistant keepers, one clerk, one gardener, six laborers, one copyist, one messenger, two painters, two carpenters, four watchmen, and one teamster.[111] Yet much of the work, of course, was performed by an unsalaried "gang of laborers" who dug fence post holes at an exhausting pace of fifteen to twenty holes per person, per day.[112] "Common laborers" performed everyday yet essential tasks, only rarely rising through the ranks to better compensated positions.[113] The zoo could not function without fences, roads, walkways, pipe networks, electricity, water pumps, and storage facilities; and these laborers deserve credit. Fence-building, in particular, was a monumental task—even worthy of its own newspaper article, reminding the public that "now everybody who rides or walks out in that vicinity will have a reminder of the existence of the park in the shape of a board fence. . . . This fence will be seven-feet high and in order to deter the small boy from trying to climb over . . . there will be two rows of barbed wire."[114] The fence that enclosed the zoo not only demarcated (and tamed) the land of the Rock Creek but also memorialized the landscape, transforming the valley into a national monument. Though much labor was devoted to placing shrubbery and planting Japanese honeysuckle, creeping junipers, birch, poplars, and thousands of willows to prevent floods, the barbed wire twisting atop the zoo's fences seemed to announce that the federal government would always keep a watchful eye and guiding hand on the park's future.[115] Yet congressmen, taxidermists, Smithsonian officials, and construction workers alike carved the zoo from a forested Rock Creek valley for purposes of both practicality and nationalism, popular zoology and science, and in so doing, they installed another Washington monument in the capital city, one surely more sprawling than the 555-foot obelisk erected on the National Mall six years before. No exact records exist regarding attendance in the early years after the National Zoo's grand opening in 1891. Nonetheless, it is known that in 1892, an average of seven thousand visitors entered the park each day.[116]

From taxidermic beginnings within the National Museum of the Smithsonian Institution, the Department of Living Animals expanded quickly as Hornaday amassed a diversity of North American lives around the Castle, drawing the excitement of legislators, bureaucrats, and children alike. Once in motion, a zoo emerged even before national infrastructure could formally establish a National Zoological Park. Zoos were always complex institutions—knotted and layered—and their infrastructures needed to be built and then networked with the surrounding city. This always took time. Cities often struggled to keep up with the needs of their zoos. They still do. As the only zoological collection in American history that required the support and funding of Congress, the National Zoo became a topic of *national* debate. These circumstances gave zoo animals extra symbolic valences, as some Southern Democrats ushered rhinoceroses and sea lions into a Civil War still boiling within the public sphere. While zoo animals were given surprising ideological lives, most Americans supported a new zoo in the nation's capital, just as they began to clamor for their own local zoological gardens. Once teams of laborers prepared the Rock Creek valley for a zoo, Washington officially gained a new tourist fixture. Today, millions of zoogoers still stroll along the Olmsted Walk that meanders over the original promenade.

CHAPTER TWO

Runaway Animals

Bear Hunt in the Capital

At daybreak on Tuesday, May 24, 1892, William Gross, a boy who lived near the zoo, was walking down Pierce's Mill Road when he stumbled upon a grizzly bear climbing a hill alongside the avenue. The boy ran. And the bear pursued.[1]

Luckily, William got away unscathed. Prior to this encounter, according to other reports, the bear had been spotted wandering down Quarry Road, over the Walbridge property, along Klingle Road, and through Harvey Page's lettuce patch. After foraging in the field, the grizzly approached Page's stable and encountered children who "ran screaming in every direction." It was this commotion that had apparently scared the bear back over Klingle Road and into Gross's walking route.[2]

The grizzly had escaped from the zoo shortly after four o'clock in the morning, when the night watchman last saw the ursine.[3] The bear pits, dug out of the soft-rocked cliffs between Ontario Avenue and Quarry Road, were supposedly "quite homelike" for bears.[4] (After all, though grizzly denning ecologies vary greatly, cave dens in North America can reach, in specific topographies, up to 79% of all den types.[5]) The zoo's rocky pits were also supposedly secure. Yet, somehow, the bear escaped.

The runaway was around three years old, "small for its age" but known for a "bad disposition." Captured near Yellowstone's Lake Hotel the previous September, the bear was extracted from Wyoming, relocated to a cage in the national park, and then shipped promptly to the National Zoo for the pleasure of Americans who could not afford the expensive animal tourism scene emerging in the American West. Now, as the bear wandered at-large in the nation's capital, alarm spread, causing "a feeling of uneasiness, especially to [parents] who had to send their children to school." Neighboring farmers immediately checked on the safety of their livestock, and at every household "the doors and windows were kept closed, for fear that the animal might appear at the wrong time and play havoc with the people."[6]

Yet when an *Evening Star* reporter reached Pierce's Mill Road, he found "no such fear" among the men tracking the bear. The zoo's first head keeper, William Blackburne (who was hired in January 1891, from the Barnum & Bailey Circus),[7] saw no need to panic, for he believed that since the bear was recently fed, he would not hunt humans for the purpose of "securing a square meal." In fact, his diet consisted mostly of roots and grasses anyway. Blackburne believed that the bear was relatively harmless and would attack only if "it was cornered or an attempt was made to recapture it."[8] The troubling fact, though, was that such an attempt was unavoidable. A grizzly bear could not be allowed to roam at liberty through Washington, DC.

Upon discovery of the escape, and despite his apparent calm, Blackburne sent immediately for a pack of search hounds to sniff out the bruin. However, for unknown reasons, the keeper could not obtain the hounds. Soon thereafter, around eleven o'clock in the morning, a party of local citizens formed, spearheaded by two unrelated men with the last name Cramer, one apparently an employee of the Rock Creek Park and the other an employee of the zoo. Citizens Coyle and Houghton were among the first to offer their support. Farmer Routt, wielding a breech-loader, also offered his. Some were mounted on horseback. And the *Star* journalist chose to observe the chase on a bicycle. Once this People's party had departed, zoo superintendent Frank Baker (who we will formally meet in the next chapter) arrived at the scene and formed his own search party, calling "almost every employee" to serve their zoo. Armed "with pick handles, pitchforks, stones," and pistols (like a Gaston-led search party in *Beauty and the Beast*), the zoo mob took off after their bear. They tracked the animal to Pierce's mill but doubled back when they lost the trail.[9]

The number of people involved in the hunt increased steadily over the course of the morning, totaling between twenty or thirty individuals.[10] Eventually, after scouring the area around the zoo, the posse located the bear. The zoo's carpenter Louis Hess and a citizen with the fitting last name Hunter surrounded the grizzly and "for the first time the infuriated

Figure 2.1. Superintendent Frank Baker headed the National Zoo from 1890 to 1916. He will be an important person to many of the stories throughout the rest of this book. Picture taken circa 1900. Courtesy of the Smithsonian Institution Archives.

animal showed fight." The bear attacked Hess, sinking teeth into his leg until Hunter managed to beat him away with a club. The bear, then, ran for the woods, and en route attacked other members of the search party. Shots reverberated through the valley. Head keeper Blackburne emptied two loads of buckshot into the bear, and Routt contributed a bullet. The grizzly collapsed and died.[11]

At first, in response to the incident, Frank Baker blamed "the insufficiency of the watch force, as their funds enabled them to employ only one watchman for night duty." Blackburne simply stated, "Animals cannot be properly cared for unless the proper quarters are provided."[12] The immediate response of both Baker and Blackburne was to frame the entire incident as a by-product of insufficient funding. The next day, the *Evening Star* provided a more detailed account of the escape:

> This morning he [Baker] hoped people understood that killing the bear was an absolute necessity in order to prevent the maiming and perhaps killing of some person. The bear, he said, had never been tamed, and but few visitors to the park had ever seen him. During the day he always remained in the cave, and at night he did his prowling. . . . A watchman was put to watch him every night and his attempts to escape were always ineffectual. After being in there seven months it was thought that escape was impossible. An examination of the cave showed that he got away over the rocky cliff, where there is about fifty-five feet of almost perpendicular rock.[13]

Baker now admitted that the bear's escape was not entirely due to a dearth of night watchmen. He agreed with Blackburne's assessment—that the bear's quarters were to blame. However, rather than vaguely suggesting that the zoo lacked the funding to build "proper quarters," Baker instead seemed to acknowledge that a flaw in the cage, an oversight on the part of the zoo, made escape possible. The cage did not properly seal off the space between the edge of the ceiling and the rock face because its designers apparently had not realized that a grizzly could scale straight up sheer cliff. (Bears of all species are incredible climbers. Even polar bears have been recently observed cliff-climbing for bird eggs to supplement their diet, now that seals are vanishing with sea ice.)[14] No formal apology or admittance of blame was ever delivered. Technological shortcomings and the mysteries of wild animals served as rhetorical exculpation. Baker's greatest worry, it seems, was that the killing of the bear may, in fact, bother the public more than his escape, so he took extra measures to emphasize that the death was regrettably unavoidable.

The following day, the lifeless grizzly was carted from the zoo to the Smithsonian, where taxidermist William Palmer examined his body. The autopsy revealed that the bear had either died from a load of shot that mangled his left eye, where some grappling penetrated into the brain, or from the bullet embedded in his side. A revolver bullet was also discovered in the bear's paw, an early attempt to mildly injure the bear so that he could be captured alive. Palmer then proceeded to cut "the porterhouse, tenderloins and round steaks" from the carcass and give them "to all applicants" hungry for bear meat.[15] Some of this flesh was even sent to the Metropolitan Club and served in the club's dining room.[16] Palmer continued to assess the body for mounting, determining "that the skin was in excellent condition, but the hair was considerably worn on the rump, the result of the animal's sitting on the rocks" in the zoo. The mounting would, of course, take several months, but eventually the bear would be preserved and displayed either in the zoo or in the National Museum.[17]

Every detail of the escape and the ensuing drama fascinated and worried the District public. On May 26, two days after the bear was killed, the *Washington Post* reported that the bear hunt "had the effect of attracting an unusual number of visitors" to the zoo, as they crowded to see the grizzly, polar, cinnamon, and black bears. The article especially conveyed the "condition of uneasiness" that weighed upon local neighborhoods. The runaway bear sparked fear, but Baker exacerbated the disquietude of residents when he admitted that "accidents are liable to occur owing to a shortage of funds" and that he was "not prepared to say what he would do should one of the big elephants or the fierce African lion get loose."[18] The media had long molded public opinion about the zoo. With the runaway bear, Washington newspapers, with the lead of Baker's public relations, tried to push anger toward Congress for failing to pass new appropriations. Unlike the grizzly, the zoo dodged a bullet. The zoo was not a danger to Washington; Congress was. The keepers did not fail to do their jobs; legislators did. And, for the most part, this spin seemed to work.

The grizzly actually became a rallying symbol for zoo boosters. One satirical dialogue, printed in the *Evening Star*, put the issue succinctly:

> "Why was it that you fled, oh Bear?
> Why did you leave the Zoo?"
> "Alas, with this economy,
> It was the only thing," quoth he,
> "That I could safely do."
> "Why were you forced to fly, oh Bear;
> What led you thus to do?"
> "Appropriations are so scant
> That I must surely come to want
> By living at the Zoo."[19]

The zoo even underwent an inspection that scrutinized the bear caves and elephant house. Inspector Harry Lanning concurred with Baker's assessment that the cliff served as the grizzly's escape route: "Any of the bears with their sharp claws could climb the inclined rocky faces to freedom."

To prevent this, he recommended that a grated iron covering be used as a complete roof, extending from the top of the fence to the cliff and fastened directly and securely into the stone. In addition, Lanning believed that the elephants could escape with ease too if they so desired. His report concluded, "I have no blame to attach to the managers of the park. They have done remarkably well with the small amounts at their disposal—better than I could have done—but I do blame Congress for its 'penny wise and pound foolish' policy."[20]

The *Washington Post* elaborated on this "policy" in a lengthy rebuke of Congress. The philippic began, "It may be necessary that human blood shall flow and the lives of a few persons be sacrificed to bring Congress to an appreciation of what it is doing." Then, the article continued on the critical problem of funding. The issue was not only about not receiving *enough* funding (although this was a serious concern—for that year, the zoo received only $29,000 of the $73,000 requested) but also about how the money appropriated was earmarked for specific uses. As the *Post* put it, "every dollar is so tied up that it must be used for a specific purpose or not at all, regardless of the welfare of the Park or public." For example, if the zoo needed to build a fence or purchase oats, money had to be preapproved by Congress for these expenses. It would make more sense, the *Post* argued, for Congress to grant a general appropriation and allow the regents of the zoo to "lay out the money in such a way as seemed best" than to let politicians decide how a zoo should be managed.[21]

From this straightforward critique, the harangue on Congress intensified. The following passage is remarkable because it still carries traces of sectionalism (written into how "types" of men are depicted, translated into the language of Gilded Age and Progressive Era politics), because it puts on display a dialectical discourse that will come to define zoo history—zoo ideology (science, education, entertainment, and Nature) versus cruelty—and because all of this arose from a dead bear:

> It may be asked, why should Congress have established a national zoölogical park? True; however great an advantage the people may derive from . . . such an institution, it was not absolutely essential to the perpetuation of our Government. . . . But it is too late to ask such questions now. Congress may vote to-morrow, if it chooses, to abolish the Park, sell the land, and give away the animals. As long, however, as it pretends to continue the Park's existence, it has no right

> to impose upon a body of humane men of science the duty of starving several hundred wild beasts, or to expose a city's population of a quarter of a million souls to the perils of having daily bear-hunts in the public streets. Let one thing or the other be done. Let us have a Government menagerie of which every taxpayer will not be ashamed, or else no menagerie at all. Let the ignorant backwoods statesmen, who set the pace for our governmental expenditure, have things all their own way, and spend the money which better men put into the Treasury exclusively on the trumpery little creeks and ponds in their own pet "deerstriks" . . . or else give the intelligent legislators a chance to redeem themselves and the Government from disgrace. We can afford, as a nation, to get on without science and without art, with little useful knowledge and with less beauty, but we cannot afford to be deliberately cruel.[22]

Frank Baker, the zoo, and the Washington press had transformed a dead bear into political capital. An incident that resulted in some frightened schoolchildren, a carpenter with a bit leg, and a dead grizzly bear produced local panic about runaway animals. This alarm was not difficult to ignite, for the plot of the dangerous "animal," the lurking Wilderness, invading the safety of civilization was etched deeply into Western society. As the historical characters and contexts changed—ancient dragons, early-modern wolves, nineteenth-century leviathans—the story and the fear produced stayed the same. The runaway bear activated this latent fear, looming just below the Euro-American surface.[23] Yet as zoo neighbors worried about which animal might next lumber onto their lawns, the District public simultaneously consumed a different plotline as they read excitedly about the bear that gained freedom despite captivity. This different zoo story filled a narrative structure that occupied a particularly sacred space in the American imagination, one that applied to all sorts of American dramas—from a group of fledgling colonies trapped in a monarchy, to Black Americans trapped in slavery, to women trapped in a patriarchal democracy, to former Confederates trapped in recession, or to industrial workers trapped in the capitalist machine. Though all Americans feared a runaway bear, they also wanted to see themselves in this bear. Zoo animals—or how Americans perceived them—were always many. And a bear could be *felt* (experienced unconsciously through the shimmer of different narrative framings) as both protagonist and antagonist. What an intriguing, exciting bear! It is no wonder that when the city of Washington read about the shards of metal buried in the lifeless brain of their feared and loved grizzly, they flocked to the zoo. The gap between the death of a grizzly bear and his emergence into cultural lives bolstered the zoo's popularity.

A dead bear in Washington meant different things to different people. As Palmer began the reconstruction of the grizzly's carcass, the zoo and the press began stuffing his symbolic body with memory and meaning. The National Zoo and all the animals therein lived and died at the center of American politics. "Griz," as the *Post* called the bear, was no exception. As demonstrated by the printed invective above, the runaway bear was quickly resurrected as a martyr for the zoo, as both a progressive lobbyist in support of "science," "art," "knowledge," and "beauty" (and the federal subsidization of these ideals) and an activist against cruelty. The legitimacy of the very idea of a "national zoo" was far from settled in 1892, so the death of an escaped grizzly bear underscored debates about government spending, congressional oversight, and the role of the state in the daily affairs of the American people while also conjuring a sectional, polarized legislature seen as divided between "ignorant" and "better" statesmen. What a political bear!

The dead bear also rose again to serve popular zoology, to promote public education, and to push for an ursine-infused popular culture. Almost two weeks after the bear hunt, an *Evening Star* article titled "Concerning Bears" hit the newsstand. This article opened with a line that captured perfectly the influence that animal escapes could have on the popular imagination. "It is safe to say that the escape and killing of the little grizzly bear at the Zoo, just outside the city, last Tuesday morning has stimulated the telling of a thousand bear stories by the people of this town." "Bear stories" had already saturated nineteenth-century nature and travel writing, natural history, and fiction, and to no surprise did the zoo's runaway stir up these stories too.[24]

"Concerning Bears" tapped into the ursine lore of American history. The author, who was a "well-known" yet unnamed naturalist, began by summarizing general data about bears, the bare necessities so to speak—range, habitat, diet, hunting behaviors, and their interactions with humans. He continued, in eight paragraphs, to tell a dramatic story about a previous encounter with a cinnamon bear along a creek in British Columbia some years past. Then, he dispelled stereotypes about the savagery of bears, emphasizing that wild ones usually "will run away from man," though qualifying that captive bears do the opposite because they "become in a certain sense fearless of man." Next, he described indigenous bear rituals in New England. This story, though filled with eighteenth-century racist tropes concerning Native Americans, sought to educate about Indian culture while also engendering respect for bears. To conclude, and transition back to the topic of the National Zoo's runaway, the naturalist drew recognition to the fact that those who had shot the zoo bear surely did not "venerate" the animal like Indians would have, thus problematically naturalizing native culture for the purpose of fetishizing the bear and subtly critiquing the handling of the zoo escape. Nonetheless, the author emphasized (playing both sides) that "it was wise and timely that it was destroyed" because it would have naturally lingered around all the local gardens and orchards.[25]

Articles like "Concerning Bears" became increasingly common in newspapers, magazines, and journals throughout the latter half of the nineteenth century and the beginning decades of the twentieth. These publications not only stood as symbols of popular zoology but also demonstrated how the zoo increased popular zoology's reach. On the surface, "Concerning Bears" appeared as an unorganized collection of ursine-related trivia compacted into three columns of newsprint. The genre of popular zoology often wove together seemingly heterogeneous information that surrounded any given nineteenth-century animal. After the rise of the public zoo movement, though, popular zoological writing often linked this data to specific animals of a local zoo. Articles like "Concerning Bears" confronted commonplace assumptions and stereotypes about wild animals. They discussed the latest scientific facts and discoveries about these animals. They told stories about animals in wild environments—sometimes true, sometimes fictional, usually somewhere in-between. Popular zoology, like "Concerning Bears," amalgamated multiple discourses into a single, new and popular, animal-centered discourse, one that described a single species in both the wild and captivity. Zoos functioned as important forums for readers of popular zoology. In May 1892, everyone in Washington had heard about the runaway grizzly bear, for as the article admitted, the incident sparked thousands of bear stories in the District. Therefore, any readers of "Concerning Bears" automatically had a frame of reference, a context. Washington readers no longer needed to imagine bears as simply chimerical symbols of far-off frontiers or long-gone wildernesses. No longer were bears completely Other for urban Americans who lived in the nation's capital. Indeed, for William Gross, and the other schoolchildren living around Rock Creek, bears were now very much "real."

When animals escaped from a zoo, they transcended far more than the bars, wires, and fences that held them in captivity. When zoo animals ran away, they burst suddenly and unpredictably into the Open, into the public imagination, where they took on multiple symbolic lives, even in and after death.[26] Animals who absconded became heroes of newsprint and gossip, publicizing and popularizing their species and the zoo that held them. When animals escaped, sometimes they were caught. Sometimes, they died. Other times, they vanished. Always, these animals resisted the illusion of a zoological park, a place where, in theory or fantasy, Nature could be reconstructed, bound, and controlled. Runaway animals gave the lie to the zoo. The institution did not create any environment that could be called Natural. In fact, the zoo represented quite the opposite, the Artificial par excellence. And runaway animals made clear that whatever we call the zoo's creation or curation, it could never be fully contained or controlled, even in the most literal sense. Animals always ran away. On another level, however, zoo escapes were integral to the zoo. Whereas zoos may have had embodied ends meant to convey a powerful human ability to manage the world, they also embodied a more amorphous desire to be near "the Wild" again, to rewind or unwind the history of the very power they sought. Escapes offered cities more than exciting dramas and cheap content for print media. They offered brief, unscripted, possibly dangerous encounters that stimulated, as we saw with the grizzly, a spectrum of emotions. In this way, escapes reanimated the emotional landscapes and rekindled the economies of desire upon which the zoo and its ideals were grafted. Indeed, if we use escapes as a metaphor, runaway animals illuminate how opposites, impossibilities, become entangled in the zoo. The zoo symbolized human control, so it required human failure (runaway animals, literal and figurative) to legitimate the need for that control. The zoo sought to create and contain Nature to provide wild encounters, yet to overcome this paradox, it required the inevitable—that wildness (agencies beyond human control) emerge, break free, and (most important) be felt by the public. This experience staged the "object of desire" (the encounter, the wild) that breathed "life" into the curation of Nature in the first place. When runaway animals—the meanings experienced on

Figure 2.2. Grizzly bear of the National Zoo, pressing snout through the bars (ca. 1900–1915). As excellent climbers, bears escaped frequently. However, zoos "strengthened" their enclosures, and breakouts had decreased markedly by 1910. Courtesy of the Smithsonian Institution Archives.

their surfaces—burst into the Open, the zoo turned briefly inside out, baring, for a fleeting moment, its psycho-ontological interior upon its *own* surface for all its zoogoers to feel (and for us to see). In this way, the zoo (like all institutions, identities, or ideologies) always gave the lie to itself. Runaway animals simply captured and embodied the "invisible" energy of this repressive work. Runaways had always existed in the heart of the zoo, even before their escapes.

Animals Everywhere!

Runaway animals demonstrated that building a secure zoo was a hard thing to do. As Renaissance and animal studies scholar Karen Raber acknowledges, "the whole ethos of enclosure . . . mitigates against random movements and nonlinear, nonhierarchized usages of space, yet the creatures that are enclosed often fail to appreciate their containment, and indeed may be inspired to yet greater efforts at violating the orderliness of their neat boxes."[27] Zoo history is unsurprisingly riddled with the inspiration of runaway animals. Washington was not the first city that observed a bear hunt. In the summer of 1890, a young black bear squeezed through the bars of his cage in the Central Park Zoo to explore the park. Before the bear could be captured, a panther reached through their bars, grabbed and then squeezed the bear into their own cage, and ripped the ursine's neck open. As a passerby explained, the panther mauled the bear's lifeless corpse into pieces. He could even "hear the crushing of the bear's bones."[28] In 1891, a younger bear climbed out of the den in the John Ball Park Zoo of Grand Rapids. The cub dashed up a nearby tree, and to force the bear from its branches, zookeepers wrapped the end of long poles with rags, caught them on fire with kerosene, and waved the lengthy torches above the cub's head. The bear was captured, and wire netting was added to the ceiling of the den.[29] A more dramatic bear hunt occurred in July 1900, when a black bear, shortly after being purchased, escaped while being led into his cage in the Bronx Zoo. The bear bit and scratched three men, one of which was Hornaday (who had by then become the director of the New York Zoological Park, the name he preferred to the too-colloquial "Bronx Zoo"), ran down the avenue, raided a local restaurant of its pies, and swam 1,500 yards up the Bronx River. A posse of policemen and zoo laborers, armed with "shotguns, pistols, lassos and clubs" pursued the escapee. Eventually, they caught up to the bear and managed to get two lassos around her neck. However, the bear was strangled to death when she refused to submit.[30] Bears frequently escaped early zoos, which often underestimated their climbing abilities.

Many kinds of animals escaped from many different zoos. When the prairie "dog-town" of the Philadelphia Zoo opened in the 1870s, a wall extending fourteen feet underground surrounded the town to contain the inhabitants' tunnels. However, the prairie dogs (who generally dig tunnels three to six feet belowground but can excavate as deep as seven to fifteen feet)[31] pushed their abilities to burrow out and colonize the front lawn of John Penn's Solitude mansion. The zoo struggled for months to capture the animals.[32] Zoo employees eventually flooded the prairie dogs' dwellings to drive them out of their holes, and one by one the animals emerged half-drowned, only to be caught by workers waiting with nets. Eventually, the prairie dogs were secured and a new exhibit with deeper walls was built.[33] In that same zoo, around 1890, a leopard also got out of her cage one night and was shot the next morning.[34]

In 1899, an ostrich ran away from the Lincoln Park Zoo, and the famous park "became a Sahara," hosting a "genuine ostrich hunt" that was reported as far away as New Haven, Connecticut. The bird sprinted out of the zoo, through the park, down the shoreline, and finally leapt from a high bridge into the Diversey Harbor Lagoon. Zoo director Cy DeVry believed undoubtedly that the ostrich committed suicide. He commented that the bird jumped from the same bridge that encouraged many humans to jump to their death and that four months earlier the ostrich had lost his mate in a railcar, while in transit to Chicago, and had been "suffering from melancholia or some kindred ailment of purely mental origin" ever since.[35] While it is possible that a statement like this one could simply represent an anthropomorphization of the ostrich—DeVry projecting the suicide stigma of the bridge onto the ostrich or perhaps "reading" the ostrich's mood through the medical narrative of melancholia (which became an accessible nationwide discourse after the Civil War)[36]—it is worth pausing to note that there are many accounts of captive animals committing suicide in the historical record. We should, of course, recognize the impossibility of ever *truly* knowing the experience of an ostrich. Yet we should also accept that *trying* to understand the ostrich experience by seeing that experience through our own might also be the first step in recognizing that ostriches and humans share a biology and being and thus are not alien to each other. Seeing the possibility of suicide across the species divide (as many biologists who study altruism and self-sacrifice in animals do)[37] frees the action, its intentionality, and its encoded "will-to-life"[38] from humanist and anthropocentric arrogance. As the work of philosopher Lori Gruen reminds us, the dangers of anthropomorphizing lie not in the extension of ourselves onto others. The dangers result from losing those others through (the extension of) ourselves. If kept open, mutable, and receptive, anthropomorphization can serve as a bridge for an "entangled empathy," a way of trying to feel the world from and through the experiences of another.[39] DeVry's acknowledgment of the ostrich's "suffering" from "melancholia" or a similar "ailment of purely mental origin" could also be an astute ethological observation. Ostriches, for much of the year, will live either alone or in a pair. Little is known about how ostriches form these pairs, but they do not appear to be based around reproduction (as are "bonded pairs," a concept common in avian ethology), because pairs can be composed of all three combinations of sexes. Pair bonds, a term created in the 1940s and infused with Western heteronormative-monogamous Cold War politics, prioritizes reproduction, and thus conceals as much as it reveals regarding the relationships among members of a species. So much is known about the "bonded pairs" of different bird species, and much is known about the behaviors of ostriches when they aggregate into a herd during mating season. Yet nearly nothing is known about ostrich pairs in the wild. Certainly, though, the bond formed between ostriches, whether cultivated on semiarid plains or in captivity, has the potential to become an intimate one.[40] No matter why the ostrich jumped from the bridge, DeVry described a bird that was running away from misery at the zoo.

In 1900, a yellow lemur of the Philadelphia Zoo snuck out of her cage when the door was left ajar and ran up a tree, across the buffalo yard, and "over the high fence to the tracks of the Pennsylvania Railroad." Three keepers, one chief gardener, and other reinforcements, holding "pitchforks, brooms and whips," chased the lemur along the tracks. Eventually, the monkey was cornered in a coal bin, and the yellow-turned-black lemur was finally captured and returned to the zoo.[41] That same year, seven bison of the Lincoln Park Zoo broke down their fence and ran amuck through a Chicago neighborhood. It took two hours for keepers to catch them.[42] Also, in 1900, a wildcat escaped from the zoo in Baltimore's Druid Hill Park.[43] Luckily, this feline was captured easily, unlike the one that was killed in a hotel after escaping from a zoo near Old Point Comfort, Virginia.[44]

In 1901, an elephant charged out of the Indianapolis Zoo, "smashed" into a barber shop, and then continued running down Market Street.[45] In 1902, a panther named Tracy escaped from the Bronx Zoo. When she was spotted one mile away, an alarm was sounded and about two hundred Bronx residents, "armed with all sorts of weapons," sought to capture the animal. When the crowd finally confronted Tracy, two pursuers stood their ground and tossed a net over the cat. She was returned to the zoo that day.[46] Two years later, eighty snakes, including "rattlers, moccasins, chasers, pinesnakes, coachwhips, copperheads, and blacksnakes," slithered cage-less around the Bronx Zoo. After three hours, seventy-seven snakes had been recovered. The remaining three were found the next day.[47] In 1906, a wolverine ran out of the same zoo and was at-large in New York for twelve hours. Hornaday, "with a squad of keepers armed with ropes, poles and nets, scoured the country far and near," but to no avail. Only when two boys caught a "queer looking dog" in Fordham Park was the wolverine finally seized and returned.[48] And in 1910, a sea lion escaped from the Walbridge Park Zoo (eventually known as the Toledo

Zoo) and swam in the Maumee River for six days until caught by a local fisherman.[49]

These are just a few of the many examples of runaway animals. Whenever a zoo was established, runaway animals followed. A leopard escaped from a zoo in Baltimore and clawed a boy in the face.[50] Two monkeys got loose in the Lincoln Park Zoo and wandered to the park police station.[51] In Venice, California, three bison broke out of a nearby zoo, ran down a nearby beach, and stampeded an automobile. Luckily, some cowboys at a nearby Wild West show managed to lasso and corral the bison.[52] Every zoo dealt with animals that wanted their freedom. And these rebellious animals captured the public imagination. As the endnotes here demonstrate, newspapers regularly printed reports about runaway animals in other cities, often hundreds or thousands of miles away. People in San Jose cared about snakes in New York, and citizens of New Orleans were interested in charging elephants in Indianapolis. By running away, zoo animals helped mold the United States into a zoo-networked nation. Also, by running away, zoo animals forced zoos to redesign enclosures to prevent future escapes. Mostly, though, by running away, zoo animals gave the public that dose of "the wild" that it craved, as well as, upon capture, the relief of control that it was programmed to want. Runaway animals accomplished all of this as they sought to escape the zoo.

Zoo animals could also run away from *within* their cages. A glance at the first few years of Philadelphia Zoo history is revealing. In 1877, a bison stampeded the fence surrounding his pasture so that he could charge the elk in the neighboring exhibit.[53] A kangaroo broke a leg attempting to jump out of the enclosure after being frightened by a nearby locomotive.[54] "Jim," the Bengal tiger, grabbed the tail of the tiger in the adjoining cage, "dragged it in until he could seize its leg," and ripped the thigh bone from its socket, killing the victim.[55] A macaque tore another primate in half with his hands and had to be quarantined in a separate cage, marked wisely with a sign warning zoogoers to keep a safe distance from the bars.[56] An elephant tried to "crush" his keeper "against the bars" of the exhibit.[57] And a young boy climbed over the railing of the open cages where the lions sunned themselves and barely made it out alive.[58] Zoo animals, kept unnaturally within zoos, resisted containment. If they could not escape, their captivity often pushed them into violence with other animals, with themselves, with their keepers, and with zoogoers. Over time, zoos, through trial and error and by communicating with one another, as an emerging and networked carceral system, reduced the frequency of these violent encounters. However, they never ceased.

In 1891, a Mexican lion held in Atlanta's Grant Park Zoo grabbed a child with a paw and bit his arm, refusing to let go until a crowd of zoogoers attacked the lion with sticks and umbrellas.[59] That same year, a keeper in the Philadelphia Zoo was attacked by a tapir when he entered the enclosure to stop two males from fighting over a female. This keeper also had to stop two elephants, with a pitchfork, from charging the brick wall surrounding the exhibit.[60] In 1904, a brown bear snagged seven-year-old Robert Mayer as he passed by the cage. The bear "sunk his teeth into the boy's leg." Robert was eventually rescued by keepers and rushed to the hospital, but the injuries proved so severe that he was "lamed in all probability for life."[61] In 1907, a keeper in the Bronx Zoo was nearly killed by deer when five (including both mule deer and black tailed deer, in a multispecies mutiny) charged him during feeding time. The deer butted the keeper with their horns and kicked him with their hooves, breaking two ribs and inflicting serious head wounds and abrasions all over his body.[62] This type of deer attack was not novel, though, for, in 1896, a keeper of the zoo in Cadwalader Park, New Jersey, was also almost gored to death inside the deer enclosure.[63]

Zoo animals ran away. They attacked. Zookeepers, zoo architects, and zoo correspondents often framed these behaviors as "unpredictable": *Zoo animals are unpredictable; they are dangerous, instinctive, wild animals.* However, given the omnipresence of runaway animals, there was something quite predictable about this so-called unpredictability. Restrained individuals always pushed against their confines. By casting behaviors that complicated the smooth functioning of the zoo as unpredictable or dangerous, zoo public relations accomplished two feats. First, they disguised their ignorance about animal behaviors behind narratives of deviant, "troublesome," animals "acting out."[64] Second, by emphasizing their inherently dangerous characters, they could animalize their inmates. Doing so helped the zoo offer the *wild* encounters that zoogoers were seeking. Doing so also helped zoogoing society construct the zoo as a "modern" correctional institute where threatening wildness could be tamed. This allowed zoogoers to see animals as innately aggressive rather than driven to "madness" by their solitary confinement.[65]

The first breakout in Washington occurred when the National Zoo was still in its gestational phase within the Department of Living Animals. In June 1889, Smithsonian workers built the puma a new cage, for the other one was "deemed insecure." However, during relocation, the puma escaped. She did not free herself from the building to roam along the Mall; instead, she prowled around the coyote pen and eventually found her way into the bear pit. The puma harassed a bear, but keeper Weeden heard the commotion, came to the rescue, hit the puma with a club, and returned the cat to the new cage.[66] A Gila monster escaped prior to shipment, and a weasel, "with powerful teeth," chewed through a box during transport (a method of escape that occurred all the time during shipment).[67]

In 1891, shortly after the National Zoo's official opening, a deer jumped over a fence and ran around the zoo until she was finally caught and returned.[68] In the winter of 1892, Blackburne told the *Washington Post*, "I have had a good many hard fights with wild animals and savage beasts, but about as close a place as I was ever in was the other day, when I walked into the deer pen." He continued to tell a good, old-fashioned zoo story:

> The wild buck came bounding at me, as I thought in play, for, you know, I'm such a good friend to all the animals that they delight to have a bout with me now and then. But, sir, the old buck was mad, and he lowered his broad horns with which to give me a toss in the air. I did not have time to jump, and one prong of the horns struck me on the arm. I saw then I had to fight, but I had nothing but my naked hands. . . . In an instant he was upon me. I seized him by the horns, and he gave me such a shove that he rammed my back against the tree, but as he did so one sharp prong pierced my leg just below the thigh. I knew I could handle him for a time with the leverage of his horns, but the question was how fast would I be exhausted by the loss of blood. In the jerks and turns, twists and pulls, I soon got the deer between myself and a sapling, and, retaining a firm hold on his long horns, I tried in every way to exhaust him, or wrench the horns off and thus wound him.

The battle continued in this manner for thirty minutes until Blackburne ripped the right antler from the deer's head. Immediately, the buck ran away in pain. Blackburne concluded this saga by emphasizing that he "never dreamed of being attacked" by a deer and then reminisced that the encounter "was the deer-est joke I have had played on me for some time."[69] Zookeeping in 1892, just like hunting, offered the perfect stage upon which to validate one's masculinity.

In this narration, Blackburne began by casting himself into the role of the fearless, masculine-yet-gentle zookeeper who could tame wild beasts. In his words, he was a "good friend to all the animals" in the zoo, yet he still had his share of "hard fights" with "savage beasts." Blackburne, in this way, opened his deer story with a familiar figure, a keeper who—as a hybrid of Noah, Daniel, and David—could lead an ark in the park, lie among the wild animals, and still stand his ground against giants when needed.[70] This character (the "lion tamer," "elephant man," "superintendent of animals," or "head keeper") embodied the story lines inscribed into both the menageries and the circuses of a previous era, as well as into the zoos of Blackburne's day.[71] The enlightened zookeeper used Reason to care for a diversity of animals. He exercised compassion to befriend animals usually imagined as ferocious. And he could subjugate animals when they threatened him, for he was used to facing "thrilling and dangerous feats" in his former life with the "wild beast show."[72] The hero of the zookeeper echoed the lesson that first reverberated throughout zoos, that the modern nation-state could simultaneously manage, nurture, and dominate Nature. Blackburne's story kept in tune with these larger tropes. However, he added two twists that hint at changing ideas about animals. First, the showdown described was not one between human and lion, elephant, bear, or other "savage beast." Instead, the battle was between human and deer, a creature usually viewed as skittish and docile. Blackburne, though, made sure to connect his experience to popular zoology, offering a zoological explanation for the deer's unexpected violence—as spring approaches, rutting male deer, for mating purposes, become territorial. Second, even though the story underscored Blackburne's strength and ingenuity, the central lesson of the zoo story taught that zoo animals asserted themselves in unpredictable ways. Within the zoo, even typically tractable deer could turn deadly.[73]

Zoo animals always had the potential to run away from the National Zoo. A mule deer escaped from an enclosure. A wildebeest did the same. A leopard sank claws into a zoogoer. A zoo worker was bit by a dingo. Another was bit by a rattlesnake.[74] One especially dramatic episode occurred in May 1904. On the fifteenth of that month, around noon, keeper

Arthur Edwards was cleaning the soiled ground of the coyote enclosure when one of the inhabitants "sprang at his throat." Edwards threw the canine to the ground, but the animal pounced again. To protect his life, Edwards covered his neck quickly with his hands. First, the coyote bit down on his right hand, breaking every bone. Then, he bit his forearm. The struggle brought Edwards down, underneath the coyote, and "finding the strength gone from his right arm, Edwards put forward his left, from which . . . the ligaments were torn horribly." He fought for his life until another keeper saw what was happening, ran into the enclosure, and beat the coyote until he released Edwards. The keeper was rushed to the hospital. His "right hand and arm were badly torn." His left hand was almost crushed. The ligaments of his left arm were ripped "out of place." And the pinky finger of his right hand was amputated. The day after the attack, a visitor reported that the coyote was simply lying in the sun as if nothing had happened.[75] Zoos made animals unpredictable, for coyotes (almost) never act aggressively toward humans in the wild.[76]

Not all runaway animals were understood to be dangerous. In the autumn of 1896, while their enclosure was undergoing renovations, five beavers escaped. The next morning, four returned by themselves to their artificially built dam, but the fifth chose to remain at-large. Zoo employees searched "up and down Rock creek, in the Zoo and outside." After a few days, the fifth beaver was "given up for lost." Six months later, though, while a group of boys from Mount Pleasant meandered through the woods, they stumbled upon a beaver dam along Rock Creek, north of the zoo, built by the long-lost beaver. The runaway had built a dam from felled beech and maple trees that grew along the creek's banks. All around the dam were logs without bark, trees partially cut, and others completely girdled. The willows, poplars, and birches in the vicinity were consumed as food. Once locals read about the discovery, visitors began arriving to the site to examine a "real" dam, the likes of which had disappeared from the mid-Atlantic almost two centuries earlier. This beaver that chose freedom encouraged both popular and scholarly zoology. As the *Evening Star* reported, a professor from the University of Pennsylvania (perhaps esteemed paleontologist Edward Drinker Cope, who wrote on extinct giant beavers in the 1880s) traveled to Washington just to study the dam, which led to a series of essays.[77]

Not all runaway animals were found. In 1899, the National Zoo received a shipment of one dozen Arctic foxes sent directly from an Alaskan island. Since no exhibit was prepared for them, they were placed temporarily in a cage located within the zoo's stables. While in this purgatory, one of the foxes managed to pry apart the bars of the cage, climb up to a windowsill, and "vault" "to his liberty."[78] A plea was published in the *Evening Star* asking neighbors to keep lookout for a dog-like animal "prying into yards or woodsheds," but the fox was never found.[79]

This was the experience for the vast majority of absconding zoo animals. Runaways rarely received headlines. Usually, when zoo inhabitants got away, they did so quietly and mysteriously, disappearing anonymously into the urban jungles that surrounded their zoo.[80] Of course, when large, charismatic animals broke out or potentially dangerous animals sneaked off (like the grizzly), they would be more likely to catch public attention. In many cases, the public was warned directly by zoo officials, either to prevent children from being devoured while walking to school or to increase the likelihood of finding their expensive investments. Since the significations of zoo escapes were essential for the maintenance of zoo ideology, sometimes reporting runaways could bolster attendance, as shown above. Yet for less exotic animals, disappearance was often just grumbled about by keepers and recorded as boring statistics in yearly reports.

The report for the fiscal year ending on June 30, 1900, stated straightforwardly that "a number of raccoons escaped from their enclosure."[81] The report for the year ending on June 30, 1904, recorded one female mule deer killed by a male and one Japanese bear that died from "shock." It also listed the deaths of two moose, an antelope, and a puma, leaving the reason "unknown."[82] The report of 1905 listed two elk and one pronghorn that died from an "accident."[83] The one of 1906 declared a bison dead from a broken spine and a zebra dead from an "accident."[84] Eventually, the yearly reports actually created entire catchall categories that could encompass vague deaths, which included animals that came up missing, died running away, or died in unspecified accidents. In the 1908 report, three animals were listed under "cause of death not found."[85] In the 1909 report, seven animals died from "accident."[86] The 1911 report stated that thirteen animals were killed by "accidents and injuries."[87] The 1912 report listed a remarkable nineteen animals dead from "Accident (fighting, killed by wild animals, etc.)."[88] Not only does this category announce an interesting register for "wild" (describing urban animals not-on-display-as-wild *as* wild for killing the supposedly "wild" captives), but the et cetera surely indicates the mundane regularity of violent deaths in the zoo. And the 1913 report revealed an absolutely astonishing thirty-two animals dead by "accidents."[89] What specifically caused these deaths will always remain a mystery, though nearly all of them were likely derivative of captivity and its discontents.

William Blackburne told the *Evening Star* in 1897 that "in order to handle animals properly, the person must understand them thoroughly. They [the animals] have much more sense and understanding . . . than they are generally given credit for. They can tell in an instant when any one is afraid of them and will take advantage of it in a moment. Their moods and tempers change just the same as do those of persons, and treatment which will be well received one day, will probably invite an attack another."[90] More than in his 1892 deer encounter, this reflection five years later strayed further from the mythology of the keeper or tamer. Zookeepers could not simply dominate or manage the animals that they "kept." They needed to "read" them, their "moods and tempers," and then engage them on their own ground, so to speak. While in an enclosure, Blackburne seemed to suggest that he needed to respect the animals as individuals, as "persons." Zookeepers could not take animals lightly or let their guard down. Yet this did not mean that they should be afraid, for if they demonstrated fear, the animals would "take advantage." Zookeepers needed to enter an enclosure displaying "nerve," for displays of blind arrogance could provoke aggressive reactions too. Keepers needed to take animals seriously, if for no other reason than their own well-being.

Blackburne never argued, by any stretch of the imagination, that keepers should treat animals as equals. In fact, he believed that control was crucial to keeping keepers safe. He elaborated in the following manner:

> If there is any familiarity to be indulged in between a keeper and an animal always let it come from the man and never from the beast. If he wants to pet an animal, all well and good, but he must never let one of them have an idea that it can presume to "jolly" a keeper. If this takes place control is lost over the animal, as it immediately imagines it is as capable of running things as its keepers. The old adage that familiarity breeds contempt is never more truly exemplified than between a keeper and an animal.[91]

Striking in this statement was Blackburne's treatment of zoo animals as opportunistic. He did not, though, cast animals as conniving, the way in which bears and wolves were frequently stereotyped at the end of the nineteenth century. Blackburne did not employ overwrought mythologies like the "cunning fox" or the "coy raccoon." The agency he attributed to animals was not the type that allowed humans to dismiss or degrade them. Instead, Blackburne depicted zoo animals as storehouses of initiative, and it was the keeper's job to do everything possible to acknowledge this potential in individual animals while minimizing the chances that they would put keepers, animals, or the public in danger. Throughout the twentieth century, zookeepers, like financiers, learned how to manage and lessen risk. Never could they subdue it. The business of zoos was always risky business, quite possibly because it was a business that "breeds contempt" into its capital.

A Truant Wolf and a Rabid Dog

On October 18, 1902, a little more than a decade after the runaway grizzly incident, another animal ran away. Once animals left their enclosures, they found themselves in surprising places. This time, prints were tracked to the steps of the District Supreme Court.[92]

At around nine thirty in the morning, Mrs. George Jackson, who lived in a small white house along Klingle Road, only a short distance from the National Zoo, was doing chores when she suddenly heard her children screaming outside. She rushed through the door and saw, across her lawn, what looked like a vicious dog on top of her three-year-old, teeth buried into the back of the child's head. Her seven-year-old daughter was attempting to pull her sibling away from the dog, but she could not outmuscle the animal. The mother sprinted toward the dog, gripped the fur, and "with a strength born of her frenzy," threw the attacker into a nearby chicken coop. The child lay upon the ground bleeding. The dog escaped through a window of the coop and vanished.[93]

Thirty minutes later, the residents of a house in Cleveland Park belonging to G. H. Powell heard the piercing cries of a bull terrier chained in the backyard. A servant saw the pet being attacked at the throat by a dog. She grasped a clothes prop and beat the attacker, who released the terrier and ran to the front of the house. Mrs. Powell and Andrew Jackson (an uncle to the siblings first attacked) ran out the front door to see what the commotion was all about. The dog tried to bite Jackson, but he, armed with a cobblestone, scared the dog off the property. Shortly thereafter, an employee of the National Zoo arrived at the Powell property and

Figure 2.3. Admiral George Dewey, pet owner. From William McKinley, *Exciting Experiences in Our Wars with Spain and the Filipinos* (Chicago: Book Publishers Union, 1899).

tried tracking the animal. About one-third of a mile down the road, the dog again attacked—this time, a collie owned by a Mr. Sharpe. The pet held her own, and the dog ran off toward a field along Woodley Road, owned by Nevada congressman Francis Newlands (next to the home of the famous admiral George Dewey, who four years earlier won the Battle of Manila Bay). A gardener by the last name Bowles, employed on the Newlands property, grabbed his revolver, shot, and then bound the still-living dog with rope. A few hours later, the keepers from the zoo arrived, seized the injured animal, and then shot the dog dead in the zoo.[94]

Soon after the incident, those who had encountered the vicious dog had come to the conclusion that the animal was indeed no dog at all but a runaway wolf from the zoo. Why else would zookeepers arrive mysteriously on the Powell property, track and chase the canine, and then take the "dog" back to the zoo to be clandestinely killed? Word spread like wildfire, and by the time the *Washington Post* caught wind of the story, it was popularly understood throughout the District that the ruthless attacks were led by a "Mad Wolf from Zoo." Somehow, a wolf had escaped and gone on a rampage against innocent citizens living on the outskirts of Washington. The nation's capital, according to the *Post*, was again in a "nervous frenzy over the possibility of the escape of other and probably more savage animals."[95] And thus goes the story that was first told. When the public first read about this event in their most trusted newspaper, there was no question that the "dog" in question was actually a runaway wolf.

However, upon further inspection, facts are never as factual as they first appear. Only one set of facts proved absolutely true—the Jackson children were in the hospital after sustaining serious injuries. The three-year-old had a "large piece of scalp" lifted from the skull, teeth marks embedded into the back of the neck, and a large wound under the right eye. The older sibling had a "badly lacerated" hand, which received three bites that penetrated to the bone. Luckily, the children, although "well bandaged," survived the attack. Other information associated with the original story became blurry as new information was released. First, the dog-wolf that was captured on the Newlands property and taken back to the zoo was indeed a wolf. Second, this wolf was not killed by the zoo but was instead "placed in one of the strong cages outside of the animal house." Mrs. Jackson went to the zoo the next day and identified the injured wolf as the animal that attacked her children. The zoo, though, claimed that Mrs. Jackson originally stated that the animal that attacked her family was wearing a collar. Mrs. Jackson responded that she had never said any such thing, insisting that the attack animal was the escaped wolf. The zoo, though, counter-insisted that the children were attacked by a wild dog, not by the wolf that, only by an unfortunate coincidence, happened to escape around the same time as the attack. As the *Washington Post* stated skeptically, "The real Hound of the Baskervilles is in town . . . a Sherlock Holmes would seem to be needed to sow the seeds of the detective instinct in the special policemen who stand about the gates of the Zoo."[96] Let us begin with the zoo's defense.

Superintendent Baker, when questioned by a reporter from the *Post*, "ridiculed the idea that the animal [wolf] had bitten the babies." To prove his point, Baker, in front of the reporter, walked into the cage holding the injured wolf and "stroked it with a broom." Baker insisted that the wolf was "gentle and quiet" and "wouldn't hurt anyone." The journalist wrote that this demonstration was "not altogether reassuring, however, as the animal was helpless" and injured from a gunshot wound. Blackburne tried to bolster Baker's defense by stating, "He [the wolf] is nothing but a baby . . . eight months old." The journalist countered, though, that this supposed "baby wolf" was oddly larger than their mother and also reported that Blackburne "bet $100 that if he were to let out either of the brother wolves, they would wander about the Zoo and make friends with any children they happened to meet." Blackburne also made a joke about how if he had to choose to be bitten by a wolf or a dog, he would prefer a wolf because the teeth would be "clean and white and free from poison [rabies]."[97]

Baker and Blackburne had their propagandizing work cut out for them. It was undeniable that the wolf had escaped from the zoo. The cage, surrounded by a chicken wire fence, possessed a floor made of two-by-fours, and it was clear that at one side of the cage this floor was "water-soaked and decayed" and had "given way under the scratchings" of the wolf. Baker and Blackburne could not deny that the wolf had escaped. Instead, they had to convince the public that even though their wolf was at-large that day, the nearby attacks were still prompted by a pet, stray, feral, or wild dog.[98] Others in the zoo came to the zoo leaders' aid. The lemonade seller and a security officer both claimed they saw a "mammoth dog" running along the cliff across from the Chevy Chase entrance. To corroborate this claim, paw prints that "were unquestionably those of an animal at least as large as a full-grown mastiff, too large, in fact, to have been made by the wolf" were discovered around the zoo. Both witnesses also claimed they saw a collar on the dog and observed this pet running "madly out Connecticut avenue toward Cleveland Park."[99] However suspect this evidence may seem on the surface, on the afternoon of October 20, the police at the Georgetown station were notified of a rabid dog loose in Cleveland Park. They sent officers to kill the animal, but the mad dog could not be found.[100] This report legitimated officially the stories told by the lemonade man and security officer. A canine suspect was indeed on the loose in Washington!

At first, it seemed as though public opinion was pitted against the zoo. What are the chances that a stray or rabid dog attacked the children living near the zoo on the same morning that a wolf escaped? This argument seemed to be only a far-fetched attempt by the zoo to either protect its reputation or sidestep a lawsuit. Certainly, this would not be the first time that Baker and Blackburne tried to shape public opinion. However, when reports about a rabid dog reached the police department two days after the wolf escape, public opinion became more mixed. William Blackburne told the *Evening Star*, in an article titled "It Was Not a Wolf," that "if we thought the wolf inflicted the wounds, we would willingly admit it, as we have no desire to conceal anything, but there is no necessity for people to become unreasonably alarmed." He admitted that the zoo has had "agents investigating the case, and we are positive that the wolf did not leave the park until 10:30, while the attacks are alleged to have taken place earlier."[101] Surely, different readers formed differing opinions about whether the zoo would "willingly admit" that an escaped animal had attacked two children. In a show of sympathy, though, Frank Baker did make a trip to the hospital to visit the maimed victims.[102]

The saga gripped the interests of the nation's capital. Some sided with the zoo. Some sided against it. On October 21, as the *Washington Post* appeared to be leveling an all-out assault on the National Zoo, the *Times* decided to weigh in on the controversy by publicizing a case in defense of the zoo. The paper reported that a few days before the attack on the Jackson children, there was reportedly a "dog fight" in the same neighborhood, in which an aggressive and anonymous dog attacked a pet bulldog. The "fight" was stopped by servants, and the aggressor supposedly ran off to Cleveland Park and bit a child. Then, the attacker was pursued and shot by neighbors of the park. The *Times* printed the following:

> A remarkable point in this case seems to be that the wolf was not recognized as such by any of his assailants, all of whom have been living in close proximity to a zoological park. . . . This, however, is not the first instance of mistaken identity, for it is a matter of recent history that a bill was rendered to the master of the local hounds for damages for the killing of a pet dog, which had been taken not only by the hunter, but by the hounds themselves, for a fox. But all this does not excuse the fact that the wolf succeeded in escaping through the carelessness of some employee of the establishment; nor does it speak well of the vigilance of the keeper that his loss was not immediately discovered and the hue and cry of "Wolf!" promptly raised.[103]

The *Times* seemed to support Baker and Blackburne's stance. It appeared unlikely that the wolf was responsible for the attack. This fact, though, should not exculpate the National Zoo from criticism, for this "establishment" allowed a wolf to escape! Whether the wolf attacked the Jackson children or not, such negligence was unacceptable. The "daughters" of the District should not be forced to play "the role of Little Red Riding Hood."[104]

Now, if the *Times* was correct, there appeared to have been two crimes committed on October 18, 1902. First, a dog attacked the Jackson children. Second, the zoo allowed a wolf to escape. However, both cases were deadlocked until evidence could be found. And that evidence finally presented itself when five or six days after the attack on the Jackson children, Gardener Bowles, the gardener who shot and injured the runaway wolf, came forth and stated that he was also bitten by Admiral George Dewey's dog.[105] The details of this story were immediately released to the public. Dewey's dog, a collie named "Prince," bit Bowles on the same day that the wolf escaped. Dewey also admitted that the dog probably had rabies, recalling that "Prince" was himself bit two weeks earlier by a strange, stray dog. Once this news went viral, "the majority of the residents of Cleveland Park" believed that the collie, not the wolf, was responsible for the atrocities. Locals armed with sticks and a "corps of sleuths" from the pound mounted a search for the dog. Admiral Dewey himself issued a death sentence for his pet, instructing his own gardener, T. W. Ogden, to kill the collie "on sight" if "it" happened to return. Dewey reportedly told Ogden to kill the dog with a brick, but lacking the courage possessed by the "hero of Manila," Ogden went to town and purchased an old Harpers Ferry musket, "loaded it with slugs, and barricaded the place."[106] The northwest corner of Washington was on watch for an enraged collie, frothing at the mouth, while waiting anxiously for the next zoo animal to run away. All the while, Dewey paid for the Jackson children and Bowles to be sent to the Pasteur Institute in Baltimore to be tested for hydrophobia (rabies).[107] And the injured wolf in the zoo still lay panting on the floor of the zoo cage, most likely soon to be "killed and stuffed."[108]

What took Gardener Bowles, the only person who had seen both the wolf and the collie, so long to come out with his story remained unclear. Bowles did say that he had "been bitten by dogs before," so maybe he thought nothing of the implications of a snapping collie.[109] Also, how the lemonade seller and zoo officer could confuse a collie (which would be of a comparable or smaller size than a wolf) with a much larger mastiff also remained unclear. Nonetheless, for the meantime, the District forgot about the wolf in the zoo and instead grew frantic about a rabid dog still at-large.[110] For the time being, even the *Washington Post* redirected its gaze from the zoo to the area around Cleveland Park. In fact, the *Post* decided to capitalize on this new phobia of hydrophobia by printing a five-column feature about rabies, which surely only heightened public fears. This article exclaimed that "all Washington . . . shares the feeling of apprehension and solicitude that exists among friends and neighbors of the victims of the mad-wolf-mad-dog invasion." The article continued to tell a story about a mad wolf that "invaded" Fort Larned, Kansas, in 1868, leading to the horrific death of a soldier stationed there.[111] In this way, the *Post* increased its readership by exploiting a long-standing fear of rabies and further frightening citizens about canines running amok in Washington. However, the possibility remained that both the wolf *and* the collie had bit innocent victims. The *Trenton Times* put the problem succinctly: "The victims of the animals' fangs cannot tell for certain whether they received a wolf bite or a dog bite."[112] It is also possible that Dewey's collie was not the only dog biting residents. Just because the admiral's pet bit the gardener did not mean that it necessarily bit the children.

The District clamored over the incident as he became the topic of conversation in every restaurant, tavern, and government building. Some citizens even amplified their voices by sending editorials to the papers. One person wrote:

> We do not quite see how this community is to get back to normal conditions until somebody solves the problem. It was bad enough to hear that a ferocious wolf had escaped from the Zoo. . . . It was still worse to read in the next day's paper that the beast had attacked a number of children and old persons as he went upon his devastating way. But we could have put up with that all right, especially after the announcement of the wolf's death [this announcement was either false or not made by a zoo official]. The incident had not been without a certain charm. It was rather pleasant than otherwise to read about the huge, gaunt monster, with bloodshot eyes and glistening fangs and lurid tongue.

This sarcastic editorialist continued with a harangue against those who dismissed the wolf as only a "skulking chicken thief" and instead chose to worry only about Dewey's dog. How can the people forget that the National Zoo allowed a wolf to escape? At the same time, however, this writer was not trying to free Admiral Dewey from blame, for he "must [also] be held accountable for everything dreadful that has happened in the neighborhood." The column served as an assault against both the zoo and the admiral. Both committed crimes. Yet the question remained—did a wolf or a dog bite the victims? The writer concluded, "We protest against injustice. Call a court of inquiry, somebody; that's the way to get to the truth!"[113]

The animals of the National Zoo always lived political lives. Sometimes this simply meant that the *idea* of zoo animals possessed political significances or symbolisms. The wolf of the National Zoo and the dog of a US admiral, though, entered directly into political and legal arenas as the wolf-dog debacle shifted from a topic rattled about in the public sphere to a formal legal case. As this transition took place, Dewey's dog seemed (although, not for certain) to have been caught and killed. One report stated that "Prince" was captured and shot on the same day as his escape, but, for some reason, this was not made known to the public or discovered by authorities until three weeks after the original incident.[114] A differing report asserted that Poundmaster Einstein, who had "endeared himself afresh to this community by a careful . . . investigation of Admiral Dewey's dog," found on November 4, two weeks after the Jackson children were attacked, that a "large dog of a dark grayish color" was recently killed by a Georgetown man. Einstein presumed, for reasons not quite clear, that this dead dog was, in fact, Dewey's runaway.[115]

The zoo wolf was killed on Wednesday, December 31, 1902, the last day of the year, over two months after the escape. Upon order of Frank Baker, the wolf was sent to the Bureau of Animal Industries, exterminated, and tested for rabies. The only reason that the zoo had kept the injured wolf alive for so long was to prove to the public that the animal was not rabid. However, "zoo authorities shrouded the death of the wolf in a veil of secrecy."[116] Immediately after Baker announced publicly that the wolf had been put to death (the announcement was made on January 2, 1903), rumors began swirling that the wolf killed was not the one who escaped. This suspicion arose after someone in the bureau leaked information that the wolf executed was dying from ingesting an object thrown into the cage—"a steel coil spring of intense power, which, once fairly in the stomach, recoiled with great force, and caused the miserable wolf considerable trouble with its digestive apparatus." The "truant wolf is still alive!" some citizens clamored. Of course, no evidence came forth to corroborate this conspiracy, and zoo authorities and members of the bureau refused to comment.[117] Baker only announced, "The wolf was a harmless, playful, innocent little animal, and there was nothing the matter with it except the gunshot wound in its hind legs. . . . It had been in considerable pain, and we kept it simply to prove that it was healthy."[118] By the beginning of 1903, two dead canine bodies represented two different suspects, but both runaways found new symbolic lives in the legal system.

Chatter about potential lawsuits commenced immediately after the wolf-dog incident. However, nothing of this sort took place until the following summer. In June 1903, one *Post* article titled "Wolf Case in Court" announced that "the wolf that escaped last fall . . . [has finally] invaded the City Hall in the guise of a damage suit." Augusta Jackson, on behalf of George Jackson, filed a lawsuit against Frank Baker for the recovery of damages to the children. The suit, presented to Justice of the Peace Mills, was for $300. More important than the amount, this lawsuit gave "judicial intellects of high caliber" the opportunity to decide, once and for all, "the mooted question whether it was the wolf or Admiral Dewey's dog" that attacked the girls and spread pandemonium. Before the case could be heard, though, Baker applied to the District Supreme Court for an appeal. Being the director of the National Zoo, Jackson's suit against him was indeed a motion against the government of the United States, and for this reason, Baker wanted the case to be heard by a higher tribunal.[119] The appeal was granted and scheduled for the upcoming autumn.[120] Runaway dogs, though, were not entirely new characters in high profile courtrooms. In *The Law Is a White Dog*, American studies scholar Colin Dayan examines the 1897 case *Sentell v. New Orleans* in which the US Supreme Court ruled that dogs must be registered, collared, leashed, and their movement controlled in order for them to be legally considered personal property. As Dayan astutely analyzes, both this ruling and the previous year's 1896 *Plessy v. Ferguson* ruling, which established the "separate but equal" cornerstone of racial segregation, sought to control the movement of two animalized others—Black people and dogs.[121] Baker's appeal, then, promised another hearing about restricting the movement of what white urbanites saw as unwanted predators in increasingly changing cities. Attorney General Philander Chase Knox appointed two highly skilled lawyers to defend Baker, to prove before the jury-to-be that it was indeed Dewey's dog that attacked the girls. The defense was preparing for a case that "will be fought to a finish on the side of the government," and the Washington public was excited to hear all the mysteries of the wolf-dog incident revealed and unraveled.[122]

The case, though, never reached the courtroom. Most likely, this indicates that either the Jackson family dropped their suit upon learning of the granted appeal or that Baker simply decided to settle with the family outside of court. Either decision would make sense. The Jackson family may have decided that time, effort, and potential court costs and lawyer fees would not be worth risking for a case taking on the US government. The kids who were attacked were undoubtedly better by the summertime of 1903. Surely, they had scars, but the wounds would not be able to be reexamined for any sort of proto-forensic analysis. Furthermore, the bodies of the suspects no longer existed (if the wolf was ever truly killed and the dog ever truly caught in the first place), for both would have been cremated by the summer. Frank Baker, for his part, may have decided to pacify the Jacksons with a settlement so that he would not attract bad publicity to the zoo, reminding the public that a wolf so easily dug out of an enclosure. All that is known is that the lawsuit, for whatever reason, fizzled out. Yet a truant wolf still managed to escape and creep to the steps of the District Supreme Court.

The mania sparked by a runaway wolf and dog sheds light upon citizens' latent fears toward wildness. In 1902, Americans recognized the agency of animals. Although zoos, for some, symbolized the domination of Nature and the subjugation of wild natures, in reality, as the citizens around Cleveland Park and as zoogoers everywhere realized, zoological parks only held wildness at bay. As feminist science studies scholar Juno Salazar Parreñas shows with orangutans held in rehabilitation centers in Borneo, animals kept in even the best sorts of zoos and sanctuaries are still animals held in arrested development, with an arrested autonomy and "thwarted freedom."[123] Ironically, the animal behaviors experienced as "wild" in zoos were symptoms of the failures of zoos, emerging within the institution's lack. This "wildness" was visceral and at hand when it broke free of captivity as animals acted out, attacked, escaped, and ran away. Zoo animals meant many different things to many different people. All zoo animals lived multiple symbolic lives simultaneously. However, when they ran away, the lives of animals, literal and metonymic, were pushed to the brink. When zoo animals ran away, they often confronted injury or death, yet at the same time, their rhetorical power magnified as zoo directors, zoogoers, journalists, and citizens harnessed the agency of animals for their own different purposes. Runaway animals revealed two important lessons. First, zoo animals possessed agency. They ran away all the time. And as they did, as they pushed against their captivity, they gave the lie to the zoo. Zoos never truly kept wild animals. The animals they tried to keep always ran away. Zoo animals, then, were only "kept" *and* only "wild" as constructed by an ideology of captivity, pressed upon their bodies. Second, zoo animals, runaway or not, became visible to the American public, or at least their surfaces did. In the last quarter of the nineteenth century, Americans invested animals with many meanings, yet no matter how symbolic animals were employed in culture and politics, the living animals represented always ran away from their representations. In some ways, this agency enhanced their utility for humans, making them stronger, more intriguing signifiers, staging a "wildness" that zoogoers (and Western cultures) were after. The agency of zoo animals also, though, limited this utility, underscoring the point that animals could never be controlled, managed, or cared for completely. Even the dog, the archetype of domestication, "man's best friend," ran away.

Zoological parks functioned as a site where all discourses about animals converged. People used their zoo as a place where they could project ideas onto animals, and thus as a place to work through beliefs, agendas, politics, anxieties, fears, and passions. A runaway bear could attract people to the zoo and scare people away. A runaway wolf could incite a contagious public fear. Fear of captive beasts and fear of rabid pets arose from the same place within human sensibilities. Fear of the one only bolstered fear of the other. No matter what specific meanings runaway animals held for specific individuals, zoological parks, in all cases, brought animals closer to humans and thus increased the number of distortions that circulated about animals.

The most surprising facet of the truant wolf debacle was that the fear of runaway animals quickly subsided. District citizens and zoogoers demonstrated liberal forgiveness. They quickly forgot about how the National Zoo had allowed a wolf to escape, and they seemed far more worried about rabies than about runaway carnivores. In fact, like zoo publics everywhere, they quickly forgot about all the runaway animals of the National Zoo, refugees lost within the dusty archives of human forgetfulness. These animals were accepted as inevitable accidents both endemic and essential to the very idea of keeping wild animals in the first place. If a city chose to build a zoo, it would occasionally serve as a stage for great escapes. Urbanites, though, surely needed these occasional dramas—these confrontations with the Other, the wild, and the exotic. For the living escapees themselves, these encounters almost always ended tragically. Yet maybe individuals preferred these ends to living in the zoo.

CHAPTER THREE

The Crossroads of Science and Popular Culture

It should come as no surprise that a zoo sponsored by the Smithsonian found a place at the center of American science. Conveniently located in the nation's capital and in the midst of a scientifically minded Smithsonian community, the National Zoological Park forged inroads into myriad branches of the life sciences. This chapter will explore some of these resulting entanglements, demonstrating the centrality of a zoo to ever-specializing and ever-professionalizing sciences. By offering scientists both a laboratory and a repository of living and dead specimens, the National Zoo contributed to these trends. However, the zoo functioned not only as a laboratory and a cabinet but also as a crossroads for science and popular culture, providing a communicative place where popular zoology could be forged by and showcased for an urbanizing United States.[1] This chapter will introduce to the history of American science little-known zoo director Frank Baker and will then turn to stories about primatology, public science, and aeronautics, respectively. The chapter will close with a reflection on animality and narrative within scientific discourses.

Frank Baker: A Superintendent of Science

As the Department of Living Animals underwent its metamorphosis into a national zoo, its marriage to science became foregrounded through congressional debates. One of the key justifications for the zoo bill was the role that a national zoo could play in advancing the scientific agenda of a modernizing nation. The National Zoo, in the eyes of the public, indeed became a symbol for *national* science. The *New York Tribune* emphasized that the zoo "originates with scientific men." The *Washington Post* claimed that "it is the duty of every Congressman who is a friend to science" to support the zoo bill. The *Boston Globe* recognized that the "object" of a national zoo would be "the advancement of science and instruction . . . of the people." The *Washington Star* predicted that a national zoo would be a "resort for scientific study and investigation." And the *Journal of Comparative Medicine and Surgery* described the zoo as an "invaluable adjunct to observation and a storehouse of material for exact work in natural science."[2] The nation's newspapers and legislators bestowed a scientific purpose on the zoo, and as we will see, despite the practical setbacks that came with zoo management, the nation's newborn zoo fulfilled its destiny.

William Hornaday was crucial to linking the mission of the department to scientific endeavors. He was also crucial to the creation of the zoo itself. Not only did he first advertise the need for a national zoo, garnering grassroots support as he traded for animals, but he also spoke to Congress directly as the principal voice of the zoo. He interviewed with journalists. He met personally the property owners of the Rock Creek valley. He communicated with zoos around the world. He contacted all types of scientists, professors, animal dealers, and naturalists concerning animal-related issues. Hornaday always served as a liaison between those inside and outside the Smithsonian. Nonetheless, in May 1890, before the zoo's official opening, Hornaday resigned suddenly due to a falling out with the Smithsonian secretary.[3] The contention between them can be attributed to several factors. First, frankly, Langley possessed a difficult demeanor that made him frustrating, sometimes impossible, to work with. And Langley constantly (always palpable in their correspondence) asserted his authority over Hornaday, even taking credit for his accomplishments. Second, Langley refused Hornaday the position of zoo director because he believed that the taxidermist, despite his experience, lacked the appropriate graduate degrees. Third, Hornaday felt exhausted and disheartened with the altercation concerning District taxes. Finally, Hornaday had received information that New York City was about to establish a new zoo. Surely, the prospects of a zoo leadership position in America's largest city would have been especially enticing for a man feeling undervalued in Washington.[4]

After Hornaday's departure, Langley gave the responsibility of managing the National Zoo to Frank Baker. Born in Pulaski, New York, in 1841, the future zoo man emerged into adulthood at the beginning of bloody civil war. He enlisted with the 37th New York Volunteers in 1861 and faced the war's worst ferocity at the battles of Fredericksburg and Chancellorsville. In 1863, when his regiment was disbanded, Baker worked in a Washington bureau where he befriended John Burroughs and Walt Whitman; and after the war, he earned an MD from Columbia University in 1880 and a PhD from Georgetown University by 1890. While at Georgetown, Baker forged quite the reputation for himself within the District's medical and scientific communities. Most famously, he worked at the bedside of James Garfield as he looked desperately for the bullet lost inside the president after he was shot at the Sixth Street Station. In 1883, Baker became an anatomy professor at Georgetown. And in the late 1880s, he established the Anthropological and Biological Societies of Washington, became a member of the famous Cosmos Club (which formed the National Geographic Society), served as an editor of *American Anthropologist*, and authored articles about human morphology. Baker's zoo career commenced in 1889 while he was simultaneously serving as vice president of the Association of American Anatomists. Langley hired him as acting manager, and in 1893, he was appointed officially as superintendent of the National Zoological Park, a position he would hold until 1916, two years before he would die at the age of seventy-seven.[5]

Zoo life did not halt Baker's scholarship. One year before the National Zoo's inauguration, he published an article on evolution titled "Ascent of Man," which he also delivered as the keynote address for the American Anthropological Association annual meeting in 1891.[6] And Baker pursued evolution again in 1899 with the publication of "Primitive Man."[7] In "Ascent of Man," he advanced Lamarckian evolutionary theory (the major competitor to Darwinian natural selection) through analysis of the limbs of humans and anthropoids.[8] While "Ascent of Man" showcased expertise in comparative anatomy, "Primitive Man" presented a timely review of anthropology's discoveries for human evolutionary history, including the famous Neanderthal (first uncovered in Belgium's Engis cave in 1829) and the newly found Pithecanthropus (a hominid discovered in Java in 1891). These articles certainly adhered uncritically to the standard Gilded Age narrative that white humanity had "climbed from darkness to light, from bestiality to civilization," a story line that also underwrote zoo-building modernization rhetoric. Within this racial hierarchy, however, Baker also sought to nuance the natural history of humankind by reading bones for the traces, "characters long undeciphered," of lost evolutionary pasts.[9]

Frank Baker's scientific research lay at the intersection of comparative anatomy, evolutionary theory, archeology, and anthropology. This cutting-edge scholarship invited job offers from the University of Michigan (tempting for Baker because of Michigan's progressive stance on the coeducation of men and women) as well as the University of Pennsylvania.[10] His own scientific interests were not related directly to zoos, but they certainly prepared him to head a scientific institution. More important than his specialized knowledge of primates (despite their prevalence in zoo enclosures) was his breadth of knowledge and range of connection. Even in the earliest years, Baker wasted no time in displaying intellectual versatility. While simultaneously overseeing the construction of the zoo, the relocation of animals from the Washington Mall to Rock Creek, and the management of a voraciously popular public attraction, he read and reviewed Professor G. E. Manigault's paper on the osteology of chinchillas.[11] He fielded questions, from government employees, concerning the current literatures of zoology and geology.[12] He advised New York publishers about which entries of an anatomy dictionary needed illustrations. (He even hinted that he would be willing to produce some of these illustrations—spinal cord, cranial nerves, hippocampus major, ovum—if the price was right.[13]) He, unsurprisingly then, also wrote and published his own anatomy textbook.[14] And he gave advice about how to organize a national anthropology conference that would promote the new fields of somatology, psychology, ethnology, and philology.[15]

Baker did not stop there. He planned experiments to test the "venomous character" of unknown reptiles.[16] He personally identified a dead beetle mailed from Texas and did the same for a coatimundi sent from New Orleans.[17] He reviewed books concerning vertebrates, embryology, and natural history for the zoo's library.[18] In the midst of the antisepsis revolution in the history of medicine, Baker developed a new "disinfecting apparatus" that worked so successfully in sterilizing the tropical house that he applied for a patent.[19] He sent a letter discussing the extensor tendons of a gibbon's hand to medical doctors, recommended special cages be purchased for an experiment that two scientists asked to conduct on the zoo's birds, encouraged the official photographing of zoo animal expressions for the "interests of science," and peer-reviewed an article concerning brain morphology.[20] He even found time to ship the feces of a zoo snake to a laboratory, to pursue his own genealogical research and trace his family to the revolutionary Vermont hero Ethan Allen, and to read Virgil's *Aeneid* with the latest scholarly annotations.[21] Frank Baker far exceeded an average comparative anatomist. He expressed a curiosity that spanned the sciences. In keeping thousands of diverse animals alive, breadth served a zoo superintendent far more than depth. In an age of scientific specialization, the National Zoo served as a site where the distinct trajectories of individual sciences collided. With a man like Baker at the helm of Washington's ark, potential chaos could be redirected into unique scientific opportunity.

Richard Garner: Zoo Experiments and the Birth of Primatology

From the very beginning, the National Zoo popularized zoology. Yet this purpose was not unique. Zoological gardens in Europe had long been built, at least in part, for the sake of science.[22] To gain support for zoos, boosters heralded these institutions as places that would support research and education, thus contributing to the advancement of knowledge. While European zoo boosters from London to Paris to Hamburg employed this rhetoric of refinement as well, the idea of "advancing knowledge" held a particular resonance in an America attempting to get past the legacy of the Civil War and subsequent Reconstruction. By "advancing knowledge" in the Gilded Age and Progressive Era, Americans strove to legitimate themselves on the world stage; and as the twentieth century approached, American intellectuals worked to create an *American* literature, *American* philosophy, and *American* science on par with Europe's. Doing so validated an American exceptionalism necessary for an imperialistic state competing with others in an age of empire.

In 1874, when the Philadelphia Zoo became the first public zoo in America, entomologist John LeConte enumerated four "events of importance to science" that occurred that year—"the appearance of the Coggia's comet," "the establishment of the Anderson Scientific School on Penikese Island," "the establishment of [public] hospitals," and "the opening of the zoological garden" of Philadelphia.[23] Even though zoos doubled as entertaining urban spectacles, the Philadelphia Zoo marked the beginning of a new type of American institution where science and entertainment could become a single pursuit. The zoo was designed, as one newspaper put it, to be a place where "children could gain knowledge of natural history . . . and fresh air and exercise."[24] A booster touted that "every man will be able to comprehend the value of a superb zoological garden as a pleasure ground, apart from its scientific importance."[25] The

National Zoo followed the precedent set almost two decades before by the Philadelphia Zoo, blending science and entertainment.

Science in action—that is, science-in-the-making, in its production—captured District attention.[26] In 1891, shortly after the National Zoo's grand opening, Professor Richard Garner carried controversial speculations about primates to the nation's zoo, believing that the new institution could serve as a personal laboratory for groundbreaking research. A local schoolteacher and armchair evolutionist living in Roanoke, Garner postulated, in the words of a Washington journalist, that monkeys possessed "articulate speech like that of human beings, differing from the latter only in development." Garner supported such claims with a series of experiments done with a graphophone,[27] Alexander Graham Bell's recent 1887 improvement on Thomas Edison's phonograph, which took the following form, as explained by the *Washington Star*:

> He has been experimenting of late with the graphophone . . . securing a "record" of squeaks and gibberings by one monkey and subsequently grinding them out through a trumpet for the benefit of another monkey, in order to observe what remarks were elicited in response. In this way he has hoped to gradually make up a lexicon of the monkey language, which would probably not be very voluminous, inasmuch as the simian vocabulary does not appear to contain more than forty words. The doctor thinks he has already secured five of them.[28]

Word of Garner's experiments spread quickly.

Eventually, Garner reached out for the support of fellow anthropologist Frank Baker. While Baker remained skeptical of Garner's claims, the acting zoo manager admitted to the *Washington Star* that "there was something" compelling or plausible "in the idea . . . of evolution backward from the language of man to the speech of monkeys." A fitting response from the anthropologist who had just published "The Ascent of Man," Baker recognized some merit in Garner's hypothesis. If evolution could explain man's body, surely it could explain man's language as well. Garner argued that the language of a monkey appeared analogous to the language of a human infant—mostly composed of vowel sounds. He suggested that the human child possessed advantages not given to monkeys: "education for" and the "hereditary experience of using" vocal organs. While their language proved to be structurally similar, human children, according to Garner, inherited greater future possibilities due to a mix of culture (education) and evolution, or, that is, both Nature and Nurture. Baker would have surely been well-versed in this sort of recapitulation theory—a biogenetic notion that the stages of development for an individual animal (in this case, a human child) resembles the stages of development in the evolution of that animal's species (in this case, a monkey). However, he doubted Garner's claims that he had successfully deciphered the monkey's "monosyllabic" words for "pain," "satisfaction," "fear," "menace," among others.[29] Through the *Star*, the public received a zoological education on evolution via Garner's theory and Baker's riposte.

Later in 1891, another article, written by Garner himself, was published in Washington. In this feature, Garner summarized the origin story of his notion of simian speech by centering it in a zoo experience seven years earlier in the Cincinnati Zoological Garden (which opened in 1875, just after the Philadelphia Zoo). Garner explained, "I was deeply impressed by the conduct of a number of monkeys caged with a savage rib-nosed mandrill, which they seemed to fear very much." He noticed that "every movement of the mandrill seemed to be closely watched by the monkeys that could see him and instantly reported to those" who could not. This observation inspired Garner's theory of monkey speech. Elsewhere, Garner remembered that one day at work, as a schoolteacher in the 1880s, he argued to his colleagues—in a faculty meeting concerning the new field of phonetics—that animals also possessed language, but he was derided for this contention. Shortly, thereafter, he began, with resolve yet great difficulty, to listen to and track by hand the utterances of monkeys. This proved mind-numbingly difficult—trying to, first, accurately listen to monkeys' sounds and, second, translate them consistently into a functional notation.[30] However, the new recording machines of the late 1880s presented new opportunities. So Garner continued, between 1890 and 1892, to observe and record monkey "specimens" in cities with traveling menageries, museums, and hand organs; aboard ships; or held as pets. Most important, though, Garner traveled to the zoos of New York, Philadelphia, and Chicago; to the Department of Living Animals in 1890; and to the National Zoological Park in 1891, to study monkey language in more "controlled" settings.[31]

Though Baker found the topic of language compelling, he only reluctantly granted Garner permission to stage experiments at the zoo, expressing skepticism about Garner's methodology and tendency toward reductivism. In a letter written to Langley, from one scientist to another, Baker expressed concern about Garner's prior use of professionally trained primates (the ones that cranked hand organs, for example) for his experiments. He also called Garner's qualifications as a scientist into question, telling Langley that Garner "was first a peddler of trees and nursery stock and afterwards a real-estate drummer. Do not understand that I consider that these occupations in any direct way detract from his merits as a citizen or diminish his general worth. I merely think that they are not exactly those that enable one to 'catch the eel of science by the tail.'"[32] Garner fell short of the standards that Baker believed should characterize a man of science. Nonetheless, he (with Langley's permission) gave Garner access to the zoo. Indeed, Baker explained to Johns Hopkins morphologist William Keith Brooks that since the "National Zoological Park is a public institution, I have thought it proper as Manager to give investigators as wide a liberty as is compatible with the health and comfort of the animals." In other words, since Baker saw no harm in Garner talking to monkeys, he felt that he had no grounds upon which to censor his research—especially since the National Zoo was a *public* zoo, a place where any American should be able to put science into action. Even though Garner's prior use of trained primates may not have passed a peer review, and even though Baker found Garner's experiments "illusory and of no real value," in the world of popular zoology and public zoos, *Garner's* peers approved: Americans were captivated by the idea of talking with monkeys.[33] As Hornaday would state from his new zoo directorship at the New York Zoological Park, "The human mind loves to contemplate . . . the apes that stand nearest to man,"[34] a sentiment that echoes many early-modern naturalists (from Linnaeus and Rousseau to Buffon and Haeckel) who pondered the location of the ape within the great chain of being.[35]

Thus, Garner conducted, before journalists and prominent scientists, a series of highly visible experiments at the National Zoo. He separated two capuchins, female and male (staging the "separate spheres" of gender that seemed to be dissolving in an urbanizing America).[36] Then, turning the crank of the graphophone, he recorded the sounds of the female and then replayed these recordings for the male: "The surprise and perplexity of the male were evident. He traced the sounds to the horn from which they came, and failing to find his mate, he thrust his hand and arm into the horn quite up to his shoulder, withdrew it, and peeped into the horn again and again. He would then retreat and again cautiously approach the horn, which he examined with evident interest." Garner repeated this procedure in reverse order, recording the male and then playing his utterances for the female, who reacted similarly. This methodology, which historian and philosopher of science Gregory Radick has carefully analyzed as the first "playback experiment," was straightforward and easy to reproduce. And scientists beginning in the 1940s and continuing into our twenty-first century will resurrect Garner's simple approach as a model for studying animal communication—from fish, insects, amphibians, and birds to the famous 1980 experiments with vervet monkeys.[37] When his National Zoo tests were completed, Garner looked for patterns and concluded that the repertoire of screams, coos, and chatterings that he recorded represented a structured verbal language. He had conducted similar experiments previously in the Lincoln Park and Cincinnati zoos, but it was within the National Zoo that "the simian tongue was reduced to record" for "the first time in the history of philology."[38] In fact, philologist E. P. Evans in 1893 believed Garner's phonograph would do for his field what the microscope was doing for medicine. Even eminent psychologists William James and Wilhelm Wundt acknowledged the promise of Garner's graphophone experiments, the former even annotating his *Principles of Psychology* with a reference to Garner's zoo recordings.[39]

In Chicago, Cincinnati, and Washington, Garner scrutinized his recordings, especially capuchin vocalizations, seeking patterns that could break their code; and in the National Zoo, he claimed to have deciphered sounds that translated to *milk-water-drink-thirst*, *eat-hunger-food*, *pain-sick*, and *menace-cry of alarm*. He also tried speaking monkey himself. From this research, which he admitted was still in a stage of infancy, Garner drew several conclusions about primate elocution. Some of these, as detailed in "Do Monkeys Express Themselves in Words?" include:

1. The simian tongue has about eight or nine sounds, which may be changed by modulation into three or four times that number.
3. The sound used most is very much like "u," or "oo" in shoot. The next one, something like "e" in "be." So far I find no "a," "i," or "o."
11. They use their lips in talking in very much the same way that men do, but seldom speak when alone or when not necessary.
14. Simians reason from cause to effect, and their reasoning differs from that of man in degree, but not in kind.

In this way, Garner demonstrated that monkeys spoke with similar sounds as humans, produced those sounds in similar ways, and, most important, articulated with Reason—the divine faculty that philosophers like René Descartes had long argued separated humans from brute creation.[40] All of these conclusions were extrapolated from the single species of monkey that Garner most cared about—the capuchin. According to Garner, the capuchin marked the "Caucasian of the monkey race," for "it" was "less vicious and more willing to treat one civilly." Also, capuchins possessed, for Garner, more refined language capabilities.[41] Despite (or probably because of) Garner's racism (which Central African historian Jeremy Rich has astutely analyzed),[42] his grander notions about language and monkeys intrigued a white public invested in zoos and hierarchies and poised to voraciously consume his 1892 book *The Speech of Monkeys*—which, from a twenty-first-century vantage, was saturated with a concoction composed of equal parts linguistic-zoological musing, evolutionist theory, pseudo-ethnography, and a noxious blend of racism and xenophobia sometimes normalized in social Darwinist rhetoric and other times laid bare through vulgar slurs and stereotypes.[43]

Garner not only fine-tuned his recording methodology within the National Zoo, but he also utilized the institution to expand his research. Using the zoo as a home base, Garner gained financial support from the Smithsonian and planned an expedition to the Gabon River, in French Congo, where he could continue his recordings with wild gorillas.[44] Henry Morton Stanley had delivered an exciting public lecture at the National Zoo a year earlier, but now the zoo had its very own explorer![45] Garner intended to take with him a selection of the most cutting-edge technologies capable of, in Donna J. Haraway's words almost a century later, the "possession, production, preservation, consumption, surveillance, appreciation, and control of nature."[46] Garner's luggage included "phonographs, telephones, photographic apparatus, an electric telegraph, and a complete taxidermist's outfit." He also prepared to take two cages—one to protect him while conducting the experiments and one to hold a living gorilla in case one could be captured alive. By attracting wild gorillas to his cage, Garner believed he could reconstruct his zoo experiments in equatorial forests, recording the sounds of gorillas that approached his cage and then replaying those recordings for others. In comparing the responses of gorillas over time, Garner hoped "to get an inkling of the meanings intended." The Gabon River expedition would simply function as a larger version of his zoo experiments, and it would, therefore, end where it started, since one goal was to "secure desirable specimens alive for the National Zoological Park." And these desired "specimens" included not only gorillas, which were quite rare in zoos prior to the twentieth century, but also other "strange beasts as yet unfamiliar to science."[47] By procuring new animals, Garner could reimburse the zoo for its support, supplying exotic-animal capital at a moment when most of the collection hailed from North America. Animals from Africa would help transition the National Zoo from an "embryo zoo" to a "modern" zoological park. Between the world wars, as historian Daniel Bender has noted, "many ships that left Asia and Africa bound for the American market were loaded with animals," but gorillas in the 1890s would surely give the National Zoo a head-start in the international zoo movement.[48] Furthermore, by donating living specimens, Garner was also doing himself a service by bringing a non-capuchin primate back to the zoo nearest his home, where he could easily orchestrate future experiments.

These plans for the expedition received two columns of newsprint. Garner's experiments, though, captured attention far beyond Washington. Sometimes, the story of Garner's relationship with the National Zoo even developed comedic dimensions. On December 13, 1891, the *Buffalo Daily Courier*, for example, published "Talking with Gorillas," which informed its readership not only that Garner planned on capturing a living gorilla for the National Zoo but also that (not-so-subtly referencing Hornaday's Smithsonian departure) "the idea is to place the animal in an official position directly under Langley, by which means it is absolutely certain any gorilla would be swearing in seven languages within six months."[49] Between 1891 and 1892, magazines and newspapers across America publicized Garner's zoo experiments and theory of simian speech. *Current Literature* published "Phonographing the Language of the Apes."[50] The *Youth's Companion* published "Monkey Talk."[51] And the *Cosmopolitan* published "Simian Speech and Simian Thought," authored by Garner himself.[52] Similarly, in New Orleans, "Monkeys Can Count: They Can Also Call for Peanuts in Their Language" headlined in the *Daily Picayune*.[53] Citizens of Salt Lake City, in their *Weekly Tribune*, perused "Richard L. Garner: Now in Africa Making Investigations in the Monkey Language."[54] And the *Wheeling Sunday Register* referenced Garner's research in an article about a lovesick monkey.[55] The experiments Garner created in the National Zoo, as he attempted to replicate them along the Gabon, aroused the sensibilities of a nation increasingly interested in popular zoology. Jules Verne in his 1901 *The Village in the Treetops* even fashioned a character (German doctor, Johausen) after Garner, observing apes from a cage.[56]

Yet Garner also raised the ire of some established men of science. Frank Baker, again in a letter to Brooks, admitted his true feelings:

> His scheme for going to Africa appears to me wild and fantastic. . . . He proposes to take a cage weighing 700 pounds with a large quantity of electric apparatuses (with which by the way he appears to be imperfectly acquainted) through a trackless jungle into the heart of gorilla country. In this cage he will establish himself, enticing to him those shy and wary animals by means of lures and then get them to speak into his phonograph so that he may record their language. This sounds more like Jules Verne than a proper scientific experiment."[57]

Even more bluntly, Baker told a Mr. Guthridge that Garner was "unsatisfactory to a marked degree."[58] And he told J. B. Skehan of the Electric Illuminating Company (which produced other recording machines) that Garner's research "was regarded as of no importance, no record of it was kept, nor has any report of it been published by the [Smithsonian] Institution."[59] Nonetheless, a public who loved *Around the World in Eighty Days* held different opinions. Certainly, Baker was quite aware that Garner's fantasies could at least function as a publicity stunt during a critical year in the financial life of the zoo. (Garner attracted so much publicity that "fan mail" meant for him even ended up on the exasperated Frank Baker's desk.[60]) The public associated Garner with the zoo that served as his storehouse, booster, advertiser, marketplace, publisher, and laboratory. And as his fame attracted ticket-purchasing zoogoers to Rock Creek, he also pulled in hefty donations from Edison, Bell, Grover Cleveland, and other rich and famous Americans.[61]

Garner embarked for Liverpool in July 1892 to gather supplies and fundraise in England. From there, in December, he left for the great lakes of Africa (the region surrounding Lake Victoria and Lake Tanganyika and an entrepôt of the British colonial network in Africa) to gather information about Central Africa's human and animal inhabitants. Months later, he finally placed his cage 2,600 miles away near the Lagune du Fernan Vaz along the Gabon Estuary (an inlet of the Gulf of Guinea), where the

A Peep at My Cage

Figure 3.1. A dramatization of Garner's cage "fieldwork" in Gabon. From Richard Lynch Garner, *Apes and Monkeys* (Boston: Ginn and Company's Athenaeum Press, 1900), 75.

largest concentration of gorillas (specifically, western lowland gorillas) still existed. When describing this outdoor laboratory, Garner emphasized the rugged masculinity that enabled him to tame the perilous wild, especially when highlighting the need for protection. Along with food and scientific equipment, Garner armed himself with "two revolvers, one magazine rifle, one air gun and hollow arrows filled with prussic acid, which he discharged with a blowgun." He also fashioned makeshift ammonia bombs that could suffocate a gorilla from ten feet away in case the above firepower was not enough to stop an angry ape.[62]

He did see, in his words, "a great many gorillas," informing the press after his return to America in 1894, "My position was such that gorillas would come very close and I could sit silently and study every detail of movement and expression." Garner also reported, confusing two types of primates, that he learned "six or eight words of the chimpanzee language" (because, apparently, all nonhuman primate species spoke a common tongue), yet conveniently he could not explain what these words were because they were "not amenable to any known etymology or orthography." He also concluded that, while far from completely deciphered, the language of monkeys contained forty or fifty words, "utterances." Finally, Garner's expedition enabled him to discover that "many things . . . in magazine articles and natural history about the gorilla are fiction."[63] What he did not tell the press (which might explain his vagueness) was that he was never actually able to get a working phonograph or graphophone (an essential component of his zoo experiments) to the French Congo! In fact, as Radick details, virtually every aspect of Garner's expedition remains foggy, especially since he did not formally document the expedition until sixteen years after its conclusion.[64] In some ways, keeping with a tradition of explorer frauds,[65] the so-called expedition could be read as a short-lived media event held on the outskirts of a Catholic mission. In other ways, however, Garner's 1896 monograph *Gorillas & Chimpanzees*, inspired in part by this trip, was one of the first books to work at dispelling popular stereotypes of "ferocious" apes.[66]

Clearly, Richard Garner could not discern simian language. Clearly, he could not reduce the complex behaviors of nonhuman primates to the signs of English or to the frameworks employed in the new scholarly studies of historical linguistics and human anthropology. Yet his experiments were very successful in drawing public attention to the *possibility* that monkeys and humans shared an even closer affinity. By traveling to Africa, Garner traveled, in the eyes of the public, to Eden, to a land before time where the origins of humanity still lived in wild jungles. Haraway, in *Primate Visions*, has importantly emphasized the Edenic dimensions of early primatology, calling attention to notions of race and gender entangled with the images of primates throughout the twentieth century.[67] Even though Garner pushes back Haraway's intellectual history of primatology to the early 1890s, his exploits does not alter her primary thesis that primates, as boundary dwellers, functioned as perfect tabula rasae upon which "White Capitalist Patriarchy" could inscribe their own pretensions and insecurities, a process that supported the emerging field of primatology, its determinisms, and its place within the sciences. Garner also underscores an important corollary evident in Haraway's classic (and a major contention in the "Science Studies" movement in which she is a major figure)—namely that "the boundary between technical and popular discourse is very fragile and permeable."[68]

Garner used the zoo to popularize primatology. He, and the science-culture hybrid he embodied, helps explain the emergence of not only this specialized science but also the growing public interest in scientific issues previously considered esoteric. Frank Baker, in his own research, for example, asked similar questions of primates as Garner did. *How did human cultures arise from primate cultures? What can humans learn about themselves by looking at their primate ancestors? How can "modern" Science conquer the previously "wild" and unknowable domain of "the animal"? How do studies of primates solidify American-male-white-scientific supremacy?* Whereas Baker encoded these questions in the language of evolutionary theory and the morphological concerns of anatomy and anthropology, Garner packaged them in rhetoric and narratives that lay audiences could understand. He fashioned seductive zoo stories from, or within, the gap—reproduced by the zoo—between the desire to control, to know, "the animal" and the desire to encounter its alterity, otherness, wildness. Garner's adventures laced these desires together for Americans following his experiments. After his return from Gabon, in Charlotte, newspaper audiences read "He Knows Six Gorilla Words."[69] In Knoxville, "Talks to Gorillas."[70] And in Topeka, Bismarck, and Trenton, citizens read about "The New Philology."[71] Even in Prescott, Arizona, newspaper-toting miners relived Garner's expedition two years after his return.[72]

For Garner himself, the years between 1891 and 1894 marked the beginning of a long career dedicated to the speech of monkeys. And throughout the 1890s and the first decade of the twentieth century, he returned to the French Congo multiple times to amass more evidence for his theories and to collect gorillas for zoos, as animal historian Tracy McDonald has shown through her moving and tragic biography of Dinah, an individual captured by Garner for the New York Zoological Park in 1914.[73] In 1900, Garner published *Apes and Monkeys*, which detailed the research he had conducted on "simian speech" thus far in his career. Garner, though, never lost sight of the general public. Indeed, in many ways he stands as an exemplary symbol of popular zoology because his entire scientific career was directly shaped by and tailored to the interests of the public. In the opening pages of *Apes and Monkeys*, Garner expressed the intentions that undergirded his science, stating, "A careful aim to avoid all technical terms and scientific phraseology has been studiously adhered to, and the subject is treated in the simplest style." Furthermore, Garner, writing in the third person, admitted (reaffirming one of Baker's original concerns), "Most of the acts related [in this book] are those of his own pets. A few of them are of apes in a wild state. The author has carefully refrained from abstruse theories or rash deductions but has sought to place the animals in the light to which their own conduct entitles them, allowing the reader to draw his own conclusions."[74] Garner not only candidly revealed the *popular* motivations of his science but also underscored two important dimensions of popular zoology. First, popular zoology must not rely upon the arcane and ostracizing languages of professional science. Second, and similarly, popular zoology should not make theory its focus. Instead, it should allow lay audiences to experience animals and "draw" their "own conclusions." This simple belief in *experience* defined the entirety of Garner's career. Prior to his first trip to Africa, he told the public, "Many who are called men of science spend their lives in acquiring knowledge of what others have done, and these men of letters are usually regarded as the high priests of science; but they are really the laymen."[75] Garner pushed Americans to experience the wild and make zoological discoveries for themselves.

In 1905, Garner built a house on the shores of the Lagune du Fernan Vaz. And between 1916 and 1919, again with full Smithsonian sponsorship, he made his last expedition. Shortly after this last return to the United States from Africa, Garner died in Chattanooga, Tennessee, five years before the Scopes Monkey Trial would erupt forty miles away in the small town of Dayton.[76] Although many of his overblown claims about simian language proved to be, at best, naively overstated, or, at worst, purposefully deceptive, and although his experiments proved to be, at best, unprofessional, or, at worst, pure profiteering, his larger project addressed questions that fascinated both scientists and laypeople alike. *How similar were humans and monkeys? What could monkeys teach us about our own natural history? Did we have language and culture from the Beginning?* Questions like these made Darwin and Lamarck relevant and digestible for Americans at the turn of the century. As American zoogoers increasingly came into contact with their living ancestors, held captive in monkey houses, they increasingly confronted the implications of this encounter. Garner created a narrative that, in some ways, bridged the divide between human and nonhuman primates by embracing the ancient dream of cross-species communication[77]—one that is still alive and well at the time of this writing, especially as displayed in the launching of the

Cetacean Translation Initiative (Project CETI), which seeks ultimately to use Garner's playback experiment to decipher the language of sperm whales.[78] In other ways, of course, for countless obvious reasons, Garner's experiment extolled and centered (white, male, American) humanity as it conceived of other primates as lesser than, worthy to be used as "subjects" (objects) of science.

By "communicating" with monkeys, Garner experimented at the fringe of humanity, in a realm Haraway labels "science fiction," "where possible worlds are constantly reinvented in the contest for very real, present worlds."[79] For American zoogoers in the 1890s, primates represented new, exotic worlds, and language functioned as a denominator through which to order humans and other primates. Despite Frank Baker's professional opinion, and despite his serious problems and biases, Garner's experiments were repeated within zoos—by herpetologist Raymond Ditmars in the Bronx Zoo in 1920 and by broadcaster Ludwig Koch and evolutionary biologist Julian Huxley in Regent's Park Zoo and Whipsnade Park Zoo in the 1930s (which led to *Animal Language*, a "sound book" that presented zoological text, pictures, and records of zoo animals).[80] Garner's research became directly important to formal primatological science by inspiring comparative psychologist, ethologist, and primatologist Robert Mearns Yerkes in the mid-1920s.[81] Yerkes, hired as a psychobiologist at Yale University in 1924, founded both the Yale Laboratories of Primate Biology and the Anthropoid Breeding and Experiment Station in Orange Park, Florida, where the primate language Yerkish was first developed and where primatology was partially legitimated as its own scientific field.[82] Furthermore, as Radick argues, the results of Garner's experiments and showmanship were not nearly as important to the history of science as the "playback" apparatus and methodology that he developed and popularized in the zoo. The "playback experiment" will become an "off-the-shelf resource, susceptible of endless variation," adaptable for laboratories or the field, and can be tracked through twentieth-century primatology and zoology, marine biology, musicology, or even (revealingly) through early anthropological studies of American Indian languages.[83] The history of language and primatology is complex. Richard Garner, though, shows us that this history, in part, finds its first chapter in the birth of the National Zoo, where primates were, for the first time, showcased before a nation and where a tree-peddler-turned-scientist first staged experiments for a zoogoing nation craving zoology.

Zoo Sciences and the Public Sphere

Beginning in 1891, Garner's graphophone experiments established the National Zoo as the perfect public laboratory, where science and entertainment met and mingled, where science could educate a nation, and where a nation could shape its science. Shortly before monkeys talked into Edisonian devices, Samuel Langley ordered an astrophysical observatory to be built in the park so that the Smithsonian's current observatory could be relocated from a ramshackle shed behind the Smithsonian Castle (where the Department of Living Animals was also located) to a more appropriate setting.[84] Like the controversies about whether a federal government should and could support a zoo, a parallel controversy ensued before Congress about whether the same federal government should and could pay for an observatory built on a site reserved for animals. One House Representative from Tennessee, Benjamin Augustine Enloe, complained that the purpose for such an out-of-place building within a zoo was solely to provide "men of science" a place to conduct research on the nation's dime. Langley responded to this criticism (upon the floor of Congress even) by emphasizing the humble nature of the observatory he intended to build. The structure would not be an ornate domed building but simply a "low shed." He explained that the seclusion of Rock Creek, "free from noise and tremor," made it an ideal site for housing instruments designed to measure earthquakes and view stars.[85] Unfortunately, for Langley, the Smithsonian had run out of political capital, and Congress denied the request. Nonetheless, the request itself underscores the scientific intentions built into the zoo from the very beginning. The National Zoo could support all sciences, even the nation's new geo- and astrophysics.[86]

Yet the zoo most often served the life sciences. Usually, scientific education about animals did not happen as part of a highly publicized experiment like Garner's, nor as part of a systematic educational program spearheaded by the zoo. Instead, popular zoology operated below perception, entangled within the zoo experience. In other words, as zoogoers engaged the zoo, as they found amusement, entertainment, and escape along its walkways, they simultaneously learned about science—or in the very least, they absorbed scientific discourses through their consumption of animals. Journalists, by weaving scientific data into their zoo stories, played an important role in this type of quiet education. As early as June 1889, the *Washington Post* published an article titled "Sunday at the Zoo," a column that would become a consistent feature. This article discussed a wide range of topics: a goat's consumption of popcorn, the pumas' hatred for small dogs, and a grizzly bear's fondness for garden-hose baths. Interwoven into this medley appeared a scientific blurb about antlers. The article raised a question frequently asked by zoogoers: "Many persons have called attention to the antlers of the elk . . . covered with grey velvety material?" The *Post* then continued to describe the sensitive nature of "antler velvet," the rubbing behaviors employed to "allay irritation," and the seasonal rhythms of antler-shedding.[87] While zookeepers could not yet conceptualize this velvet as a growth hormone, articles like this one placed "antlers" into the realm of popular zoology. Maybe most important, the column taught readers that an "ideal" zoogoer should approach the zoo through observation and questioning, situating the zoo experience within a zoological tradition of observation stretching back to Aristotle.[88]

One year later, the *Evening Star* tackled zoology more directly. In "This Is Our Own Park," the newspaper explained that one of the primary functions of the public "playground" was to conduct scientific experiments that could examine "the habits of beasts and birds and the limits of species." The *Star* then explored some topics that sat at these limits. "A rough definition" of "species," according to the paper, should make clear that individuals *of the same species* must be able to produce fertile offspring capable of reproduction. Accordingly, the dog and the wolf as well as the horse and the ass were considered members of different species (the first two are now understood as members of *Canis lupus*). However, as the article emphasized, the species of many known animals remains a mystery. "Take the case of the puma," the article posed, "which is found, with variations in shape . . . all the way from the Far North to the other end of South America. Are these different pumas different species or different varieties?" The National Zoo, the *Star* contended, would be the perfect laboratory to settle such philosophical-scientific debates: "The park will afford naturalists opportunities for adding to the present knowledge of zoology." In so doing, a leading Washington newspaper educated the public about systematics (a branch of biology devoted to classification), which consumed biologists and zoologists around the Atlantic world.[89]

Zoogoing publics read about all sorts of scientific issues in newspaper articles that were, on their surfaces, devoted mostly to entertainment, leisure, and publicity. In 1899, for example, when the National Zoo began a large-scale purchasing campaign to bring legions of new and exotic animals to Rock Creek, a torrent of articles flooded the national public sphere, from New Orleans to Grand Rapids, advertising this grand expansion.[90] While boosters tried to seduce their readers with superficial prose devoted to spectacle, they also taught zoological knowledges. One article (with the title "Uncle Sam, His Show" and the subtitle "It Will Undoubtedly be the 'Greatest on Earth'") discussed in detail the new acquisitions. First, the National Zoo hoped to obtain a gorilla, which would create quite the fanfare, since "the gorilla has never yet been brought to the New World."[91] (Garner never did bring one back to Washington.) When hyping this potential addition, the article detailed the gorilla's range, habitat, defensive habits, and early natural history (beginning, of course, in antiquity with Admiral Hanno of Carthage) while also emphasizing the dangers and dramas of capturing a gorilla that are essential to a good zoo story.[92] Second, the zoo planned to acquire chimpanzees, baboons, and an assortment of monkeys. In similar fashion, the article provided data about their range and defensive habits, even detailing the physical characteristics of a particular species of baboon called the hamadryas while also calling attention to the occult powers of baboons. "Uncle Sam, His Show" blended scientific information with advertising. And after forming these hybridized descriptions of animals, fusing entertainment and science, the feature listed other specimens that the zoo wanted—giraffes, wart hogs, manatees, zorillas, spectral tarsiers, and tamarau. The National Zoo's expansion, the article concluded, would "out-Noah Noah," attracting throngs of zoogoers who would become acquainted with a long register of unheard-of creatures.[93]

Zoo stories used science to legitimate the zoo. And they played an important role in structuring the experience of zoogoers, allowing them to feel as though they were taking part in a refined "civic zoology."[94] The zoo stories that weaved science and entertainment added texture to the spectacles of circuses and menageries. Zoo stories made both science and entertainment more appetizing for zoogoing publics. And they continue to do so today in the twenty-first century. Zoo stories changed with the times, but the idea of popularizing science and disguising spectacle through subtle "edutainment" (a portmanteau first used by Walt Disney in 1954) has remained a central mission of zoos and aquariums.[95]

Throughout the National Zoo's first decade, entertaining articles about science consistently brushed against page-turning fingertips. In the winter of 1891, a bison calf was born, the first "zoo baby" of its kind in Washington, DC.[96] In an early iteration of cuteness, an article titled "An Odd Infantile Pet" christened this calf "the idol of the children" and sought to bring kids and their parents to the zoo.[97] The bison, "Baby May Weeden," received her surname in honor of keeper W. C. Weeden, the department's first keeper, and her first name was bestowed upon her by a child:

> One day last week a pretty girl, who has taken a lively interest in the baby buffalo, went down to the pen on the Mall and christened the baby May Weeden. The young lady herself is named May. . . . The pretty girl stood beside the pen with the gloved fingers of one hand hooked in the wires of the enclosure, while the little brown calf stuck its wet, black nose out to her inquiringly, and the old cow watched rather sullenly, and the young lady said: "I christen you May Weeden."

The article continued describing this endearing encounter of entangled cross-species innocence with thick anthropomorphisms and gendered, classed, romantic imagery, portraying Baby May's ears flopping "back and forth," describing the manner in which she sniffed and licked the hands of children, and generally emphasizing her "wonderfully spry" character. Although clearly giving precedence to entertainment, the article nonetheless nodded to zoology. What began as a maudlin story of a little girl and a neotenous baby bison concluded as commentary on the "doctrine of affinities," which consumed every evolutionist and morphologist of the day. The last two paragraphs explored the common descent of ruminants and the maturation process of bison as they grow from infancy to adulthood and offered a brief comparative analysis between the evolution of ruminants and the evolution of species descended from the wild ass—horses, mules, and zebras.[98] Entertainment and science went hand in hand.

In July 1900, a three-page feature informed the public about the diversity of animals housed, including the largely unheard-of tayra, a marten-like mammal of South America. The feature announced that the zoo offered formal classes in natural history and art.[99] That summer, zoogoers even had the opportunity to observe famous wildlife artist Ernest Thompson Seton at work drawing animals from a temporary cabin built by the zoo solely to support his art and ecological studies, the most fascinating of which explored antelopes' uses of flatulence and tail-signaling for communication.[100] In November 1900, through an article tellingly titled "Some Queer Freaks Among the Cubans," readers learned not only racist adjectives by which to think of the Caribbean subaltern but also about the habits of iguanas, the habitats of bats and trogons (birds in the same order as quetzals), and the natural history, habits, habitats, anatomy, range, speciation, and diets of hutlas (a tree-climbing, ratlike rodent).[101] In 1901, District citizens became educated on the range, plumage, and taxonomy of the king vulture.[102] And they learned about the nocturnality of many species, from hyenas to wolves.[103] In 1902, readers examined in precise detail the characteristics of the world's largest carnivore, the Kodiak bear of Alaska, which was for the first time acquired by the zoo.[104] And on the same day, the public also read an entire column about the habits of musk ox.[105] In all of these writings and countless others, journalists reported gripping stories about the National Zoo that presented scientific information about zoo-animal protagonists. Journalists treated the zoo, like the National Theatre, as a regular source of stories, an important site in the cultural fabric of the District.

As evidenced above, science and entertainment comingled in print culture. Sometimes newspapers favored science. Other times, entertainment. Rarely did articles fuse both together in equal proportions. In fact, in most cases, especially in the 1890s, entertainment usually trumped scientific concerns. These priorities could be seen in the headlines that introduced readers to a good story rather than an informative tract. 1892: "Battled With a Deer: Keeper Blackburne's Bloody Fight with a Vicious Buck."[106] 1892: "Christening at the Zoo: Pet Names Bestowed Upon Animals in the Collection."[107] 1892: "Fraternalism at the Zoo: Native Animals Make Friends with the Imported Ones."[108] 1893: "Midnight Out at the Zoo: Sights and Sounds after Gates Are Closed to the Public."[109] 1896: "Tiger That Eats Rats: Curious Tastes Developed by Ex-Man Eater."[110] On the surface, articles with these titles seemed to be intellectually hollow, casting the zoo as a lowbrow cultural theater that staged dramas performed by a multitudinous cast of anthropomorphized nonhuman actors. However, no matter how colloquially comical or lively a zoo headline appeared to be, science often made subtle and surprising appearances within the plots and scripts of zoo stories.

After the turn of the century, articles about the happenings of the National Zoo moved away from the amusing headlines that signaled comedic plots and more frequently presented technical headlines that signaled reports meant to inform and educate. Issues of science moved from the margins to the text itself. Headlines and bylines speak for themselves. 1902: "Sea Water for the Zoo: Vessels Bring It in Tanks from Off the Coast of Virginia."[111] 1902: "Arctic Foxes at Zoo: Novel Experiments with Fur-Bearing Animals."[112] 1903: "Patients at the Zoo: How the Sick Animals Are Cared For: A Full Record Kept."[113] 1903: "Painting Animals' Eyes: One of the Most Difficult Tasks of the Government Taxidermist."[114] 1903: "Meat Contract for Zoo Animals: Clerk Tells Why the Lowest Bidder Didn't Get It."[115] 1907: "Big Lion Chloroformed At the National 'Zoo' to End His Suffering."[116] At no single point in zoo history did technical headlines replace colloquial headlines. Yet the rising rate at which technical and scientific headlines appeared in newspapers marked a transformation in the intellectual palate of Washington zoogoers. These zoogoers were, by the turn of the century, interested in the intricacies of zoo animals—their bodies, diets, and deaths, topics that often required some type of scientific, medical, or technological vocabulary. The National Zoo certainly amused zoogoers, but it educated them as well, even (as the titles reveal) on the deficiencies of the zoo, topics that zoos will increasingly bury from public scrutiny by World War II.

Throughout the first two decades of the zoo's existence, correspondence riddled with questions about science inundated the Holt Mansion, where the administrative offices were housed. The zoo not only schooled zoogoers in discourses of experiment and observation but also taught the specialized languages of the professionalizing sciences. In 1903, a *Washington Post* byline described the National Zoo as "A Place to Improve Your Pronunciation." This article quoted keeper Blackburne, who contended that the zoo made the schoolchildren of Washington "quicker, shrewder, brighter, and greater sticklers for proper pronunciation." The "Washington youngster has even taken the shine off the Boston urchin," raised in the Puritan epicenter of American education. Blackburne relayed the following story regarding a Himalayan bear:

> We received a very fine Himalayan bear. . . . A little girl who, after gazing at Mr. Bear for some few minutes, turned to interrogate me as to its species. 'It's Himaalayeyun bear,' I replied. The words had hardly left my mouth when the little miss, with a look half scornful, half pitying, replied: 'Oh, don't say Himaalayeyun; don't you know better?

> That isn't right; it's Heemahlahyahn.' Since which time I have been careful to say 'Heemahlahyahn.'

Blackburne similarly told a story about a "bright-eyed tot" ("well posted on zoology" and "forward with his French") who informed him that the sacred cow of India should be pronounced "zayboo" and not "zeebyew." This child also corrected Blackburne in his use of the word "buffalo" when referencing American bison, for the name "buffalo" was only supposed to refer to water buffalos, like the ones of the Philippines, housed, the child knew, in the kangaroo exhibit.[117] These musings about pronunciation advertised the zoo as an educational medium of science for the emerging middle classes, a place where children could practice fancy zoological elocution.[118]

Few anecdotes from this section would have been recognized as Science to professional scientists. The zoo stories that blended scientific information with age old tropes and myths about animals would never have constituted a scientific education in the eyes of any university zoologist. (In many ways, even these scholars were infants. In the academic year of 1890–91, just as the Department of Living Animals was becoming the National Zoo, Harvard University had just opened its Zoölogical Department, the first in the United States.)[119] Void of specialized scientific languages, lacking the structure of scientific method, missing the veneer of scientific neutrality, and created for the obvious purpose of entertainment, zoo stories have been overlooked, then and now, for their cultural influence. Zoo stories popularized science, translating the erudition of laboratories into the everyday conversations of zoogoers. If, according to critical theorist Nancy Fraser, the "public sphere" is "a theater in modern societies in which political participation is enacted through the medium of talk," then this study expands the conversations of the public sphere to encompass science. The zoo served as a "theater" for "scientific participation," ushering everyday zoogoers into conversations about animals previously reserved for naturalists and scientists.[120] In the space of the zoo, zoogoers talked about the animals they encountered. Surely, the discourse of popular zoology often consciously entertained more than it educated, yet subconsciously, and in the aggregate, through the personal experience of curiosity, the objects of lighthearted amusement could become the objects (and sometimes subjects) of serious interest. Just as literary scholars have critiqued commonplace notions of highbrow versus lowbrow as overly simplistic, similar challenges should be leveled at the way science is and was perceived. A child trying to properly pronounce "Himalayan" or zoogoers pondering antler velvet constitutes Science as much as an astronomer experimenting with flight—a topic to which we will now turn.

Samuel Langley: Zoo Experiments and the Birth of Aviation

Samuel Pierpont Langley served as the third secretary of the Smithsonian Institution, from 1887 to 1906. When it came to the National Zoo, Langley usually played the role of overseer, delegating the day-to-day responsibilities of zoo governance to Frank Baker. Nonetheless, Langley still let his power be known by sending Baker frequent directives about issues of public visibility. And while he typically allowed Baker to manage the zoo as he saw fit, Langley would occasionally fall into binges of micromanagement, something for which he became notorious throughout the Smithsonian.

As secretary, Langley still pursued his own scientific interests. Before the Smithsonian, he worked at the Harvard Observatory, Naval Academy, Western University of Pennsylvania, and Allegheny Observatory.[121] Shortly after his arrival to the District, Langley advanced the new line of research for which he would be remembered—aeronautics. On May 6, 1896, he flew the Aerodrome No. 5, an aircraft equipped with a one-horsepower steam engine, for three-fourths of a mile, exceeding any distance reached by previous experimental aircrafts. After this flight, with support of the US Army, Langley developed the Great Aerodrome (also known as Aerodrome A). Piloted by Charles Manly, this flying machine failed to actually fly. It crashed into the Potomac, before crowds of onlookers, on October 7 and again on December 8, 1903. Nine days after this second failed flight attempt, the Wright Brothers famously achieved their four successful flights near Kitty Hawk, North Carolina. However, despite failed experiments, Langley represents an important figure in the history of aeronautics, and his Great Aerodrome, despite its crash landings, marked an important, high-profile flight attempt. In addition to his Aerodrome experiments, Langley also published several seminal articles on aerodynamics.[122]

On the surface, Samuel Langley's involvement in science seemed far removed from the life sciences. Aeronautics, the scientific study of human flight through the atmosphere, fused the sciences of physics, meteorology, and aerodynamics with engineering. Nonetheless, upon close inspection,

Langley rooted many of his aeronautic experiments in the life sciences, and he took advantage of the zoo at his disposal. Since these connections do not usually get showcased in early-twentieth-century flight experiments or their historical accounts, we will need, for the next few pages, to zoom in to isolate important details that quickly get scrubbed from the surface of science.

On March 14, 1901, Langley informed Baker that he wanted to measure the wings of six different birds—the condor, golden eagle, Egyptian crane, Jamaica buzzard, albatross, and frigate bird. He hoped Baker would "prepare a table . . . of columns to be filled in later with the weight of the bird; the wing surface, divided by weight, showing how much goes to the pound; and the distance from tip to tip of the wings as extended in flight." Langley asked Baker to procure any of the above species not currently held in the zoo, suggesting that the frigate bird could be found in Santiago's harbor, the albatross "in the neighborhood of New Zealand" or "on the Island Laysan [Kauō], one of the Hawaiian group," and the buzzard (also known as a "John crow") from Jamaica. In addition, the Smithsonian secretary added that he was "particularly interested in getting the *largest* specimens of birds," and he stressed that he wanted all the measurements taken on living, not dead, birds.[123] In response, Baker informed Langley that the zoo possessed only the golden eagle but that the other desired birds could probably be purchased from Carl Hagenbeck, the famous Hamburg animal dealer. Regarding the "John crow," Baker suggested that perhaps a Turkey buzzard would suffice since they belonged to the same species.[124]

Five weeks later, Baker had acquired a condor and Jamaica buzzard, and he immediately sent the requested measurements, taken by head keeper Blackburne, to Langley. The condor weighed eighteen pounds and eight ounces, with a nine-foot, two-and-a-half-inch wingspan. The golden eagle weighed twelve pounds, with a six-foot, ten-inch wingspan. And the "John crow" weighed three pounds and twelve ounces, with a four-foot, nine-inch wingspan. In addition, in regard to finding a frigate bird and an albatross, Baker informed Langley that he had not only sent a letter to Hagenbeck but also sent one to the Commanding Office in Santiago, Cuba (which employed a prominent animal dealer who had previously sent the zoo a pair of flamingoes) as well as to New Zealand's fisheries commissioner.[125] The National Zoo, then, served not only as a storehouse of living animals for scientific study but also as a purchaser for that storehouse, one that utilized the expanding imperial and diplomatic reach of the US government, as well as the economic reach of the international animal trade.[126] Popular zoology ushered the public into science and science into the public, but it also depended upon a global market economy, increasingly driven by big business (like Hagenbeck's) and big government. The animals supplied to zoos were often purchased from the same dealers who supplied menageries, circuses, pet stores, and laboratories.

The zoo also provided Langley with laboratory assistants. Baker mentioned that Blackburne did not "attempt the calculation of wing surface to weight on account of the considerable variation in the shape of the wing."[127] After hearing this, Langley responded with a four-page instruction manual about the appropriate procedure for calculating wing surface. First, he described the basic method that Blackburne should employ:

> The determination of wing surface may be made far more accurately by placing the bird, alive or dead, on a sheet of coarse white or brown paper wide enough to receive the shadow of the body and tail of the bird and at least one of the wings. Let the shadow be taken near noon, when a pencil can readily be run around it and a correct outline traced. The wing should be supported by the tip so as to be a trifle turned up as during flight, when the shadow is taken. Afterward the sheet of paper may be ruled up roughly into inch squares and the number of whole squares may be taken and also the number of broken squares (which may generally be averaged a half a square each). Hence we get the area of the wing and thence of the two wings.[128]

This process seemed quite rudimentary. *Stand a large bird on a piece of paper at high noon. Spread the wings to their fullest extent. Trace the shadow. Rule paper into squares. Repeat.* Langley continued to elaborate on how to assess wing curvature.

"The curvature," he explained, "should be referred to a plane parallel to that of the level sheet of paper the bird is standing on." He continued with an example:

> Suppose, for instance, the bird was facing the south, and was consequently under the sun at noon when it is highest; then we may suppose two north and south sections to be made through the wing, one

quite close to the body (1), the other at some point near the middle of the wing (2). In many specimens the section nearest the body will have the chord of its arc pointing slightly downward, as in this illustration, (C) while the chord near the middle of the wing may be pointing horizontally or a little upward, (D). This may be not always so.[129]

This process for calculating wing surface area and curvature was quite mechanistic. The avian individuals described would be stood upon paper, manipulated, encouraged to hold still, and then reduced to rough pencil sketchings, delineated with bisections, a perfect (dis)embodiment and twentieth-century version of the Cartesian *bête-machine*. It is almost hard to believe that the birds in this description would be *living* birds! The "soaring birds" that captivated Langley were definitely not the same ones that captivated zoogoers. Langley's birds looked like the one below—pinned and opened for study.[130]

Blackburne and Baker, though, knew personally how *living* Langley's machines were, for they were the ones chosen to pose living birds of prey on a single sheet of paper, with wings extended and presumably flapping, long enough to get an accurate tracing that would meet the secretary's expectations. While no record exists concerning how this task was accomplished, it is clear that the condor gave Blackburne a difficult time, because Langley concluded his letter, "I am very sorry to hear of the annoyance that I have been to the condor in these measurements, but I hope that useful bird will feel persuaded that it is all in the cause of science."[131] Whether Langley's concern about the condor was genuine, or whether he was using "condor" as an interspecific euphemism for "Blackburne," remains unclear; however, no matter Langley's true feelings, the condor must have, one can imagine, resisted in some way to being manipulated and modeled upon stationary as hands pulled open her wings.

These measurements of living zoo birds helped Langley pursue flight. To accompany the instructional letter discussed above, Langley also sent Baker and Blackburne a copy of a missive he had sent a year earlier to Robert Ridgeway, the curator of birds at the US National Museum, which revealed the true purpose for his study. Langley informed Baker and Blackburne, "I may mention that the line AB in the drawing enclosed . . . is supposed to represent the trace of a plane passing through the center of gravity of the bird when its wings are extended in the attitude of flight." While these drawings, Langley admitted, would be difficult to

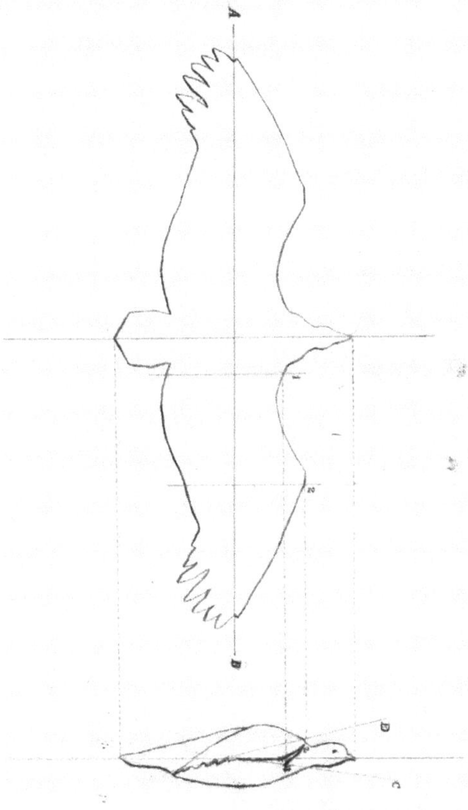

Figure 3.2. Soaring Bird Sketch. A model sketch that Langley sent to Baker and Blackburne. Courtesy of the Smithsonian Institution Archives.

obtain from a living bird, he wanted Blackburne to "familiarize himself" with them.[132] More important, the images of a soaring "John crow" [below] reveal how the birds of the National Zoo could be helpful to a scientist interested in aeronautics, aerodynamics, and the engineering of flying machines.[133] With the exception of German aviation pioneer Otto Lilienthal (specifically through his seminal 1889 *Birdflight as the Basis of Aviation*), few experimenters of human flight gave birds close, serious, scholarly attention.[134] Rather, birds often just functioned as rhetorical backdrop, sources of inspiration, over which humanist stories of striving and invention could be told by those, like the Wright Brothers, seeking to achieve flight.[135] Yet as Leonardo da Vinci recognized in 1505, "birds are instruments working according to mathematical law."[136] Langley, four centuries later, just wanted to discover their equation.

In studying these living birds, Langley was most interested in their

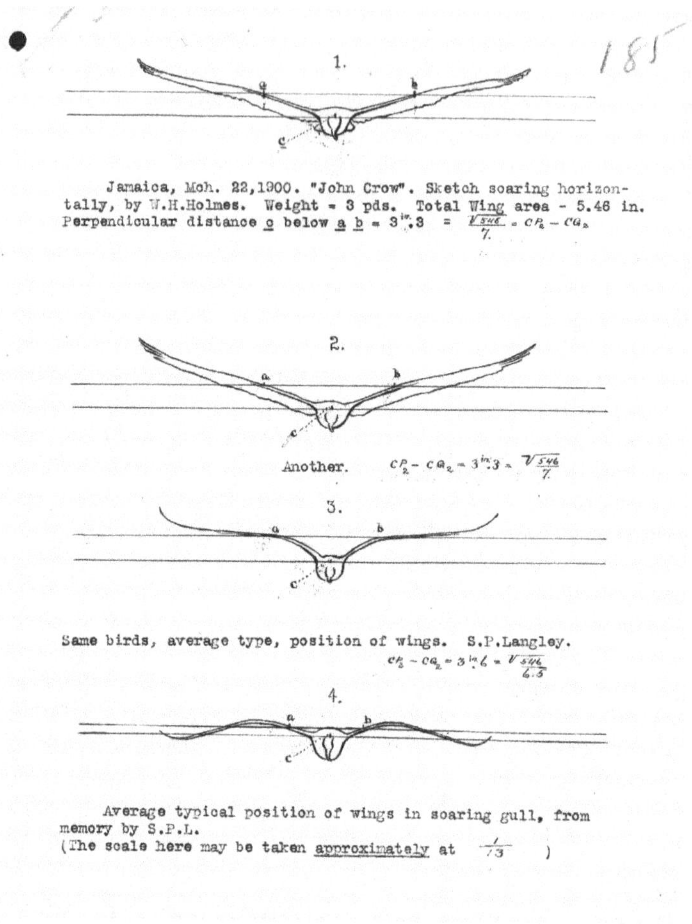

Figure 3.3. Sketch of soaring John crow and seagull sent by Langley to Ridgeway. Courtesy of the Smithsonian Institution Archives.

centers of gravity and pressure. As he explained, all birds possess one center of gravity, but this center could be "considered with reference to its position on the horizontal plane of the bird with wings extended" or it could be considered "with reference to a vertical plane" when the bird's wings are closed. In other words, the center of gravity "would have a certain position when the bird was at rest, and another position when it was soaring and the wings were above the body." The center of gravity while at rest was labeled "C" in Figure 3.2. Points "A" and "B" in Figure 3.3 represented the two centers of pressure present on a soaring bird, one under each wing. At these points "all the efforts of the upper pressure of the air may be supposed to be centered." Langley also thoroughly summarized the methods used to find the center of gravity and centers of pressure, which, ironically, could only be done on a recently deceased bird. To find the center of gravity, a dead bird's wings needed to be spread and held in position with wire. The bird then needed to be suspended from the ceiling by the tip of a wing so that the calculator could note where this line would theoretically pass through the body. The same process would be repeated after hanging the bird from the ceiling by either head or tail. The point where these two imaginary lines crossed, within the body of the bird between the breasts and mantle, would theoretically be very close to the individual's center of gravity. The center of pressure for one of the bird's wings could be found by accurately tracing its outline on thick paper, cutting it out, and then balancing it on the tip of a pencil! The point at which a single wing could lay horizontally on the end of a pencil would be very close to the center of pressure for that wing, and the second wing should have a corresponding center of pressure (assuming that wings were always perfectly symmetrical). These centers of pressure could be corroborated by cutting the wings from the dead bird's body and balancing these as well.[137]

The purpose of explaining, in such detail, Langley's methods of analyzing bird anatomy is to underscore four important points about the avian foundations of his aeronautical experiments. First, it is clear that Samuel Langley thought about birds as machines that achieved flight by neutralizing the opposing forces of downward gravitational pull and upward air pressure. By thinking about birds in this way, Langley hoped to replicate their physical structure in his designs for flying machines, thus rooting his aeronautical technology in the evolutionary "technology" of soaring birds. Second, to properly assess this evolutionary "technology," Langley needed both dead and living birds. From living birds, Langley could find weight (something that could not be calculated from a dead bird once fluids were lost and decomposition set in); wing surface, divided by weight, creating a surface-per-pound ratio; and distance of *actual* wingspan, as a living bird spread wings in life, not as they were artificially spread by humans after death. Modeling living birds enabled Langley to "check" many of the figures extrapolated from dead birds. Third, comparing how Langley thought about birds with how he thought about flying machines will inevitably bring parallels to light, showing the influence of the life sciences on a branch of science (aeronautics) usually considered as distinct from those sciences focused on organisms. Fourth, Langley's conflation of birds with flying machines, and vice versa, shows how deeply Descartes's mechanistic perspectives on animal automata had infused the mentalité of science over the preceding three centuries.[138]

While Langley's interests in zoo birds materialized in 1901, he had thought about birds before. According to his memoirs, Langley "often stated that even as a boy he was interested in watching the motions of hawks and buzzards, and he wondered by what mysterious power birds so much heavier than the air could maintain themselves in space and could move about at will without apparent movements of their wings." In 1886, while listening to a paper delivered at the American Association for the Advancement of Science, this foundational curiosity developed into a professional interest when "it seemed to him that prevailing theories as to how birds fly were not based on sound facts, and he resolved . . . to ascertain . . . what mechanical power was required to sustain a weight in air and make it move at a given speed."[139] From the beginning (at least according to how the man narrativized his own life), Langley's interest in human flight centered on living birds.

In 1891, through publication of *Experiments in Aerodynamics*, Langley first declared that "mechanical flight is possible with engines we now possess."[140] While this is not the place to detail Langley's physics, mathematics, or proposals concerning the actual engineering of engine-powered flying machines, it is worth noting the bird references perched in this foundational text. In his third chapter, titled "The Suspended Plane," Langley challenged what he saw as a widely held misconception about the physics of avian flight, namely, that "the energy expended [by flying birds] increases with the velocity attained." This assumption, first forged by the French engineer and physicist Claude-Louis Navier in 1832 while studying the flight of a swallow, wrongly likened avian motion to pedalistic animal motion, advancing the view that "there exists the same ratio between the efforts necessary for simple suspension and for rapid flight as exists for terrestrial animals between the effort required for standing upright and that required for running."[141] After "proving" this mathematically, Navier's assertion led to the long-standing opinion in the scientific community that "mechanical flight is practically impossible."[142] However, through *Experiments*, Langley blamed Navier for faulty differential equations, figuring wrong constants, and assuming that the physics of flight for all birds (from swallows to eagles) could be computed in exactly the same way.[143] Essentially, Langley accused Navier of generalizing about all birds, ignoring their differences, and imposing mathematical formulas used for calculating terrestrial animal work and motion on analysis of a bird's work and motion through air. In other words, the class of animals labeled "birds" should not be thought about monolithically. By challenging Navier's conclusions, Langley opened a previously closed space regarding the possibility of human flight. It is fitting, then, that in *Experiments in Aerodynamics*, Langley used "bird" and "flying machine" interchangeably when he argued the opposite of the Navier thesis—that "effort . . . of a bird or flying machine in the air is greatest when it is at rest relatively to the air, and diminishes with the horizontal speed which it attains."[144] While Langley's direct discussion of birds in *Experiments* occupied limited page space within the book, birds lay at the core of this work's central thesis that human flight may indeed be possible. Langley's revision of aeronautical physics depended upon viewing birds differently (as different both from each other and from land-based animals) and flight differently (as different from land-based locomotion). Unsurprisingly, *Experiments in Aerodynamics* led Langley to experiments with birds themselves.

Two years after asserting the possibility of human flight, Langley published "The Internal Work of the Wind," which examined directly the flying behaviors of birds. In fact, the paper opened with the following observation: "It has long been observed that certain species of birds maintain themselves indefinitely in the air by 'soaring,' without any flapping of the wing, or any motion other than a slight rocking of the body; and this, although the body in question is many hundred times denser than the air in which it seems to float with an undulating movement, as on the waves of an invisible stream."[145] In this paper, Langley offered an explanation for "soaring," which challenged Navier's claims resulting from an overemphasis on flapping. Langley concluded that wind was inherently "irregular," and that, in the concise words of future Smithsonian secretary Charles Walcott, "these irregularities might be a source of power and might account for the ability of certain birds to soar with outstretched, unflapping wings."[146] This important conclusion set the precedent of taking soaring birds seriously in the study of aeronautics, for if the wind provided these birds with "invisible support," then it could potentially provide the same support for a "future aerodrome." Langley recognized, though, that a successful flying machine would need a "mechanical brain"

Figure 3.4. Langley photographing a buzzard "seeing the wind" from the roof of the Arts and Industries Building, circa 1901. Courtesy of the Smithsonian Institution Archives.

that replicated birds' abilities to "see the wind," constantly exploiting its ever-changing irregularities.[147]

As Langley wrote these two classics, he worked on designs for his future aerodrome. Between 1891 and 1895, Langley built four models. In 1896, his fifth model became the world's first successful motorized flying machine, achieving a distance of three thousand feet.[148] While quite the accomplishment, Langley recognized that such a feat was far from the daily successes of the planet's albatross, seagull, and buzzard populations. To further his research, he needed to take his work one step further. And he decided that he should not depend upon the *idea* of these "soaring birds" but should study the actual, living soaring birds themselves. Enter the National Zoo.

Langley's examination of zoo birds did not end with the measuring instructions he sent to Baker and Blackburne. Between 1901 and 1903, Langley again asked them for a favor, this time requesting photographs of buzzards in flight.[149] Since zoo buzzards could not "soar" in their cages, Baker decided to lure "wild" buzzards from outside the zoo to the park's rock face promontory by exposing carrion. Baker realized, however, that there was no need for bait because hawks and vultures were already "quite frequently seen flying about over" the birdcages within the zoo, eyeing the smaller captive birds as potential (yet unattainable) prey. The entire zoo itself (filled with captive and feral animals, booming rodent populations, and carrion for the carnivores) functioned as bait, operating as a peculiar medium between the encaged and the wild. And to photograph wild buzzards, Baker even had spectacular, fifty-foot-tall scaffoldings built.[150] In this way, the zoo served as a malleable laboratory that could alter its infrastructure according to the needs of the experiment.[151]

While the towers were in construction, Langley, unsurprisingly, sent Baker further instructions concerning their design. He requested that their roofs be raised eighteen inches higher, that a weathercock (of very precise dimensions) be placed on top of the western tower, and that a compass card be attached to the base of the weathercock. He even included a sketch in case Baker found his directions confusing.[152] To conduct these "photographic experiments," Baker hired local photographer E. L. Geyer at a rate of eighty dollars per month to assist National Museum photographer T. W. Smillie.[153] And after he had electrical lines routed to the towers for powering a helio-photographic apparatus, Baker had the western tower fitted with a chronograph (a type of stopwatch), protected by the elements with a specially made wooden covering, and immediately ordered a chronometer (an instrument for measuring time accurately despite motion), two anastigmatic photographic objectives, and all the parts needed for a fully functional camera.[154] Baker and Blackburne were not just building simple towers; they were designing a well-equipped photographic eye alongside the birdcage. As we saw with Garner, studying animals during the Progressive Era often involved "capturing" their images and sounds with the new technics of a new century.[155]

Yet the primary purpose for all this equipment was to take pictures of buzzards circling overhead. To follow their movements, Langley had the camera attached to the wheel work of gunstocks (a common practice in early photography), enabling it to rotate a full 360°.[156] Also, the cameras in both towers were connected by an electric wire that enabled them to fire simultaneously, producing two images taken from two different angles at the exact same moment.[157] From atop these towers, Langley studied birds in flight. In the words of Rene Bache, who published a paper about these experiments:

Figure 3.5. One of the bird towers erected in the zoo for the purpose of photographing wild buzzards. Courtesy of the Smithsonian Institution Archives.

> These soaring birds point their heads against the wind, spread their wings in a certain way and are lifted . . . straight up into the air without any exertion on their part. They can progress against the wind by falling forward and slightly downward. When they go with the wind it holds them up by catching in the hollows beneath their wings. To an observer on the ground these birds appear to sail through the air for miles without moving a wing.

These were the birds of Langley's childhood. (Nature's greatest fliers expended almost no energy in flight. Why should humans not be able to do the same?) A buzzard "is sustained by the wind in just the same manner as a boy's kite is upheld in the air." While watching the buzzards soaring, Langley focused specifically on the angles of their wings in relation to the wind, for he had long realized, since he conceptualized "The Internal Work of the Wind," that the question of flight is only a question of angles, of geometry, of mechanics.[158]

As Langley pursued his buzzard research, he simultaneously built the Great Aerodrome. Concerned with "how air currents act in supporting plane surfaces," Langley worked on designing artificial planes that could replicate the wings of buzzards. Yet balancing artificial wings was no easy task. Ironically, the condors caged next to the towers provided Langley with another insight. Since these birds were indeed prisoners and sadly lacked the space for a running start into flight, they would instead simply drop from the top of the cage and "soar" to the bottom. This observation sparked the idea that if his Great Aerodrome could not produce the force needed to push itself into the air, it may be able to create enough force by being launched (dropped) from a height.[159] In this way, Langley designed his Aerodrome with lessons learned from both wild and caged birds, *both* within and above the National Zoo.

Langley's zoo experiments attracted scientific interest in flying machines and buzzards—even years later in 1908. C. William Beebe of the New York Zoological Park, for example, wrote to A. B. Baker (assistant superintendent of the National Zoo) to ask whether he could have ten wild buzzards captured and sent to New York to support an experiment in acclimatization.[160] Unfortunately, Beebe learned that the laws of the District, Maryland, and Virginia prohibited the capture of turkey buzzards.[161] To circumvent legal constraints, Baker directed the request to T. Gilbert Pearson, American conservationist and founder of the National Association of Audubon Societies in Greensboro, North Carolina, where animal protection laws were more lax.[162] Pearson promised them the desired buzzards, claiming that local animal collector E. F. Basnight was trustworthy and could obtain the specimens.[163] Six years after Langley's buzzard experiments (even two years after his own death), zoo men elsewhere still considered the National Zoo the perfect place to purchase buzzards for experimental purposes. Langley's photography, aeronautic pursuits, and observation towers gave the zoo a reputation. The National Zoo functioned not just as a laboratory itself but also as a weigh station for science, a connective medium that processed orders for living

Figure 3.6. The Flight Cage was one of several birdcages of the National Zoo. Courtesy of the Smithsonian Institution Archives.

material, tethering the world of science to the populist world of animal collecting while navigating the legal labyrinths that began to structure conservation politics in the twentieth century.

Langley's research on buzzards lasted until his death in 1906. For him, the National Zoo offered a place to conduct low-cost research when getting to "the field" would be too demanding, a function that zoos still perform for life scientists in the twenty-first century. The zoo also offered Langley a convenient and timely opportunity to think through human aviation by studying individuals who had long possessed flight technology. Langley infused his aeronautics, aerodynamics, physics, and meteorology with life science. In so doing, he should also be understood as an important figure in the early history of biomimetics, a branch of engineering and architecture that looks to plants, animals, and ecological systems for technological inspiration.[164]

Failed Experiments and the Animality of Science

Let us take a moment and consider Samuel Langley and Richard Garner alongside each other. On the surface, their experiments—in aeronautics

Figure 3.7. Langley's Aerodrome crashed in the Potomac, piloted by Charles Manly. Sueddeutsche Zeitung Photo / Alamy Stock Photo

and primatology—were staged in, and for, quite different realms of scientific thought. Yet looking closely at their experiments, at their embeddedness in the zoological park, reveals a common structure. Of course, it should be no surprise that both primatology and aeronautics were scientific endeavors born in the public sphere, for Science and Society have always been closely intertwined. Science is always a public endeavor. And Progressive Era Americans were especially fascinated by the idea of flying while they wrestled over the implications of Darwin's theory of evolution. It should also be no surprise that these experiments were built upon animals, real and symbolic. Western science had long exploited animals as windows into, and as building blocks of, "the human." For example, the entire corpus of Galenic anatomy was extrapolated from dissections of monkeys and pigs, not humans.[165] The standardized measurement by which to assess the power of steam engines at the end of the industrial revolution was indeed *horse*power (which was itself an ironic name for a standard since different horses do different work at different rates). The history of science is filled with examples of discoveries, achievements, and inventions—symbols of human exceptionalism—composed of, by, or with animals. Captive animals, active and latent, riddle the history of Science.

Science was indeed built upon animals with, and in, the public, but Langley and Garner together should also direct our attention to a larger purpose of Science, with a capital *S*, regarding ideas about humans, animals, and agency. Modern science utilized real animals to conquer "the animal"—primate language, avian flight, and other endeavors attributed to nonhumans—for the purpose of lifting "the Human." Exploring "the problem of the animal" has been an important task of Continental philosophy, animal rights philosophy, critical animal studies, and posthumanism over the last several decades. This "problem of the animal" explores questions like (as philosopher Thomas Nagel put it in 1979) *What Is It Like to Be a Bat?* or (as Jacques Derrida put it in 1997) *What is to follow if I am an animal too?*[166] Animal studies has been interrogating the substrate of Western philosophy by asking: *What constitutes "the human"?, Where does "the animal" end and "the human" begin?, Are the common denominators of Western philosophy—language, "the individual," spirit, consciousness, Reason, Being, technology, causality, creativity, religion—truly "human," originating with humans?,* and *If these denominators are not "human," if they can be found across the worlds of animals, what does this mean for "human" societies, their assumptions and ethics?* Garner and Langley were two scientists seeking to understand animals in order to coopt their abilities for the purpose of scientific knowledge and technological achievement. Langley and Garner both staged their experiments upon the boundary separating "the human" from "the animal," a fault line ("the problem") undergirding the annals of Western knowledge. Primate language and avian flight marked two nonhuman attributes ("achievements") that *composed* the perceived "gap" between "the animal" and "the human." In fantasizing about the abilities of the Other, "the animal," Garner and Langley strove to expand the territory of "the human" by colonizing that which is categorically not-human. Nonetheless, when the "real animals" (living and breathing primates and birds) inevitably ran away from the expectations of (living and breathing) human scientists, Science was born, by writing the shortcomings of humans into a progressive, teleological narrative structure, tethered to a forward movement toward Progress, all too familiar in Science while, at the same time, erasing the resistances of animals and downplaying the human failures entangled with those animals, tangible and metonymic. The experiments of both Langley and Garner lay the structure of science bare.

Langley's flying machine, the Aerodrome A, the one inspired by zoo birds, reached nowhere, especially not the heavens. It crashed into the Potomac—not just once, but twice!

Richard Garner failed too. Not only, as Frank Baker recognized, were both Garner's scientific method and critical thinking flawed, not only was Garner clearly more of an entertainer than a scientist, and not only was his time in Africa driven more by the "male imperialistic project" than by the pursuit of knowledge, but he failed to accomplish his task—to discover a discernible language of nonhuman primates.

However, failed zoo experiments can speak volumes about animals commodified.[167] Through failed experiments, the animals usually silenced by Science can be heard chirping through the fissures of triumphalist facades. Failed experiments can reveal the animality of Science usually made invisible. In *The Nature of the Beasts*, environmental historian Ian Jared Miller has presented a study of Tokyo's Ueno Park Imperial Zoological Garden, the first modern zoo in Japan, founded in 1882. Miller's "principal argument holds that when the Ueno Zoo made caged animals into representatives of pristine 'nature' it instituted a separation between people and the natural world, introducing a break between humanity and animality that recast Japanese—together with Westerners—as the rational masters of a new natural history based on Linnaean nomenclature and the tenets of evolutionary theory." He assigns the label of "ecological modernity" to this "doubled process of intellectual separation and social transformation."[168] Similarly, the "beasts" of the National Zoo—capuchins, gorillas, condors, golden eagles, Egyptian cranes, Jamaica buzzards, albatross, frigate birds, and turkey buzzards—forced to participate in Garner's and Langley's experiments, embodied Americans' removal from Nature. Indeed, they embodied the same "modern" doubled process of separation and transformation. Both Garner and Langley sought to extract knowledge from primates and birds, respectively, in order to further the ends of an imperialistic America. Seen together, their equation was simple. *Conquer animal language and flight. Then, employ both for human-American triumphalism.*

Both Garner and Langley tried to force their animals to play along in their master narratives. Garner shoved graphophones into the faces of capuchins and gorillas. Langley forced condors to hold still on paper so that they could be traced. Both Garner and Langley tried so hard to get their animals to follow methodological scripts.

Yet, zoo animals always run away from expectations. Living animals run away from their symbolic counterparts. The zoological park was a place filled with surprises. Bruno Latour has famously exclaimed that "we have never been modern," and the living animals of the supposedly "modern zoo" give evidence to this proposition, pointing us toward the deception, the lie, the hollow core, of "modernity."[169]

The primate and avian individuals above were surely used consciously by Garner and Langley as symbols of modernity and human sovereignty. Most of these animals were, after all, stolen from faraway ecosystems and held captive in Washington, DC. Then, they were enlisted in the enlightened projects of anthropology and aviation. However, these animals lived multiple lives and possessed multiple ontologies. If these captive animals, collectively, marked the rise of a looming "modernity," a separating from Nature, these same living and breathing zoo animals, when viewed from their dark-uncanny sides, from their unseen and un-storied lives, marked the impossibility, the end, of this same modernity.[170] Humans could never separate from Nature, and zoos could never display wild animals. The consistent "reaffirmation" of "Nature" and "wildness," as well "human" and "animal," in the space of the zoo always required their abjection, their "necessary disappearance."[171] Historical zoo animals always told postmodern lessons from within an Anthropocene.

Garner and Langley looked toward "the animal" to advance science and to become "modern," but in so doing (in attempting to traverse a perceived yet nonexistent chasm) they both saw more complexity in their subjects than scientists of previous generations would have seen. Garner recognized that primates possessed language. Langley recognized that the category "bird" housed tremendous diversity. And he wished to "see the wind" too. Both Garner and Langley, despite their commitment to their projects, recognized that animals had agency. In fact, it was this agency that they wanted to make their own. Garner and Langley realized that animals had faces, even if they failed to discern their features.[172] Nonetheless, they failed in their experiments because "real" animals and "real" animal abilities resisted imperialism. At the dawn of the twentieth century, animals were coming into view but were still not seen.

Zoo animals, here, should be read as embodied texts symbolizing the "modernity" that their literal bodies ran away from. As these animals became subjects in/through popular zoology, they prepared zoogoers to think about animals in new ways. By not living up to expectations, by refusing to follow "modern scripts," by being more successful than human scientists, by becoming elusive, by becoming-more-than-zoo-animals, by being themselves despite the violence of captivity, inhabitants within the zoo directed attention to the fault lines that lay beneath an outmoded humanistic tradition. By opening up a zoo, the historian uncovers a crossroads along which science and popular culture collided, producing a popular zoology that continued to open new knowledges through the continued concealing of animals.

CHAPTER FOUR

Animal Activism and the Zoo-Networked Nation

As wildlife conservation, animal rights, and vegan movements expand in the first quarter of the twenty-first century, the history of animal activism is now more important than ever.[1] By focusing on the zoo, this chapter argues that concern for animals was more widespread in the decades surrounding 1900 than is usually understood.[2] Zoos clearly functioned as sites of violence, but they simultaneously functioned as incubators of compassion. Zoos served as spaces in which common zoogoers could employ the anti-cruelty discourses that they absorbed outside the zoo's fences. And beyond those fences, as cultural historian Janet Davis has demonstrated, the "gospel of kindness" was remaking how both animals and humans were viewed during America's Gilded Age and Progressive Era.[3]

This chapter is composed of four sections. The first presents a case study of two elephants, Dunk and Gold Dust, and shows how zoogoers' concern for their well-being resulted in formal activism that extended all the way to Congress. The second section takes a wider look at zoo activism in Washington and argues that most zoogoers, knowingly or not, participated in conversations about animal welfare. The third section demonstrates how early zoos became networked through the spread of animal activism. The final section concludes by pondering larger trends for the intellectual history of "animal rights."

Dunk, Gold Dust, and Elephant Activism

Dunk, a large male Asiatic elephant, arrived at the capital in the spring of 1891 as a gift presented by James Cooper, owner of the Adam Forepaugh

Show. A few years earlier, Forepaugh had donated the skeleton of an elephant named Romeo to the Smithsonian so that he could be enshrined in the National Museum. Now, the show was giving a *living* elephant to adorn its new National Zoo.[4] The transaction received public fanfare and was treated as a grand philanthropic gesture: "Mr. Cooper gives this present . . . because it was the policy of Mr. Adam Forepaugh to encourage such institutions." The great showman also bestowed elephants Tippo Saib and Bolivar to zoos in Central Park and Philadelphia, respectively.[5] Yet beneath generous packaging, the Dunk donation proved to be more pragmatic than altruistic. The four-ton, almost-nine-foot-tall pachyderm had a long history of charging other male circus elephants, which is unsurprising, since large, male mammals often cannot be kept in containment together.[6] Cooper exclaimed that he hoped his actions would prompt other enlightened citizens to make similar donations for the "benefit to students and scientists" and to preserve animals since "civilization was driving from the face of the earth the larger members of the animal kingdom." Despite this talk, in reality, Cooper was simply discarding an individual he viewed as a troublesome brute, displacing risk from his business onto the Smithsonian in the name of patriotism.[7]

Not only did Cooper donate Dunk, but he also loaned the National Zoo two other elephants.[8] One named Gold Dust also possessed what he perceived as an irascible character, having previously killed at least one of his keepers.[9] For Cooper, the National Zoo served as an ideal correctional institute for fractious elephants. If animals were costly and difficult to control, he gave them away, as a "most generous gift." If he thought elephants could be reformed, he would "lend" them to the zoo, hoping that a zoo hiatus would serve as rehabilitation. No matter the motives behind the Dunk transaction, the gift garnered publicity because, in the words of historian Elizabeth Hanson, "a zoo was not a zoo unless it had an elephant."[10] To embrace the fanfare, William Blackburne, head keeper, led a three-elephant parade, before throngs of District children, escorting the new zoo residents "along New Jersey Avenue to S street, to U, to 13th and out on the hill and thence by way of the old quarry road down into the valley of Rock Creek." Sadly, though, upon arrival, the roof of the elephant's makeshift enclosure was still unfinished. Although the floor and the granite posts (buried eight feet underground and fixed with an iron ring to which the pachyderms' feet would be chained) were complete, the octagonal, pagoda-like building itself was literally being built around its

Figure 4.1. Indian elephants Dunk (*right*) and Gold Dust (*left*) posing with head keeper William Blackburne. Note Dunk's shortened tusks and the chain that wrapped around both their necks and bound them together. Courtesy of the Smithsonian Institution Archives.

new inhabitants.[11] The National Zoo was not ready to keep elephants. Within days, citizens objected. Even the *Evening Star* contended that "unless the S.P.C.A. [Society for the Prevention of Cruelty to Animals] prods Uncle Sam's conscience and urges more humane treatment of these captives," Dunk and Gold Dust "will be cadavers for the dust."[12] Concerns like these would reverberate around the National Zoo for years.

Most District citizens, of course, were more concerned with the dangers elephants represented than with their well-being. Society had long depicted captive elephants as savage mammoths precariously subdued.[13] And elephants had long been constructed as antagonists in the timeless "civilization" origin story about the eternal struggle between man and behemoth,[14] a story that popularized elephants, framed them as "wild," and thus bolstered their ticket-selling potential.[15] Shortly after Dunk and Gold Dust's arrival, Tippo Saib (a Forepaugh elephant donated "benevolently" to the Central Park menagerie) seized national headlines. "Tip," while charging the wooden wall partitioning his exhibit from the neighboring elephant exhibit, attacked the keeper trying to calm him. The "rogue" "seized" the keeper with his trunk, thrusted at him with shortened tusks, and then "threw him violently on the ground," knocking him unconscious. Tip then tried to crush the keeper, but fortunately the man's limp body was just out of the reach granted by the short chain secured to Tip's leg. The keeper survived, and the story riddled headlines as Tip almost added "another human victim . . . to the record of eight keepers that he had already killed." Newspaper columns especially emphasized the episode's aftermath—when the "wicked" Tippo Saib became "chained fore and aft," a chain tied to each tusk and linked to a thick fetter that wrapped tightly around the "body of the beast."[16]

As Washington zoogoers visited Dunk and Gold Dust, they (like the American public at-large) became well versed in the supposed perils posed by captive elephants. They not only read about the continued aberrancy of Tip (who was reported on for months), but they also read articles about how Miss Fanchon, a Boston elephant, escaped her stable and ended up in the kitchen of the zookeeper's wife; about how Royal, an elephant employed in the El Dorado grounds of Weehawken Heights, New Jersey, went on a destructive escapade around the gardens when frightened by a firecracker; and about how Chief, the man-killing elephant of Cincinnati, was put to death by the zoo, receiving a total of twenty-four bullets.[17] Zoo elephants, like all zoo animals, pushed against the confines of captivity, but the sheer size of these expressions helped zoos frame them, and zoogoers understand them, as indicative of an underlying "deviant" character specific to elephants.

Dunk and Gold Dust were no exception. In fact, the *Washington Post* declared Gold Dust the "meanest elephant on earth . . . mortally injuring more men than any other elephant in the country."[18] On the last day of October 1891, Gold Dust attacked his keeper, Charles Lewis, knocking him to the ground with his muscular trunk. He then created quite the spectacle for onlookers, charging around the perimeter of his exhibit until he was subdued by a "judicious use of pitchforks" and a "liberal supply of ropes and chains." Keepers forced Gold Dust to the ground, securing him until "the only things free were his ears, his trunk and his tail, and all these waved furiously until . . . the sharp-pointed prongs were felt."[19] Gold Dust required force to be contained. "His trunk was raised against every man," as an 1893 article claimed.[20] And even Dunk, the more passive of the two despite his troublesome record with the Forepaugh Show, occasionally challenged the zoo. In July 1898, Dunk "pulled up the stake to which he was chained and succeeded in making his escape." After frightening zoogoers and destroying some trees, Dunk submitted to the keepers in pursuit, one of whom was armed and ready with a rifle. The elephant was quickly returned to his exhibit and "chained fore and aft," as if an enemy vessel like the *Reina Mercedes*.[21]

Even as captivity forced elephants to play out their roles as the "wild" antagonists of zoo dramas, concern for their well-being still increased throughout the last decade of the nineteenth century. On June 3, 1891, the *Evening Star* printed a column titled "More Room for the Elephants." The author, writing under the anonym "G," wrote with conviction and precision, clearly acquainted with the behaviors of the species:

> Some better provision should be made for the elephants. . . . I saw them Saturday—a warm day—as they stood sweltering, chained like malefactors to the floor of the house that shelters them and excludes them from the air. It occurs to me that an acre or two should be fenced off for them, with access to the creek. The fence need not be expensive, though it should be strong. Wooden beams buried in the ground and connected by a single iron rope, such as is used in connection with derricks . . . would be all-sufficient. Elephants are not addicted to jumping fences nor to crawling under bars, so that one iron rope, an inch in thickness, would hold them. The posts might be twenty or thirty feet apart and should be scotched by a prop on the outer side. The high, natural wall of stone on the opposite side of the creek would answer in place of a fence on that side. . . . Elephants are fond of water, and it would be delightful to see them revel in the creek. They would keep themselves clean and would save the attendant all the trouble of carrying them water. *Their present situation is wretched and no humane person can see them with pleasure.*[22]

This elephant advocate, likely an American Society for the Prevention of Cruelty to Animals (ASPCA) member but possibly an employee in the zoo, knew the habits of elephants, the specifics of enclosures, and the layout of the zoo. Rather than simply admonishing the zoo, the writer suggested a simple way to redesign the enclosure to allow for movement, strategically using a fence and the cliffs along the creek to create a controlled space over which the pachyderms could wander. The reform would allow Dunk and Gold Dust to live an unchained life and would at least permit them to clean themselves, an important activity for all pachyderms. (One study has shown that wild Asian elephants devote at least 44% of their waking day to feeding, 38% to walking, and 8% to drinking, bathing, mudding, and other activities based around water).[23] Alas, the author's plan was not put into action. In fact, the zoo made no significant modifications to the elephants' living conditions for a decade.

On April 16, 1902—eleven years after the zoo received Dunk (Gold Dust died in 1898)—the *Post* reignited concern.[24] Two days earlier, after deliberations about Dunk's well-being, the executive committee of the Washington Humane Society decided to forward a protest letter to the District's board of commissioners. This letter, printed verbatim in the paper, attacked the National Zoo for its oppression of Dunk, who served as the "living monument of the cruelty that can be thoughtlessly committed by a rich and civilized government." The article described Dunk's exhibit—which was "hastily prepared for temporary occupancy" and "falling into decay"—as excessively cruel. Dunk still lived in the same dilapidated house originally prepared for him! The enclosure still could barely "keep the temperature in winter as high as forty degrees," yet in the summer, the house was simply scorching, "unbearable," and teeming with flies and rats. Of even greater concern was the manner by which Dunk was tethered to this environment. For eleven years, "fastened by the leg with a chain not three feet long," Dunk received little exercise, only able to pace around a circle with a six-foot diameter. In fact, according to the editorial, because of his deviant stigma, the National Zoo only exercised him after hours, when the park was "cleared of people." This allowed the "whole force of the establishment" to be present for his walks, "to guard against his escape." The author indeed wrote about Dunk as if he was serving a prison sentence. While captive animals frequently tested the chains of confinement, the perceived magnitude of Dunk as a deadly, charismatic megafauna, held precariously at the center of urban popular culture, made his agency unforgivable. Dunk was an elephant and could be trusted only if shackled to America's growing penal psychology.[25]

The Humane Society contended that "under the influence of such confinement, it would not be strange if he were as vicious as he is said to be." And the Humane Society was right. Through systematic and systemic neglect, zoos created the dangerous elephants they feared. By naturalizing this danger, a zoogoing society justified (and even erased) the violence committed in and by the zoo while simultaneously constructing their captives as "wild," in need of civilization. The Society recognized some of these hypocrisies and expressed the following to the zoo's commissioners:

> It is a matter of indifference to the Humane Society whether the government keeps an elephant on exhibition or not, or, indeed, whether it maintains its Zoological Park. But *if* it sees fit to do either, the thing ought to be done well, and no such cruelty continued as has existed in the case of this animal [Dunk] for so long. We are not insensible to the interest and enjoyment which such a park affords to the inhabitants, especially the children. . . . But it would be far better to have no animals at all, or only stuffed specimens, than to keep them throughout changing seasons under improper conditions, and at the cost of their discomfort or suffering.

These activists wanted the zoo and public to know that the Humane Society was not a collective of radicals opposed to the convention of animal display. Their demands were moderate, because, in the words of animal historian Diane Beers, "a more moderate posture generated considerable public favor and culminated in promising, moderate victories."[26] In fact, the Humane Society agreed with the very rhetoric (interest and enjoyment) that zoo boosters themselves so frequently employed. However, the Society emphasized that *if* the US government "is going to keep live animals for the amusement or instruction of citizens, they should be kept in a proper manner, and nothing savoring of neglect or cruelty should be allowed to exist in such a public undertaking." Further, Humane Society members requested that Congress approve $20,000 to build a pachyderm house, asking legislators directly to consider Dunk's well-being. If Congress could not improve the living conditions of its nation's elephant, Dunk should either be donated to another zoo "where proper accommodations already existed" or "he should be humanely killed,"[27] a stance that the Society held toward abused horses, dogs, and cats too.[28]

Petitions for Dunk did not just come from the local Humane Society. "Mrs. Clark," a concerned citizen, published "A Plea for Dunk" in the *Evening Star*, demanding justice:

> All other animals . . . are provided with comfortable quarters, inasmuch as they have freedom of action and are capable of moving around in the spaces, cages, or enclosures allotted to them; whereas the

> elephant, the largest animal in size, and in intelligence as well, is given the smallest space possible, and burdened with heavy chains which only allow him to take one step forward or backward, restraining any other motion.

Like other advocates for elephants, Mrs. Clark called attention to the horrific chains that limited Dunk's mobility, his natural "freedom," a trait that even Buffon associated with elephants as early as 1812.[29] These binds were unimaginably constraining, since Asian elephants traverse, in the wild, spaces up to two hundred thousand acres (and African elephants, ten to twenty times this).[30] In addition to the immorality of denying Dunk's "freedom of action," for Mrs. Clark, Dunk's intellectual ability made him even more in need of humane treatment. She continued to explain that "the intelligence of the elephant is classed as being superior to that of other animals, in some instances almost bordering upon human intellect and reasoning power," those divine metrics of the Renaissance so deeply subsumed in Western anthropocentrism. Possessed with a more humanlike intelligence, therefore, the elephant, "patient by nature," becomes, according to Clark, "vicious and vindictive, and seeks revenge by killing his keepers, as happened but recently in New York, as well as in other instances" (and as demonstrated in chapter 2).[31]

Like the Humane Society's editorial, Mrs. Clark emphasized that the wicked reputation of elephants did not truly reflect their "nature" and instead should be seen as a product or symptom of captivity. Whereas the Humane Society blamed the conditions of confinement for elephants' behavior, Mrs. Clark took this reasoning one step further by extending intentionality to elephants. Elephants became "vicious and vindictive" *because* of their intellectual ability. An elephant is not *essentially* vengeful. Rather, an elephant "seeks" revenge. For Mrs. Clark, inhumane captivity provided the reason for Dunk's resistance only because the highly intelligent Dunk was aware of its injustice. And because Dunk was aware of his chains and the living conditions that made him miserable, he possessed an even greater claim to the humane concerns of the National Zoo, the District of Columbia, and the body politic. Dunk was "not an object of interest, but of pity," and the National Zoo needed to provide him with an enclosure that *he* would find suitable.[32] Mrs. Clark was thinking ahead of her time, for elephants are indeed intelligent.[33] They express grief, compassion, cooperation. They use tools. They engage in verbal and nonverbal communication. They forge complex and expressive societies and pass on traditions to their offspring. We have only begun to glimpse the outer "surface" of elephant life and culture.[34]

The above story about Dunk between 1891 and 1902 illuminates three important lessons about the role that zoos played in the formation of animal activism at the turn of the century. First, the story of Dunk shows how activists directly influenced national politics, making fiscal claims of Congress for the well-being of an elephant. Although Congress did not approve the Humane Society's request for $20,000, they did allocate half that amount for building a new elephant house,[35] which was completed at the beginning of 1903 and allowed Dunk to move about without chains, in and outside his building. His house, a thirty-five-by-sixty-five-foot "barn-like structure of brick," greatly increased his space, and connected to this structure was now a seventy-nine-by-ninety-six-foot outdoor yard, equipped with a bathing pool six feet deep and twenty feet in diameter.[36] While Dunk's new home was not built around Rock Creek itself, as the 1891 editorialist had recommended, his exhibit did free him from chains and allowed him to "enjoy one of the greatest delights of jungle life, the bath"—assuming that the pleasures of standing in knee-high water, in a small and stagnant swimming pool, were comparable to the joys of wading into forest lakes and mountain streams.[37] Nonetheless, after complaining for more than a decade, animal activists eventually steered national policy. While this kind of direct appeal to Congress by outspoken zoogoers was the exception, not the rule, these advocates commonly influenced policy at the local, zoo level. Animal activists made demands on the behalf of zoo animals, and throughout the first two decades of the National Zoo's existence, these demands pushed the zoo to marginally improve the conditions of containment.

Second, the Dunk story demonstrates how the National Zoo created meanings about animals while also providing space for their destabilization. This simple lesson echoes central themes of this book: In zoos, symbolic animals, despite their captivity, ran wildly and unpredictably through American culture, and popular zoology helped Americans make sense of their new experiences with captive animals. The zoo was not structurally unique in this way. All institutions, cultural arenas, and discourses are always bifurcated—splitting subjects, meanings, and histories as (even before) they are constituted within their domains. The zoo (and its histories), however, like so many other topics, is often "read" from

Figure 4.2. After eleven years of chained captivity, Dunk wanders along the perimeter of the outdoor yard of his new enclosure. Courtesy of the Smithsonian Institution Archives.

single interpretive vantages, rather than from the invisible, erased, disavowed, and "disguised" perspectives with which they are dialectically entangled.[38] This simple hermeneutic striving toward a single meaning spawns so many of the "gaps" between binaries like "wild" and "captive," "Nature" and "civilization," "human" and "animal."[39] Yet while this lesson may seem abstract when spelled out, it feels straightforward when zoo elephants are given histories unmoored from crass assumptions about animals. Put simply, Dunk shows us how cruelty and compassion became historically conjoined in the National Zoo. As environmental historian Jason Colby puts it regarding orcas, "close contact spurs curiosity and affection."[40] Yet Dunk shows us that no matter what he meant to humans, he himself had very little say about the trajectory of his life as a captive.[41]

When zoos acquired elephants, they acquired not only their literal bodies but also their symbolic bodies. Dunk and Gold Dust captured public attention because they were deadly symbols of wild Nature. By acquiring elephants, the institution legitimated itself as a modern zoo. And by magnifying these animals' treacherous reputations, zoos advertised their ability to keep, contain, and pacify the danger that exemplified that which was bestial, unpredictable, and anti-modern. Such a rhetorical-semiotic project in Washington, DC, where urbanites had been voraciously consuming narratives of progress, refinement, control, and nationalism for decades, placed the National Zoo among the grandest in the world. However, within the space of the zoo, the narratives associated with animals changed quickly. Zoo workers, zoogoers, zoologists, and activists alike fashioned zoo stories about the animals displayed. And some of these plots assisted animal activists in encouraging publics to view animals as more than Other, as sentient individuals deserving of moral consideration—an ethical position usually associated with the anti-cruelty movements decrying the plights of working animals and pets. Yet even when the stories told of zoo animals did not lead to activist ends, they always problematized "the animal" as a singular, monolithic entity. Zoos placed an Animal on display for a public to see, and in the process of seeing, these publics ripped the Animal into myriad pieces. These were always (re)assembled by zoogoers in multifarious ways. Nonetheless, the zoo gave zoogoers an opportunity to make up their own minds about the animals they were amused by, and in so doing, collapsed, on a popular stage, that Enlightened and entrenched hierarchy separating humans and animals. Dunk's intelligence, after all, "bordered upon human intellect and reasoning power."

Last, the story of Dunk and his advocates directs attention to a component of early animal activism largely ignored by historians—the animal mind. As zoo stories challenged long-held assumptions about animals, a specific conversation about the intelligence of animals emerged from evolutionary theory, comparative psychology, philosophy, and the life sciences.[42] In the 1890s and the first decade of the twentieth century, zoogoers began to ponder the cognitive abilities of different species. *What do animals think? What do they feel? Can they speak? Can they dream?*[43] These musings, no matter where they led, delved into the interiority of the animals who told so many stories to zoogoers. By thinking about animal minds, zoogoers and scientists alike tried to find a place for animals in relation to humans. They hoped to tether "the (destabilized) animal" to some sort of truth. Of course, these projects often failed, and the truths they sought proved impartial, incomplete, or even false. Nonetheless, by pondering animal minds, zoogoers expressed a desire to make sense of what did not make sense—that beings behind bars could demand so much from their onlookers.

Figure 4.3. Dunk lived in his new enclosure for nearly fourteen years. After falling and breaking his shoulder on March 30, 1917, William Blackburne shot him, a standard euthanizing practice. Dunk and Blackburne knew each other for twenty-six years. Courtesy of the Smithsonian Institution Archives.

Zoo Activism in Washington

At the beginning of 1889, as the zoo was still in its embryonic stage as the Department of Living Animals, the *Washington Star* printed "Cruelty to Animals at the 'Zoo,'" with the subtitle "Visitors Who Torture the Brutes Will Be Taken to the Police Court." Apparently, spitting tobacco juice at monkeys occurred so frequently that W. C. Weeden, the head watchman (who was hired for the sole purpose of "compelling the small-boy element . . . to depart after reasonable time"), arrested several culprits. The youth were usually released with a simple warning, but the problem grew to such serious proportions that Hornaday began sending zoo-spitters directly to the fourth precinct station, where they were ticketed five, and eventually twenty, dollars.[44] There was something about animals caged that compelled some onlookers to submit to sadistic desires. Hornaday told the *Washington Post* that "it is strange how much deviltry one person can hatch, and their achievements will hardly be balanced by the good deeds of the ninety-nine persons who are above torturing dumb brutes." During the zoo's first summer, one visitor amused himself by coaxing his mastiff close to the cages so that he would snap at the inhabitants. The pet even harassed a young antelope donated by California Senator Leland Stanford, and "the animal, not knowing that he was safe from the dog, beat his head against the fence," suffered a concussion, and died. Hornaday promised that in case of a repeat offense, the man "will find himself under lock and key at the station-house." That same week, multiple zoogoers fed smoldering cigars to the bison to watch their frenzied reactions upon swallowing.[45]

Sometimes abuse provoked retaliation. On June 8, 1890, when a group of children "amused themselves" by again teasing the bison, one of the cows charged and tore through the fence. She bolted into a crowd of zoogoers and headed horn-first toward a group of girls that included three-year-old Rosie Fillins. As two of the older girls "ran shrieking in opposite directions," little Rosie stood frozen "almost paralyzed by fear." To save the helpless Rosie, Annie Howard, "a colored girl, employed as a nursemaid" and clothed in a "bright red jersey," ran toward the petrified three-year-old and "caught the infuriated animal's eye." Annie tried to dodge the bison, but she was "struck full in the small of the back and lifted bodily into the air." After falling to the ground, the bison jabbed Annie in the arm and tossed her again. Before striking a third time, some men armed with sticks drove the animal back into the enclosure. Annie was "so badly bruised she could scarcely walk," but she became "quite a heroine for her presence of mind in saving the little [white] girl's life." The incident terrified the other children in the crowd, and their parents unsurprisingly demanded that the zoo make the bison enclosure "more secure."[46] Teased or not, zoo animals should not be able to get out of their enclosure. Even though such escapes were inevitable, the parents of stampeded children rarely accepted this inevitability.

The above incident hit the *Washington Post* the following day. However, despite the clear and present danger of runaway bison, the newspaper surprisingly never published an editorial that blamed the zoo for the incident or demanded that a stronger fence be placed around the enclosure. Instead, three days later, the public read: "Don't Bother the Animals." This article opened with the following assertion: "It is not in human nature to bear teasing. At least there is a limit to its unresisting endurance, and animal nature is the same." After this comparison of human and animal character, the author argued that the "buffalo cow who broke out of her pen . . . is entitled to public sympathy, for all the evidence shows that her ladyship had been taunted and tantalized until forbearance was no longer a virtue." The author rightly framed the bison as victim, believing that she expressed virtue through her choice of resistance over stoic passivity. Furthermore, the journalist contended that the strength of the perimeter was not the primary issue because the "mischief is on the outside, not on the inside." If zoo "authorities are at fault at all it is in not properly policing the menagerie, so that its four-footed wonders may not be everlastingly poked at, prodded and provoked in cyclones of wrath [instigated] by the . . . small boy."[47] Despite some hyperbole, the above zoo story directed attention to a serious problem, the mistreatment of animals by entertainment-seeking crowds. Runaway zoogoers were on the loose.

Zoogoers frequently enacted violence against zoo animals. This simple and sad truth of zoo history has been well-documented by scholars across interdisciplinary zoo studies. In the big picture, ecocritic Randy Malamud got it right in *Reading Zoos* when he stated simply that "the zoo is a locus of pain."[48] While zoo violence sometimes remained hidden from the public gaze, this violence appeared raw and straightforwardly in newspapers. On August 6, 1890, James Doyle was fined five dollars for spitting tobacco juice into the eye of a bear.[49] On November 13, 1893, when the zoo sponsored a "Receiving Day" that attracted fifteen thousand visitors, the *Evening Star* commented on the multitude of "urchins" armed with "sticks and missiles of all sorts."[50] On May 26, 1895, a man accidentally poisoned a Diana monkey when he handed him a laurel sprig (deadly to primates) as incentive to continue the "funny antics" that were entertaining his small son.[51] Five months later, construction workers building an intercepting sewer unintentionally killed a sea lion when they mistakenly exploded six sticks of dynamite.[52] And by the end of 1913, violence in the park had become so endemic that a zoo employee kept a list of the "lawless, troublesome boys who have been repeatedly warned by the Sergeant of Watch." The next time Eugine Kilby, Darrell or Lowell Bancroft, Morris Johnson, Carrol Tiffany, Jas Beagle, or Harry Johnson were caught in acts of "disorderly conduct," they were to be arrested immediately.[53]

Newspapers reported intentional acts of violence, but they also directed attention to situational violence and systemic neglect. For example, as temperatures reached beyond 100° Fahrenheit during the scorching July of 1892, newspapers concerned themselves not with the dangers of zoogoing publics but with the dangers of extreme heat. Zoo employees spent "the greater part of their time with sprinkling pots wetting the interior of the cages, dens, and enclosures." Dunk and Gold Dust were even taken midday to bathe in Rock Creek.[54] The public worried especially about the polar bears, moose, elk, and wolves.[55] And this concern was not unwarranted, because zoo animals died frequently of overheating, as a cinnamon bear and a black bear did in the summer of 1893.[56] During hot days, the public wanted to know that their favorite polar bears were receiving daily rations of ice blocks.[57] During the cold days of winter, they wanted to be reassured that the elephants' building would be heated, that the llamas could take shelter in a warm stable, and that the alligators would have a place near the stoves. William Blackburne himself worried about the monkeys, for, as he reported to the *Evening Star*, "they suffer terribly from the slightest exposure, and they go into a consumption which takes them off in a hurry" (a topic we will explore in chapter 6).[58] As the wheel of the mid-Atlantic seasons turned, readers worried about how weather affected the well-being of captive animals.

There was not always a clear antagonist to blame for an animal's misfortune or maltreatment. Instead, the zoo as an institution proved culpable, structurally fostering, enabling, or encouraging of harmful behaviors toward its inmate population. In fact, much of this structural violence lay below perception, hidden within or behind the gaps produced by seeing captive animals as wild. Visitors, for example, often saw nothing wrong with feeding the animals, and they frequently tossed peanuts into the elephant enclosure and candies into the monkey cages. Encouraging such deviance, zoo animals universally learned the art of begging, manifested, of course, in species-specific ways. Bison sulked near their fences and chimpanzees reached their outstretched palms into crowds of snacking onlookers. Frank Baker knew that public feeding always harmed the animals. In a letter to Buffalo mayor Conrad Diehl concerning his city's new zoo, Baker admitted that "confinement, even under the most favorable conditions, *will impair their* [the animals'] *health to some extent* . . . [especially] if indiscriminate feeding by the public is allowed."[59] Baker failed to emphasize, however, how difficult it was to police the errant food-tossing ways of human herds. Head keeper Blackburne knew that this not only caused serious health problems for the animals but also provoked fights as they competed for food and attention.[60] Surely most zoogoers knew that feeding the animals was against the rules. Nonetheless, newspaper articles with bylines like "Small Boy and His Sister Amuse Themselves

Figure 4.4. Alligators crowded into their enclosure in the Carnivora House (which was built in 1892). More unnatural space will be provided in 1931, with the construction of the Reptile House. Photo, circa 1900. Courtesy of the Smithsonian Institution Archives.

Feeding Animals and Gathering Chestnuts" romanticized images of Dunk "opening his mouth for peanuts."[61] Zoogoers wanted to make the animals happy. Or maybe they experienced a deep joy "giving" to "the animal," in crossing the species line through a middle ground of broken bread, or in manufacturing a feeling of self-importance in the eye of the Other. (A moment ago, I offered my dog Harlan a piece of my frosted pastry. He wagged intensely. And I felt happy.) No matter *why* zoogoers offered food to outstretched hands and tongues, the problem of public feeding underscores one way in which the very space of the public zoo subsidized a type of systemic neglect.

Zoogoers did not just poke, prod, and feed the animals. They also occasionally advocated on their behalf. In the words of Tappan Adney, writing about the National Zoo for the ASPCA's magazine *Our Animal Friends*, "there are some persons to whom the sight of any animal in captivity can afford no pleasure." And these individuals made their voices heard.[62] On February 1, 1905, Mrs. Bertha Mulhall wrote to Baker asking

Figure 4.5. Superintendent Frank Baker, dressed for the occasion, bottle-feeding a baby camel and demonstrating how zoo animals should be fed in controlled and careful fashion. Images like these also prefigure the craze for "cuteness" that will begin to sweep zoos in the 1920s and after. Courtesy of the Smithsonian Institution Archives.

bluntly, "Is there any reason for keeping the coyotes and a black wolf in small boxes? Of course, you know they are natives of our north-western plains and run over great spaces. It seems unnecessarily cruel to keep the poor creatures confined in such a small space."[63] Baker replied promptly, assuring Mrs. Mulhall that the coyotes would only be held in wooden boxes until the new animal house was finished being built. Baker added (drawing from dog-keeping discourses) that the "general health" and "personal appearance" of the coyotes were "decidedly better in the small quarters" (despite that the home range of transient coyotes can reach upward of forty-one to fifty-five square miles).[64] These coyotes worried other citizens too. And many of them (unsurprisingly in light of the important scholarship of feminist theorist Carol Adams and historians Janet Davis, Moira Ferguson, Susan Pearson, and Frederick Brown) were women of the District who still did not have the right to vote. Most likely, the women advocating for zoo animals were also advocating for children, laborers, immigrants, African Americans, and women who, like animals, were caged by the patriarchy of Progressive Era America.[65] Mrs. William Anderson Miller also communicated with Baker about the "horrors" of the coyotes "shut in those small boxes."[66] On October 8, 1907, the National Zoo received a letter by Mrs. Julia Langdon Barber that extended the worry about the coyotes to all the zoo's animals, lamenting their "pitiful" housing. "If Zoos must exist," Barber concluded, "then this one at the Nation's Capital *should* be a model for others to copy."[67] The National Zoo, according to Barber, failed pathetically to meet the very standards that justified its own existence. The National Zoo could not be a zoo on a hill if the animals were shoved into cramped quarters. Zoogoers of all sorts wrote for zoo animals of all kinds. Miss Mannie Boyd Miller, for example, asked Baker to "try to make the poor lives of those Eskimo dogs a little less miserable." Specifically, she worried about the "foul greenish material" in their eyes.[68] And Colonel C. A. Williams worried about the rat nests snugged along the bears' den because once the weather turned cold, the rats would inevitably relocate inside and relentlessly gnaw at the sleeping ursines.[69]

The zoo also received correspondence sent on the behalf of animals *not* held in the zoo, and one strange incident involving a runaway pet cat shows how the zoo functioned as a centrifuge for animal activist language. In 1909, Ann Raub typed a long letter to Charles Walcott, secretary of the Smithsonian, in "vigorous protest against the wanton killing . . . of her beautiful and valuable pet cat at the hands of the clerk of the National Zoological Park," Frank Baker himself! According to Raub, whose house bordered the Park, her cat wandered into the zoo, and an authorized zoo employee shot him dead while he was resting under a tree—information she gained "upon inquiry." Raub also asserted that zoo employees had treated her rudely, telling her different stories about where her cat was buried. One worker guffawed that her cat was tossed upon a pile of shit (manure) and then pecked apart by buzzards.[70] Baker did not deny that the cat was killed. In a letter to Walcott, he expressed that he did not know how the zoo could be expected to keep squirrels, rabbits, and birds (the typical animal ornaments of urban parks) without exterminating stray cats. He added, in typical gendered fashion, that Raub was "most exasperating and hysterical" (as an aside in the history of sexism, "hysteria," by the way, comes from the ancient Greek word for "uterus").[71] Raub countered that the zoo had no "right to kill a harmless and gentle pet." Her "well-fed pets would not harm the squirrels," and she suggested that the zoo was fully invested in "common sense discrimination" by determining which species of felines should be killed, for the zoo itself kept plenty of cats. She concluded her obloquy by explaining how a few weeks earlier she had discovered a dead cat with a bullet wound lying on her property, who must have been shot in the zoo as well. She wrote, "This I presume happens more than once and the poor animals crawl off and suffer long before they die. I really feel this whole matter should be reported to the Humane Society."[72] Raub presumed correctly. According to a letter Baker sent to Walcott two years later, many neighborhood cats (feral, pets, and in-between) died in the National Zoo.[73] Yet, while Raub advocated for these cats, others complained about them. Georgia Senator Augustus Octavius Bacon sent the secretary a letter, typed on official congressional letterhead, urging the Smithsonian to kill stray felines.[74] While the zoo placed some species or individuals on a pedestal, it simultaneously functioned as an invisible "sacrifice zone" for others (borrowing animal historian Andrew Robichaud's notion for urban spaces reserved for slaughterhouses, canneries, hog ranches, blubber processing, etc.).[75] Raub saw this invisibility through her own grief. Even though many city-dwellers saw cats as vermin, she never gave up her fight. The entire episode began again five years later when she suspected that another cat had been killed by the zoo.[76]

Raub's letter called out the zoo as an institution of hypocrisy and cruelty—hypocritical for caring about some felines while shooting others, cruel for the indifferent way it treated neighboring residents who lost pets to zoo-sponsored rifles. Despite Raub's specific criticisms, she clearly connected her complaints to larger circulating conversations about the ethical integrity of zoos. Her description about the "wanton killing" and "discrimination" of the zoo suggests a larger context. And indeed many famous writers from the middle of the nineteenth century to the middle of the twentieth, from John James Audubon, Rainer Maria Rilke, and Thomas Hardy to Virginia Woolf, Marianne Moore, and Sylvia Plath, criticized zoos as places defined by, in poet Leigh Hunt's words, "lingering deaths."[77] The shooting of the housecat, for Raub, was not a lone incident. This was the standard behavior of an unethical institution that "papered over" the paradoxes of what animal historian Susan Nance has called "animal modernity," where consumers value individual animal celebrities so that they can ignore the welfare of animals in general.[78]

Other Washington citizens further developed this line of attack, not just encouraging the zoo to change *how* it kept animals but instead calling the very idea of the zoo into question. On July 28, 1910, Miss M. Gunderson mailed Baker a scathing letter packaged with a book titled *The New Ethics*, authored by the then-famous (yet now-understudied) vegetarian sentiocentric philosopher John Howard Moore, who developed the doctrine of "Universal Kinship" in 1906.[79] Gunderson's letter carried a radical tone and philosophical depth lacking in the letters of most concerned zoogoers. Excerpted below, Gunderson's letter reveals a discourse of animal activism usually associated with the era after Peter Singer's 1975 *Animal Liberation*. In clear chirography, after the greeting, "To you, the superintendent of a not very honorable institution," Gunderson blistered:

> By what "right," by what demon's right, I ask, do you condemn your victims to their cages? You point to the "tiger's tooth and claw" but are you tigers—you gentle men? If the ethics of the jungle is the only ethics comprehensible to you who call yourselves men, even *gentle* men and civilized—well then, live up at least to the standards of that ethics; do not fall below it. If the tigers and the lions in your pitiless claws were as swiftly killed as [the] more merciful tiger and lion kill their prey, they . . . [would be] better off . . . When you drag them from distant places, capture, transport, and imprison them, you engage in deeds of such moral infamy as no animal but man, the king of bandits, has ever stooped to.[80]

Gunderson elaborated for another half-page about how Baker deserved to be eaten by a tiger, and then she transitioned to the topic of "animal rights," a term used rarely in 1910.

While a few European philosophers like Jean-Jacques Rousseau and Immanuel Kant problematized the category of "the animal" in the eighteenth century, and while social reformists focused increasing attention on the welfare of animals throughout the nineteenth century, there had been little talk about "animal rights," in particular, throughout the early history of animal advocacy.[81] Rousseau briefly thought about the idea of "natural rights" in the context of the animal, and utilitarian Jeremy Bentham (opposed to the concept of "natural rights") famously argued that humans, in a just world, did indeed have responsibilities to animals.

The coupling of "animals" and "rights," though, mainly occurred in the occasional radical treatise, written in the framework of the rising British-inspired vegetarian movement. Ironically, we also see this coupling in Thomas Taylor's 1792 parody *A Vindication of the Rights of Brutes* (which mocked Mary Wollstonecraft's feminist classic *A Vindication of the Rights of Woman*). And we can also find (as historian Chien-hui Li has recently shown) the "rights" of animals discussed in the political-philosophic writings of Humphry Primatt, Thomas Young, William Drummond, and Thomas Erskine.[82] Discussions about "animals" and "rights," however, nearly vanishes by the 1830s, subsumed by the Christian discourses of "kindness" and "mercy." Nonetheless, "rights" language as it relates to animals was briefly reinvigorated by ethicist and vegetarian Henry Salt's 1892 publication of *Animals' Rights: Considered in Relation to Social Progress*, which recontextualized "rights" around Darwinian evolution instead of seventeenth-century natural theology.[83] Miss M. Gunderson, ten years before the passing of the Nineteenth Amendment and inspired by Moore, employed Salt's revived "rights" in a new, zoo way.

She continued her letter by attacking Baker's character:

> Have other animals no rights that you can . . . ignore the simplest plea for justice? "Naturalists" pointing to "nature red in tooth and claw" assert that "man need not be kinder than God"; but why then do you fly into fits when bombs are dropped beneath buildings & ban anarchists from the bomb as well as the gods when they choose to hurl their devastating thunderbolts at helpless man and his conceit, or when, there's an earthquake shock, they cause cities teeming with human life to be destroyed without the slightest regard to the "rights" of screaming man or to the "sanctity" of human life? The creatures in the jungle have ample share in the pains of life; but that certainly is not the slightest reason why you should further oppress them. "That there is pain and evil is no rule, That I should make it greater like a fool."[84]

Gunderson condemned both Baker and the zoo. In her usage of the Darwinian imagery of "nature red in tooth and claw," Gunderson posited two larger contentions about human morality. First, she claimed that acknowledgment of a violent Darwinian world could not justify human violence against animals. Second, she added that any person who used such logic to "oppress" animals had no right to speak of (secular) "rights" or (religious) "sanctity." For Gunderson, there was simply no apology for oppression in the modern world.

She then transitioned her soliloquy from gods, anarchists, and men to animals themselves, claiming that "there is suffering enough among the inhabitants of the wildwoods . . . but in spite of struggle and hardships, intermingled with hopes and joys, they are happier there by far in their native haunts than in the claws of pitiless man." Concerning "rights," she exclaimed, "Theirs is the *right* [emphasis added] to breathe the air of freedom in the forests of their fathers. Their home is there, their heart is there, their happiness is there, where they wish to live and love and struggle for their own sake and their dear ones." Despite her anthropomorphizations, or perhaps through them, Gunderson contended that animals had, beyond a shadow of a doubt, the "right" to live in their natural habitats.[85] Elephants had the right to live an elephant life; a fox had the right to live a fox life.

Gunderson concluded her letter by critiquing Western philosophy in a way that anticipates the rise of critical animal studies almost a century later.

> Are you Cartesians? Have the revelations of Darwin and Copernicus no significance to you? Are you "king by divine right?" Are you in medieval darkness? Think you still that the heavens revolve around you, that gods take interest in your affairs, that you were made for eternal glory, and tiger and lion and eagle and deer . . . were made simply to suffer? . . . If you belong to that class of benighted "naturalists" who fail to see the kinship between yourself and your prey, if you "think" that you were made for other worlds but that lion and tiger and elephant and deer and bear and ape were made for this, well then give to the creature whose home and heart and all is here, their share in the world to which they belong. If you belong to fairer worlds why not depart from this one, and relieve our world of a curse, the sooner the better! Surely, you are not making fit "preparations." . . . If, on the other hand, the delusion is not cherished by you, and you know that in all probability the only world to which man and other creatures of the earth have access, is this pitiful world . . . why then do you deny to your fellow mortals their simplest right in this world? They are not machines, but our brothers, whether we are too blind and bigoted to

see it or not and whether we will it or not. . . . Shame on the garden where lone and tortured and outraged captives pine their gloomy lives away to furnish degraded pastime for thoughtless onlookers who here learn anew the lesson of cruelty alone.[86]

In closing, Gunderson predicted that the "shameful garden of cruelty and wrong will no longer exist . . . because the more enlightened children of the coming day" will not support "twentieth century barbarisms."[87]

In this passage, Gunderson attacked religion and leveled something akin to a death threat toward Baker, welcoming the superintendent's own death because his passing would "relieve our world of a curse." "The sooner the better," she cried! Nonetheless, despite the violent rhetoric (which almost certainly was underwritten by the energy of the socialist movement, which had been uniting workers, between 1900 and 1910, against "unfettered capitalism"), Gunderson used this passage to advance radical claims about animals, elaborating on the idea that animals possessed "rights" of their own.[88] First, in asking the question, "Are you Cartesians?" Gunderson not only revealed that she was speaking to an audience larger than just the recipient of her letter (notice the plural!) but also showcased her knowledge of Western philosophy. Philosophers and naturalists in the eighteenth and nineteenth centuries credited René Descartes with formulating the predominant opinion about animals for the Enlightenment—namely, in the words of rationalist philosopher Nicolas Malebranche, that "they eat without pleasure, cry without pain, grow without knowing it; they desire nothing, fear nothing, know nothing." Descartes believed that while animals constituted living, organic beings, they possessed no capacity for reason and could not feel pain; animals were simply "automata," living machines.[89] By asking Baker to reflect upon his Cartesian assumptions, Gunderson approached "the animal" in a manner similar to the posthumanists of the last decades of the twentieth century, who quite often began their critique with the anthropocentrism of Continental philosophy. Gunderson's referencing of Descartes, and then Copernicus's heliocentrism, demonstrates that zoo criticizers could and did employ advanced philosophical reasoning to attack the convention of animal captivity. It also demonstrates that the zoo, for some, stood as a symbol of the narcissism of Western science. Specifically, by calling Baker a Cartesian and a "benighted naturalist," Gunderson challenged what she viewed as an arrogant inability to empathize with zoo animals,

Baker's "fellow mortals." Baker failed to see the "kinship" between himself and his prey.

After berating Baker for acting as if the heavens revolved around him, Gunderson expanded her notion of the "kinship" between humans and animals. "They are not machines," she asserted, referring to the Cartesian paradigm, "but [are] our brothers." Gunderson advocated for zoo animals in a different way than most activists concerned about cruelty. Many advocates for animals believed that acting cruelly toward animals would cause humans to act cruelly toward other humans, especially toward children since they were culturally conjoined with animals, as Pearson has powerfully written.[90] Other activists took this reasoning one step further, believing that animals themselves deserved, as living beings, to be treated kindly. Rarely, though, did activists explain exactly *why* animals required humanity's moral consideration. Gunderson claimed that Baker (and thus zoos themselves) denied animals from "their simplest right in this world"—"access" to the world, however "pitiful" it may be. Animals, a category of which humans were members, all had the right to "their share in the world to which they belong." Zoos fundamentally denied animals these rights, and therefore could be viewed as nothing other than institutions of "torture" and "cruelty." Zoo animals deserved to be in the world, an idea that captured Gunderson in a literal way and that prefigures the logics and languages developed by animal rights and animal studies theorists as diverse as Peter Singer, Tom Regan, and Giorgio Agamben.[91] Animals should not be held in zoological parks; they should be free in "the open," in their natural environments.

Few zoogoers wrote letters to Baker like the one presented above. Gunderson's invective reveals the discursive ceiling of popular animal activism in 1910. Most who advocated for zoo animals did so in a simple way, calling attention to a structural defect in a given enclosure or to a zoogoer's misdeed. Yet occasionally, activists looked beyond the particular cruelties of the zoo and, instead, attacked the very foundation that supported those cruelties. Critiques produced by these reformers expressed the potential of activism at the beginning of the twentieth century. They gestured toward a type of philosophical questioning that would gain ground after World War II and eventually usher "animal rights," as an established idea, into the popular lexicon of Americans. Through the penned words of Miss M. Gunderson, we see how a contentious discourse based around "animal rights" first emerged in the public sphere.

This should not suggest that all of Washington advocated for zoo animals. Some individuals complained about the animals themselves, not the conditions in which they lived. Dr. Cecil French, for example, mailed a letter directly to Baker that bemoaned the sea lions' "tormenting barking" that kept him awake at night in his nearby residence. French suggested that if the zoo "shut them up tight inside a building, be it ever so small, they will mind their own business at night and never utter a sound."[92] William Bride also complained to Baker about the sea lions, making sure to emphasize that he "appreciated the park very much" and that his "objections" were not those of a "crank." Bride had even donated some red foxes a few years earlier, but no matter how much he supported the zoo, barking in the early morning hours disrupted his beauty rest.[93] Baker admitted to Bride, "To be entirely frank . . . I do not know of any practicable way of quieting this animal, but the matter will be kept in mind. . . . By watching his habits, we may arrive at some means of controlling him."[94] In response, Bride informed Baker that he "notices that the animal stops his barking for an hour or so shortly after six in the morning," and while he "frankly knows nothing of the habits or characteristics of the animal" (pinniped vocalizing-language systems are only just now coming into view[95]), he suggested that maybe feeding the sea lions an hour earlier might ameliorate their appetite and lessen the ruckus.[96] In this advice, Bride formed hypotheses about zoo animals while also gently criticizing them. Even through its annoyances, the zoo could challenge neighbors to think critically about animals. Most who complained, however, were not so thoughtful. Disgruntled neighbors rarely worried about the welfare or "rights" of the animals that annoyed them.

Throughout his career as superintendent, Frank Baker's office was flooded with complaints against animals, zoogoers, zoo employees (himself included), and the zoo as an institution. While these letters differed, many centered on issues of well-being and justice. Whether zoogoers were angry about how their zoogoing peers treated animals, furious about the authorized killing of stray cats, disturbed at the zoo's violation of the "rights" of animals, or simply annoyed by the noises of the park, complainers necessarily had to think about well-being and justice in terms of zoo animals. Even those who poked, prodded, and teased the animals, who seemingly rejected any idea of an ethical claim made by the animals they objectified, had to first (at least subconsciously) come to terms with the individuals they enacted violence upon. Only after *seeing* "the animal" *seeing* them could choosing to poke, prod, and tease be worthwhile, producing all the potential sadistic pleasures, all the doses of oxytocin, that subjugating fostered for deviants and transgressors. Zoos forced all to ask this question: *How should I treat the animal before me?* No matter how zoogoers answered, its asking put ethical questions on display in the public sphere.

As early as 1892, the National Zoo requested that Congress pass an appropriation bill to increase the amount allocated for the maintenance of the park. To justify the request, zoo proponents emphasized that funds were desperately needed to hire watchmen. An unnamed zoo official told the *Evening Star* that not only did the zoo require them to protect children from dangerous animals, but they were also needed "to insure the safety of animals constantly endangered by malicious persons." In this article (like the one about Annie and the runaway bison), zoo animals were depicted as "helpless" and vulnerable to human cruelty. Animals and humans both required protection within the zoo. The article simultaneously advocated for Dunk and Gold Dust, calling attention to their inadequate quarters, "a shell of pine wood" in which "they would perish in a single winter's night should the fire kept in a crude stove be overlooked." Within one column of newsprint, the author married two different denunciations of cruelty—violence exerted by zoogoers onto animals and neglect on the part of the zoo itself. Then, the activist-author argued that both types of cruelty were derivative of the zoo's financial need, and, therefore, responsibility for correcting these moral failings fell upon Congress. Through this logic, animal activists, both inside and outside the zoo, cloaked the languages of associations like the ASPCA and the Humane Society in the more palatable, republican discourse of civic duty.[97]

Even though Dunk's direct claim on Congress proved to be an outlier, the conversations about animal well-being within the National Zoo cannot be understood as removed from national politics, since those politics literally sustained the zoo, and thus the lives in the zoo. When zoogoers deliberated about the welfare of zoo animals and the ethical integrity of the National Zoo, they also engaged the larger political discourses that structured that zoo. The National Zoological Park and its animals stood as symbols of government and progressivism. In this way, Gunderson's diatribe against Baker could be read as commentary on the hypocrisies of the Progressive movement, patriarchy, and science writ large. Similarly, Raub's attack on Baker's cat-extermination policies could be read as a radical appeal for the sanctity of animal life or as a conservative response to the government's intrusion into the wealthy lives of nearby property owners. The zoo and popular zoology linked politics and culture, entertainment and ethics, humans and animals. As these linkages were made, zoo animals acquired uncounted new symbolisms, pressed upon their bodies, concealing the lives within them.

Animal Activism Beyond Washington: Philadelphia and New York

The National Zoo was not the first to provide Americans with opportunities to exert violence upon animals, nor was it the first to provide animal activists with a platform upon which to criticize this behavior. Almost twenty years earlier, activist voices reverberated around the Philadelphia Zoo too. Within weeks of its grand opening in 1874, one journalist described the violence emerging:

> Men irritate the wild cat and the leopard with their canes; young women titillate the lynx with their parasols; energetic old women jab the opossums with their umbrellas; mischievous boys harass the wolves with their sticks, enrage the monkeys with torpedoes and drug the elephant with tobacco, while one man was detected the other day holding the lighted end of a cigar to the nose of the raccoon.[98]

The Philadelphia Zoo offered opportunities for zoogoers of all social rank, even the cane- and parasol-toting upper class, to exploit animals.[99] While the zoo officially prohibited violence, this sort of perverse behavior was irresistible for visitors who needed to bolster their self-esteem by asserting dominance over the kings of jungles. In a rapidly urbanizing world, maybe zoo animals served as tabula rasae upon which unseen individuals could etch their names, as they retreated from the bustling multitudes into the illusory escape that the zoo hoped to be. Many journalists saw through the zoo's facades and criticized the recklessness of the nation's first zoogoers. The above journalist even called upon the ASPCA to assist the Philadelphia Zoo in keeping its animal collection safe from poking publics. Yet, of course, the rhetoric of animal activism within the first zoo attracted mockery. For example, one article titled "More Foolish Legislation Coming" detailed a legislative bill that sought to "prohibit living animals from being fed to snakes," specifically referring to the rats and mice consumed by the serpent collection. The author criticized the representative who presented the proposal for wasting the taxpayers' time and then sarcastically warned his readers to be wary of future bills that might "prevent the keeping of turkeys because they catch and eat live grasshoppers."[100] For some, protecting zoo animals was laughable from the beginning, but others recognized that zoo animals needed advocates. Between 1874 and the 1890s, public critiques of zoos became commonplace. While there is no way to quantify the rate at which zoo animals across the United States were mocked, spat at, and targeted by food-slingers, the records of zoos nationwide show that public violence toward captive animals was universal, inscribed into the very essence of the zoo. Two decades into the twenty-first century, zoos still place signs around their grounds to remind visitors how to treat and not treat the animals. Still, some zoogoers yell, laugh, and jeer at the animals—pounding on the glass—as they try to forge connection.

On August 27, 1894, the *World* (owned by Hungarian-American publishing tycoon Joseph Pulitzer, known for financing the future namesake book prizes and for forging "yellow journalism" with others like William Randolph Hearst of the *New York Journal*) published an expose titled "Sufferings of Central Park Animals: A Legalized Institution of Torture."[101] The article was crafted by Elizabeth Jane Cochrane, who under the pseudonym "Nellie Bly" became famous in the late 1880s for circumnavigating the globe in record time (modeling her trip off Jules Verne protagonist Phileas Fogg's eighty-day trip) and for feigning insanity in order to study the workings of another institution of imprisonment, the asylum.[102] Wherever *The World* traveled, so did Bly's columns. "Nellie Bly" had become a household name by the 1890s, and in 1894 she cast her critical gaze upon the Central Park Zoo.[103]

Cochrane's article presented a jeremiad against the zoo. Bly began by wryly informing her readers that she visited the zoo frequently because it "affords such a splendid opportunity to study man, *his* civilization, *his* humanity and *his* peculiarities." Anthropocentric. Self-referential. Narcissistic. Like e. e. cummings after her, Bly depicted the zoo as a collection of mirrors.[104] She then underscored a hypocrisy lurking beneath the zoo and the city at-large: "One must remember that the same city which supports the menagerie supports a Humane Society; the one to prevent cruelty to animals, the other to promote it." She dedicated more than one full column to the abhorrent conditions of the bird house, where the inhabitants constantly died from sickness. And like other zoo critics, she described a robin, owl, and cockatoo pelted with pebbles, paper wads, and chewing tobacco, respectively.[105]

Of course, in standard muckraking style, Bly embraced hyperbole, even employing personification when quoting a robin red-breast who apparently told her:

> I am sick, and I have lost hope. I have not had a clean bath this summer, and my longing for freedom has mastered every other feeling. I want to sit in the sun; it never reaches me here; I want to breathe the pure air instead of this horrible stench; I want to see God's sky . . . Oh, I want to be free! What have I done to man that I should be kept prisoner in this unclean cage?[106]

Through anthropomorphized soliloquy, Bly not only employed the ancient convention of extending human voice to nonhumans, but she used this convention to communicate the activist message that zoos were institutions of "cruel imprisonment." By speaking for animals, Bly gave them a "voice" that her readers would be able to hear. She made anthropomorphic animals demand ethical consideration. She did not have to convince her readers that they—those *other* living beings *over there*—deserved her readers' concern. Instead, the animals of Central Park became literary characters speaking on their own behalf, ventriloquizing the complaints of animal activists. Bly's moralizing would have been palpable to her readers, who may have rolled their eyes at musings like these: "What was it to them [the birds] if the sun shone without, if the air was balmy and the trees green? What if other birds flew from tree to tree and sang and mated and built nests? God had intended that they should do likewise, but man, civilized man, had changed God's plan." No matter how Bly's readers felt about her "yellow" journalistic style, they at least entertained the idea that, first, zoos harmed the animals they kept and, second, those animals deserved moral consideration and possibly even freedom.

Bly elaborated on the horrors of the zoo. She described "torture in the monkey house," where zoogoers threw peanuts, "wastepaper," "pieces of string," "ends of cigarettes," and "lighted matches" into the cages; where a man handed his cigar to a monkey; and where another spat tobacco juice into a lemur's eyes. Bly detailed the misery of the elephants, "fastened to the wall with a three-foot chain," noting their cramped conditions and their inability to bathe, for they were "covered with a coating of filth." Even though she acknowledged that it was "needless to specify each animal in the park to tell a tale of civilized cruelty," she concluded by telling stories about polar bears and Alaskan dogs suffering heat strokes, Angora sheep lacking water and grass, foxes and wolves crammed into small pens, and bison covered in grime. All the zoo's animals were "badly fed, wretchedly watered, miserably housed, and shamefully neglected"—"to tell the case of one is to tell it of all."[107]

Depicting the Central Park Zoo as a "legalized institution of torture" created a framework through which the public could view *all* zoos. Less than one month after the Bly feature, the *World* printed "Why These Cruelties?," which presented three individuals' responses to Cochrane's expose, those of George Bird Grinnell (editor of *Forest and Stream*), an anonymous author writing under "Suffering and Discomfort," and Mrs. E. C. Halcott. To soften Bly's critique, Grinnell defended the general integrity of the zoo, for he had long been a proponent of the American zoo movement. However, agreeing with Bly, he emphasized that zoos must keep their animals healthy. And the animals' "surroundings" needed to "resemble [the environments] they are accustomed in a state of nature." They also needed the "food best suited to them," and "every attention should be paid to relieving them from the ailments *which must necessarily follow the unnatural conditions of their existence in captivity* [emphasis added]." Grinnell then gave examples about how to make the animals' enclosures and diets more "natural." Zoos, in Grinnell's estimation, were confronted with a gargantuan task. "Suffering and Discomfort" criticized the Central Park Zoo in a different manner. Rather than blame the zoo as a faceless and abstract institution, this author pointed a condemnatory finger at the inexperience of the new superintendent who replaced former director William Conklin, who had previously donated animals to both Hornaday's Department of Living Animals and the National Zoo. According to this writer, the departure of Conklin marked the downfall of the zoo because his replacement had "no more fitness than a man . . . of Tammany Hall." Mrs. E. C. Halcott, the last editorialist, took an approach more decisive than both Grinnell and "Suffering," stating simply that "it is time that all zoological gardens were done away with."[108]

American zoos were closely interconnected. As zoogoers critiqued their own city's zoo, they frequently thought about zoos elsewhere. Sometimes one zoo stood as a symbol for all zoos. Other times, zoos were compared and contrasted. Editorials about the atrocities in Central Park continued to hit New York presses. Bly even published a follow-up feature that described Illinois Governor John Peter Altgeld's tour of Central Park's zoo. In her standard style, she informed New Yorkers about the governor's horrified reaction to the zoo, who apparently exclaimed, "What neglect and what useless cruelty!"[109] (We will see in chapter 6, however, that the acclaimed Lincoln Park Zoo housed its own fair share of suffering.)

The National Zoo never figured far in the minds of zoogoers around the nation. Shortly after Altgeld toured Central Park, Frank Baker made the trip from Washington to assess the zoo himself. Zoo directors commonly monitored each other regarding the welfare of animals, establishing a peer-review system of sorts, and zoos around the nation looked frequently to Baker for approval, advice, and legitimation. During the inspection, a *Globe* journalist interviewed Baker about his thoughts, and one month after the first Bly malediction, the *Globe* printed an article titled "Poor Beasts, If You Knew!" that contrasted the tarnished image of the Central Park Zoo with the pristine image of the National Zoological Park. Despite the activism that had surrounded the National Zoo since its foundation, the author depicted it as the archetype of a zoo done right. The byline of the article described the difference between these two institutions as "the Difference Between the Bowels of Compassion and the Bowels of Tammany." A second byline declared that "Grass, Trees, and Air Are Theirs." And a third exclaimed, speaking directly to the animals themselves, that "Everything That New York Denies You Washington Gives Its Captives" (language that cannot but remind us of the vile antebellum apologetics employed by those posturing as "good" slave owners, replicated unconsciously here by a journalist attempting to seize the always-fractured voice of moral rightness).[110]

According to this *Globe* article, which was clearly orchestrated by Baker for advertising purposes, the National Zoo was "not a stuffy, cramped, ill-smelling menagerie, where dumb brutes suffer that the public may be amused." The journalist depicted the National Zoo as a paragon of virtue, stating (wrongly) that "no one there is crying out for another Henry Bergh [founder of the ASPCA]."[111] When asked for his opinion about the state of the Central Park Zoo, Baker replied simply, "I will do anything I can. I have taken much interest in the article published by The World. The principle upon which we work [in Washington] is simply to place our animals . . . under the conditions peculiar to them in their wild state," assembling the wild veneer so important to zoo ideology. Rather than discussing how to improve the Central Park Zoo, the article devoted two full columns to the greatness of the National Zoo—the spacious and well-ventilated enclosures, the abundance of water, the cleanliness of the park, and the advanced medical care. However, Baker cloaked the zoo's self-promotion in an activist guise, for the article concluded with the only other passage that concerned improving the Central Park Zoo: "I would suggest that perhaps some of the most important obstacles of Central Park may be in the difficulty of securing proper keepers. It is only possible to do this where the full power of dismissals and changes is absolutely vested in the superintendent, unaffected by any political influence."[112] Baker's boosterism was casuistical. Even though the conditions of the National Zoo were, in fact and in the aggregate, far better than the conditions of the Central Park Zoo, their animals still, as we have seen, suffered daily. Baker surely could not be honest about the weaknesses of his zoo, nor would he want to call attention to the growing animal activist campaigns for fear of diminishing the reputations of the National Zoo and zoos everywhere.

Yet Baker's honesty is beside the point. Baker traveled to New York as an ironic animal activist, as a symbol of the noble National Zoo that embodied the new national conservationism. Usually, animal activists served as bad publicity for zoos. In the day-to-day operations of a zoo, activists were viewed, by the superintendents that could never live up to activist standards, as nuisances. However, by inspecting the Central Park Zoo, Baker became "activist" himself, coopting animal activism for zoo-business purposes. By playing the role of zoo critic, Baker drew a distinction between a "good zoo" and a "bad zoo," extolling the National Zoo, in particular, and the *idea* of the zoological park, in general, while disguising the broad critiques of zoos, as well as the wild-captive paradox at the core of zoo ideology, within or *beyond* his established moral dichotomy.[113] Baker may not have told the *Globe* reporter about the chains that tethered Dunk and Gold Dust to their exhibit, but he did successfully laud, and therefore construct, the characteristics that defined the ideal zoo. Of course, these ideals always remained objects to strive for rather than ends obtained. By speaking about these ideals in a forum as public as the *Globe*, Baker made the boosterist rhetoric spoken in the halls of Congress a few years earlier relevant for America's largest metropolis. The

Central Park Zoo, Baker argued, needed to look to the National Zoo for salvation.

Zoos and Shifting Attitudes Toward Animals: Larger Trends

The National Zoo was one institution in an expansive zoo network. The lessons learned about animal activism surrounding the National Zoo apply to all American zoos, and these lessons are simple and threefold. First, the case study of Dunk and Gold Dust makes the simple point that zoos did indeed attract the attention of the growing ranks of animal activists emerging in the Gilded Age and Progressive Era. Second, newspaper articles about and letters of complaint written to the National Zoo demonstrate that zoogoers (even if not consciously participating in formal activism) often concerned themselves with the welfare of zoo animals. This broadens the category of "activist" to include informal advocates of animals and stretches the category of "activism" to move beyond politics and policy and into the realm of discourse. Third, animal activists and animal activism pervaded the American zoo movement. Just as zoos were networked together, so were their accompanying activists.

This chapter did not make a case for causation. It did not contend that the twentieth-century animal rights movement arose out of (reactions to) the zoo movement. Animal rights and zoos both have long histories that each trace back to the eighteenth century. However, animal rights (originating in natural philosophy, utilitarianism, and radical vegetarianism) and zoos (first forged as imperial symbols of the nation-state) have very different, even opposed, creation stories. While this chapter did not narrate their creation nor argue for causation, it sought to locate an undocumented context. Most everyday zoogoers did not consider themselves to be activists. Most were not enrolled in philanthropic and political organizations for the protection of animals. Most, unlike Miss M. Gunderson, did not develop a consistent philosophy concerning the "nature" of rights, the universality of emotion, the interspecific experiences of pain and pleasure, or the implications of evolutionary theory. Zoogoers, like most of us, rarely stepped out of their experiences to see the big picture. But zoogoers encountered animals. They looked through the bars of cages, over the walls of enclosures, and between the interstices of fences into myriad eyes looking back, making humans aware of themselves. These gazes from across the species divide, this seeing of the Other,

Figure 4.6. Mandrill reaching, looking, through the bars of captivity. Courtesy of the Smithsonian Institution Archives.

required response because they also carried, as art critic John Berger has recognized, a remainder, a mystery—a message, a promise, a secret, from across the "abyss of noncomprehension."[114] Yet these responses, these negotiations, could lead zoogoers in many directions, toward, for example, retaliation or embracing. Humans enacted violence *and* extended compassion toward zoo animals. No matter how zoogoers engaged "the animal," these interactions, at the largest scale, created a new discourse of animal welfare as it also created a new popular zoology.

CHAPTER FIVE

Zoo Conservation and Its Discontents: Chasing Bighorn Sheep

This chapter presents the story of the National Zoo's pursuit, between 1898 and 1902, of a single species. The story of William Hornaday and early bison conservation is well known,[1] so this chapter will follow the lesser-studied attempt to conserve bighorn sheep. By looking closely at a case study, we will see elusive herds confounding their aspiring captors. We will see how zoos always existed in the gap between animals and humans' expectations of animals. And we will see how popular zoological knowledge was assembled from the ground up. The history of North American mountain sheep begins on the shrubbed tundra of Pleistocene Beringia, upon the land bridge that connected the Americas to Eurasia thirteen thousand years ago. Even though these mountain sheep diverged from their Eurasian ancestors six hundred thousand years ago, when archaic humans (*Homo heidelbergensis*) shared the earth with *erectus* and *neanderthalensis*, they would not become "American" sheep until they crossed the Bering Sea and stepped hoof in northern Canada.[2] This diaspora migrated southward and quickly colonized the North American continent from northern Canada to northern Mexico. Today, zoologists generally recognize three species as offspring of the Ice Age newcomers—*Ovis canadensis* (bighorns), *Ovis dalli* (white, horned sheep of Alaska and northwestern Canada), and *Ovis nivicola* (Siberian snow sheep).[3] Of these three, it is *Ovis canadensis* who sheeped most of North America. This species varies greatly from place to place, yet most scholars acknowledge three subspecies: Rocky Mountain bighorn, Sierra Nevada bighorn, and desert bighorn.[4]

Millennia before Europeans sailed westward across the Atlantic, humans and sheep met in the mountains of western North America. These interactions were surely dynamic, different in the Bitterroot Mountains than in the Black Hills or San Juan Mountains, different in the Wasatch Range than on Mount Elbert or along the Baja peninsula. It is, therefore, difficult to generalize about the role of sheep in human cultures, and humans in sheep cultures, prior to European colonization. Indigenous societies and First Nations forged dynamic and diverse relations with sheep. The Hopi, for example, believe Pangwu (mountain sheep) brings the rain and grows the grasses.[5] The Apsaalooka (Crow) believe bighorns give humans wisdom, keen vision, and careful balance.[6] It can be assumed generally that one of the most developed sheep-human "contact zones" existed in the American southwest, where Puebloan peoples had long flourished, where sheep and human populations were greater, and where sheep habitats existed at more accessible altitudes.[7] Nonetheless, mountain sheep were always meaningful to humans in places where both lived, as evidenced by petroglyphs and pictographs of bighorns embedded on boulders across California and Utah and by their ocher portraits painted upon cave walls in Baja California, Arizona, and New Mexico.[8]

Even though sheep and humans long shared the continent, European colonization unsurprisingly brought quick declines in sheep populations. Mountain sheep from the seventeenth through the nineteenth centuries (like so many New World animals) were moved along a rhetorical spectrum, in the production of their alterity, from mysterious "exotica" to "marvelous possession" to commodity.[9] In 1697, a Spanish missionary in California first noted a "remarkable species of sheep" he encountered near Sonora. Five years later, Father Francisco María Piccolo also described this mysterious animal as having the characteristics of a stag, ram, oxen, and deer. And he gave them the name "tayé," probably derived from the verb "tallar," to carve out, since sheep may appear to carve themselves out of sheer rock.[10] In 1800, Duncan McGillivray, a Scottish explorer with the fur-trading North West Company, encountered the same animal near Calgary that "appeared to be a compound of the deer and the sheep, having the body and the hair of the first, with the head and horns of the last."[11] McGillivray's "discovery" was quickly communicated to the Western world and coronated "Belier de Montagne" by leading naturalist Etienne Geoffroy St. Hilaire in 1803.[12] Two years later, in 1805, Lewis and Clark met mountain sheep along the badlands of the Upper Missouri River, representatives of the newly stolen lands of the Louisiana

Purchase.[13] It did not take long for these new national possessions to become international commodities.

Robert Jameson, University of Edinburgh naturalist and future instructor of Charles Darwin, proclaimed that upon "examining the fleece, I was particularly struck with its uncommon fineness; and it occurred to me, that an animal inhabiting the temperate regions of the Rocky Mountains, with so valuable a fleece, might be easily procured, and readily introduced into this country, and form a valuable addition to our wool-bearing animals."[14] The proposed transatlantic relocation never happened, but North American mountain sheep fascinated Europeans—especially their wool, horns, and mutton. British investors especially saw naturalizing New World animals like bighorns or alpacas as a potential asset to Great Britain's textile industry, which was beginning to seize global cloth markets from India and China. They believed the Scottish Highlands, Shetland Islands, Wales, the Cheviot Hills, and the mountains of Kerry and Wicklow to be well-suited for these high-altitude coat-bearers.[15] Even though a century of scientists quibbled over classification, no one who had laid eyes on North American mountain sheep questioned their uniqueness or utility. They stood atop Nature's throne.

An 1858 article in *Spirit of the Times* described the desirable qualities of mountain sheep. First, "the coating of this sheep is more cervine than ovine . . . [like] that of elk, but coarser." Second, "their flesh has the flavor of the finest mountain mutton." Finally, "the horns of the male measure about three feet from base to tip . . . they have at their base a diameter of five or six inches, commonly weighing eleven or more pounds."[16] (In fact, these horns, which males use in spectacular and powerful head-butting contests to determine status, territory, and mates, can weigh thirty pounds each.[17]) Not only did mountain sheep, for nineteenth-century Americans, make fine coats, delicious dinners, and spectacular antler-trophies—of course precisely measured per hunters' preoccupation with size—they also were challenging to hunt and, therefore, became both the sportsman's ultimate test and prize.[18] This commodification of mountain sheep spawned a voluminous popular zoological literature. Titles alone tell the story. 1877: "Big Horns, or Rocky Mountain Sheep, Surprised," *Frank Leslie's Popular Monthly*.[19] 1881: "Sheep-Hunting in the Mountains," *Littell's Living Age*.[20] 1884: "Stalking the Mountain Sheep," *Forest and Stream*.[21] 1891: "Capturing Rocky Mountain Sheep," *Overland Monthly and Out West Magazine*.[22] Even poet and nature writer Mary

Figure 5.1. *The Mountaineers*, painted by Carl Rungius, 1912. Courtesy of the Buffalo Bill Center of the West.

Austin wrote verses for mountain sheep: *The shepherd tends his foolish flocks / Along the mountain's hem; / But free and far the wild sheep are, / And God doth shepherd them.*[23]

Despite these sensational headlines, bighorn populations plummeted over the nineteenth century. By 1895, Hornaday, who had collected a specimen for the Department of Living Animals six years earlier,[24] lamented the demise of the continent's mountain sheep: "Formerly the Mountain Sheep inhabited all the wild mountain-ranges of the West, from central Arizona to the Arctic Circle . . . but in the United States this much-prized animal has been so persistently hunted and shot that now it is found in few localities only, and those are very hard to reach."[25] Others joined the dirge. German painter Carl Rungius, for example, inspired greatly by Hornaday, memorialized mountain sheep through his painting "The Mountaineers: On Wilcox Pass," where bighorns stood proudly atop the Alberta Rockies, looking outward at the human viewer as if demanding a reason for the intrusion onto their mountainscape.[26]

Rungius's work might be best understood as an epitaph, an artistic preservation of a species that he was certain would soon be gone forever.

Bringing Sheep to Washington: The "Hunt" Begins

Late in the year of 1898, Frank Baker decided that the National Zoo needed to keep the promise made almost a decade earlier in the halls of Congress, namely to preserve the animals of North America. This mission could not be complete, he believed, until he acquired Rocky Mountain sheep for the zoo. By the end of the nineteenth century, acquiring a zoological specimen seemed to be a straightforward task—locate a collector, place an order, and wait a few weeks. Baker knew this process quite well. He had placed hundreds of orders, resulting in the successful procurement of alligators, beavers, bison, and elk. Indeed, nearly all of the zoo's inhabitants (who were not donated or loaned to the zoo) were acquired in this way. Yet, even though a national network of "specimen dealers" had emerged after the Civil War to meet Americans' growing demand for natural history, some specimens were more difficult to obtain than others.[27]

On March 1, 1899, Baker made his order through the botanist, naturalist, and University of Montana professor Morton Elrod.[28] The telegram to Elrod read: "I wish to obtain immediately a young Rocky Mountain ram alive. Can you furnish one, or inform me where to get one, and price? Answer collect."[29] The superintendent assumed that locating a sheep would be an easy assignment. To cover his bases, three days later, Baker sent an identical telegram to hunter R. W. Rock of Lake, Idaho.[30] Even after following up with Elrod with a longer, typed letter, Baker heard only silence in return.[31] He then contacted hunter and collector Vic Smith, of Anaconda, Montana, and hotel owner Leonard M. Cotton, of New Castle, Colorado, in hopes that they would know the whereabouts of sheep.[32] Both agreed to help the National Zoo, and Samuel Langley even sent Cotton the legal papers for a formal agreement.[33] Yet the year ended, three months of a new century passed, and Baker was still without even a prospect of bighorns. The hunt for sheep was on.

Secretary Langley and Superintendent Baker pursued different courses of action. Langley rushed down the paved paths of government. In February 1900, he contacted Colorado Senator Edward Wolcott for help. The senator promised to talk to his business partner who he believed would "succeed in convincing" the permit-granting authorities "that every citizen of Colorado is interested in the success" of the National Zoo. In April, Wolcott even appealed directly to the governor and asked outright for "some kind of official sanction" granted to the zoo for the capture of two or three pairs of sheep.[34] Langley also telegrammed New York governor Theodore Roosevelt (who owned a ranch in North Dakota) for further recommendations of hunters who could catch these animals.[35] The future president responded the same day with two names.[36]

Baker, relying on America's trading networks, continued down the tangled footpaths of hunters, trappers, and collectors. He waited for good news from Elrod, Rock, Smith, and Cotton, yet once Langley began appealing to government officials, Baker decided it was time to hurtle himself down the skein of trails that tied the North American West together. Throughout the month of April (and to emphasize the zoo director's thoroughness), Baker reached out to Owen Swift of Edlaw, Perry Foust of Stevensville, Frank Parks of Ketchum, Frank Robinson of Libby, J. Blizzard of Eureka, H. Weeman of Midland, John Goff of Meeker, W. H. Root of Laramie, Frank Hays of Aspen, Howard Eaton of Medora, George L. Smith and Nims Ferguson of Glenwood Springs, and Tazewell Woody of Mammoth Hot Springs.[37] He scoured Colorado, Montana, Idaho, Wyoming, and North Dakota in the hope of catching Rocky Mountain sheep. All the while, Langley grew increasingly impatient.[38] When he grouched at Baker for still not having acquired a single sheep, Baker responded, "During the last two years no chance has been allowed to slip for obtaining these animals. The failure does not arise from lack of effort but from the fact that the sheep are exceedingly wary and almost impossible to capture. But very few have ever been in captivity, and those were caught by accident rather than by design."[39] Rocky Mountain sheep, just like the American bison, had been pushed by a westbound nation to the edge of extinction, victims of bullets, disease, and dislocation. Despite their vulnerability to a settler nation, catching one alive still proved to be a nearly impossible task for white Americans. As Roosevelt acknowledged in 1886, "There is no animal in the world superior in climbing. . . . The perpendicular bounds it can make are truly startling."[40] Baker would have to find the perfect man for the job.

"Buffalo" Jones and the Colorado Expedition of 1900

One day in April, that man arrived at the National Zoo for an interview with Baker. By 1900, Charles Jesse "Buffalo" Jones had acquired quite the reputation as a frontiersman, hunter, rancher, and cowboy. As a child, Jones had likely met Abraham Lincoln when his father hired the future president for his legal advice. Jones knew personally American legends like Wyatt Earp, Pat Garrett, and Buffalo Bill Cody. He founded Garden City, Kansas, and served as its first mayor. And most significantly for Baker's purpose, Jones, alongside the National Zoo, fought to prevent the extinction of bison.[41] A fitting symbol of a tangled, turn-of-the-century West, Jones was quite the socialite for a frontiersman, quite the statesman for a cowboy. And as he and Baker meandered along the walkways of the National Zoo, sharing stories about wildlife, Jones convinced the superintendent that mountain sheep could not be obtained by the casual enlistment of individual hunters. They were simply too elusive. Their capture would require an expedition, an organized campaign.[42] Less than a week later, the Smithsonian had organized such a venture, and as Jones's grandiose signature was still drying on the contract, he headed west.[43]

One month later, Jones scrawled a letter from Red Stone, Colorado:

> Dear Friend: I came down from the mountains yesterday quite ill, but am all right today. . . . There are no young lambs yet, except one or two pair and they are in the most difficult parts of abrupt cliffs. I am after long ropes so as to let one of my party down to search for them. I feel sure they are there as the doe keeps near and is very watchful [and] besides she has a regular path leading to the cliffs. We have located five bands ranging from seven to thirty. . . . I am not disturbing them just yet. . . . I feel confident of securing some sheep, yet it is one of the most difficult and hazardous undertakings I ever experienced. . . . Don't be disappointed even should I fail altogether, yet if there is such a thing as getting them will do it.[44]

There *was* such a thing. Before this letter even reached Baker's desk, Langley received news via telegram that Jones had indeed captured his first lamb.[45] And shortly after receipt of the letter, Baker received word of a second lamb.[46] Maybe mountain sheep were not so tricky after all.[47] Regardless, Jones took the lambs, with their soft and light-colored coats, to Topeka for safekeeping and went back to the mountains in search of more.[48] At this rate, "Buffalo" Jones would have three pairs of sheep in no time. Yet the assignment was still arduous. "You have no idea the difficulty in capturing the lambs," Jones wrote. "They go into the most

Figure 5.2. Charles Jesse "Buffalo" Jones. Courtesy of the Kansas State Historical Society.

difficult cliffs and my fingers are covered with blisters . . . endeavoring to find their lair."[49]

Another letter, scribbled frustratingly, June 19, 1900, from Topeka:

> I stayed in the mountains as long as there was a hope of capturing a lamb. Even a week longer. As they are as nimble as a fox when a week old. The old ewes take them to the highest and roughest peakes as soon as they can travel. And even drop them far above where the old bucks + young sheep range. People see sheep on the side hills + think it easy to get a lamb but *they know nothing about the habit of the animals.*[50]

Jones had gotten two lambs in two months' time, yet acquiring more was simply exasperating. As head of the Colorado expedition, Jones "had men in all parts of Colorado, endeavoring to capture lambs . . . but not one of them ever saw the tail of a single lamb." Mountain sheep were impossible to catch! And they were impossible to hold on to, for Jones even complained that the two lambs under his possession kept escaping because they "scale everything they come to."[51] One of the lambs could even "run up the side of the barn."[52]

In Topeka, the two lambs were "having a hot time of it," for a prairie summer was much hotter than a mountain summer.[53] Animal dealers knew little about caring for mountain sheep except that they rarely stayed alive at low altitudes, frequently succumbing to apoplexy.[54] Jones reported in July that one of the sheep looked as if he was approaching death.[55] He recovered, but Jones continued to stay "with them all the time" to ensure that they would live to see their days in the National Zoo.[56] He monitored their diet closely, feeding them milk three times every day, but "dared not risk strangers to feed them as sour milk would be dangerous."[57] This intensive nurturing quickly grew expensive and time-consuming, especially when Jones had private business to attend to in Nebraska.[58] So once the weather began to cool in September, he wasted no time in shipping the lambs to Washington. They were both in excellent health, and he felt confident that "they would go through all right."[59]

A little more than a week later, Baker received news that the lambs had died in transit.[60] After hearing word about these deaths from his employer, Jones responded, embarrassed, "Your message reached me here. I am almost paralyzed by it." He apologized profusely, feeling guilty that he rushed the needy animals off to Washington. "I cannot express my regrets at the news."[61] "I regret my time was so taken that I could not have stayed with them."[62] The cause of death remained a mystery. The Texan who shipped the lambs reported that they succumbed to a cough (possibly tuberculosis).[63] Baker suspected food poisoning.[64] Jones, after examining the bodies, suspected that one lamb died *naturally* and the other became so lonesome that he beat himself to death upon the side of the pen.[65] Baker remained skeptical. More than one year later, another collector made reference to a pair of sheep caught by Jones who had succumbed to "injudicious handling."[66] Apparently, information about the fate of captured sheep traveled quickly across prairie towns increasingly conversant in popular zoology. No matter how these sheep died, however, the National Zoo's first two lambs resisted captivity, despite dying eight to thirteen years before they would have in the wild.

Though death-as-resistance is well-established in cultural histories concerning humans, some historians might bristle at the assertion that a mountain sheep could practice resistance through and in death. Some might argue that reading a sheep's death as suicide would be, as Jones suggested, at best, without empirical evidence and, at worst, a base projective anthropomorphization. They might add that it would further be impossible to argue that animals died, like humans, for a Cause. Nonetheless, while individual animals may not have been able to consciously resist by dying, by *choosing* to end one's life (though this is certainly not out of the realm of possibility), symbolic animals undoubtedly resisted, both human practices and their own living/biologic counterparts, through death. From the perspective of an animal collector or zoo superintendent, animals dying *functioned* as a form of resistance. When viewed collectively, animals dying could stop both the zoo and the business of animal collecting in their tracks, making the most essential tasks—procuring and keeping animals, making captives appear wild—impossible. Concerning the specific contention of Buffalo Jones that a sheep could die from loneliness: Bored, miserable, lethargic, and unhappy animals are more likely to die than wild, active animals. Animals, like humans, experience emotion, including various types of melancholia. Elephants do mourning marches. Dogs and horses grieve the death of their owners, a trope long recognized in literature and folklore. And all sorts of nonhuman animals experience sadness over the loss of other animal companions and mates. According to neuroscientist Keith Kendrick, sheep have impeccable visual and spatial memories, especially when it comes to recognizing the faces of others; and animal husbandry scholar John Webster has shown that sheep are also deeply emotional, affected in the present by past events, and especially prone to depression when experiencing stress, characterized by a drooping head and the avoidance of positive outcomes.[67] As Donna J. Haraway has noted, there is something about truly realizing that we give captive/pet dogs Prozac that "warps" our "sense of reality in the way that makes one see what was hidden before."[68] Seeing suicide or depression across the human-animal divide might do the same. It seems likely—even probable—that lonely, captive animals would have less "will to live" than lively, satisfied animals, and, therefore, the death of a lonely sheep, while being transported through hot train yards, could surely be interpreted as some sort of conscious resistance to a human endeavor. This line of reasoning may flirt with the impossible question about whether "resistance" requires "intentionality," yet locating forethought, reason, and soundness in and around anyone's suicide is always a difficult task. Of course, historians can never "prove" the intentions of these lambs—can never nail the jelly to the wall, as intellectual historian Peter Novick once said—but, as for all historical subjects, context speaks volumes.[69] And animals should not be held to different evaluative standards than humans, regarding agency and intentionality, just to make us feel more comfortable in and confident about the "human" side of our fictive binary. Jones was possibly correct. Maybe the sheep—who kept escaping from the Kansas barn—chose death. However, it is also likely that Jones was merely displacing blame onto the victim, for it is most probable that the sheep died from the horrific conditions of the train car, from lack of food or water, from a dearth of fresh air, from intense jostling, or from claustrophobia or extreme heat or simply sheer panic.

Intensification in the Colorado Rockies, 1901

Even after the Smithsonian forged a contract with "Buffalo" Jones, Frank Baker continued to fashion an extensive mountain-sheep community across the Colorado Rockies. Clearly, catching bighorns was no simple assignment, and the zoo director built a safety network of potential collectors. Even though, in 1900, to minimize financial risk, the Smithsonian refused to sign more than one contract, the National Zoo still agreed to purchase sheep from individuals should they happen to catch any on their own. Most of these contacts stretched across Colorado—Edlaw, Aspen, Manitou, Granite, Paonia, Denver, Leadville, and Eagle.[70] Others hailed from Wyoming, Montana, New Mexico, Arizona, and California.[71] None caught sheep, however. By the late autumn, Baker still believed that mountain sheep would be captured most easily in Colorado, and on the doorstep of the 1901 winter, he prepared to rely on more than one celebrity cowboy to bring the animals to Washington. That year, after lobbying for budgetary approval, Baker sent two contracted men to the mountains—hunters A. S. Sharp and Samuel V. Moninger.[72]

Before they could catch any sheep, though, they needed permits from the state of Colorado. On May 25, 1900, President William McKinley had signed the nation's first wildlife statute into law. In addition to protecting avian populations and regulating the importation of exotic animals, the

Lacey Act, Baker knew, forbade the "shipping of sheep . . . through other states" if those sheep were obtained in violation of state game laws.[73] The previous year, Colorado governor Charles Thomas approved unquestioningly a permit for "Buffalo" Jones to catch three pairs of sheep. The new governor, James Orman, however, responded tersely to the petition for permits in a year of heightened wildlife regulation: "Do not care to issue further permits at present." Enraged at the curt dismissal by a freshman governor and frustrated that the authority of state governments could attenuate the *national* sovereignty supposedly wielded by the National Zoo, Langley pleaded with Colorado Senator Henry Teller to offer assistance.[74] Sharp and Moninger needed permits promptly because it was decided, after Jones's spring and summer failings, that the best time to capture sheep would be during the deep snows of January, February, and March. Adult sheep were simply too nimble and "too wary to be trapped by the hunter." The only hope would be to corner young sheep between mountain drifts since they cannot generally move through deep snow.[75] Langley festered with impatience. He even told Moninger not to worry about a permit and to just collect for himself![76] Finally, after unceasing political pressure from Teller, Orman capitulated, granting permits to both Sharp and Moninger to "kill for your Institution six Mountain Sheep each."[77] "Kill" was then crossed out, and "capture" penciled in.

Sharp and Moninger hit the mountaintops. Moninger located a large band of forty-five sheep right away. The problem was that they were "pretty well snowed in . . . [in] one of the roughest canyons in the state" and cornered by two or three mountain lions. Moninger planned to kill the lions to ensure his own safety and then use muzzled dogs to run down the sheep.[78] This failed. Though Moninger would eventually gun down the predators successfully, he didn't realize that mastiffs can't catch sheep on rocky cliffs. Sheep had evolved for millennia to outmaneuver agile cats. Canines had no chance.[79] Later that winter, Moninger tracked down a different herd. Herds of wild sheep and goats are wary, ready to run their vertical axes if a predator appears. Yet they are not skittish; they are curious, amicable, and rarely see calmly moving humans as threats.[80] As anthropologist Brian Fagan acknowledges, twelve thousand years ago, "an easy familiarity" existed between humans and mouflon.[81] Since, then, Moninger could crawl ("undetected") within one hundred feet of the herd, he planned to trap them in a makeshift enclosure built on the mountainside. He failed again.[82] Sharp had no more luck than Moninger.

He discovered a "flock of fine sheep" outside Leadville, Colorado, and had "four men . . . after them steady." However, a blistery storm drove them from the mountains. Sharp remained optimistic: "If there is such a thing as getting them we will have them. They are one of the hardest animals to capture."[83] Yet, despite doing "all in his power," he came up empty-handed.[84] Indeed, Sharp called his task to catch Rocky Mountain sheep "the most difficult job I ever undertook."[85] The zoo's two hunters failed.

Luckily, Baker had prepared for the worst. Even though "Buffalo" Jones was no longer on contract, Baker promised that the Smithsonian would purchase any sheep Jones caught as a freelance collector. Jones was supposedly still on their trail, yet apparently the famed cowboy spent more time over a desk writing *Buffalo Jones' Forty Years of Adventure* than he spent tracking sheep.[86] Baker, frustrated, ordered the book and turned toward another freelancer.[87] Trapper and trail guide J. C. Cosley emerged from Baker's mountain sheep network exuding confidence. "I am a man who is thoroughly aquainted [*sic*] with the mountains," Cosley proclaimed to Baker in December.[88] In March, Cosley informed Baker that he had already captured two sheep in January but released them when Baker did not reply to either his letter or telegram.[89] Cosley supplied Baker with a vivid, five-page account of their capture.[90] The zoo superintendent, however, remained unbelieving; the story seemed like a tall tale to him. And its teller repeatedly proved to be unreliable. Nonetheless, Baker's growing frustration with catching sheep may have made him skeptical of any success story. In addition, it was also true that, like sheep, mail frequently became lost in the mountains.

Spring came again to Colorado. The deep snows began to recede. And still, no sheep. Baker decided that it was time to expand the search. The sheep decided to climb elevation as they transitioned from their winter to their summer ranges.[91]

Exploring the West, 1901–1902

Just as Baker began to widen his field of vision, a sheep emerged from an unlikely source. Baker learned that E. I. Shafer of Eckert, Colorado, possessed a sheep that he was willing to sell. The individual had been in captivity in western Colorado for nine months and was a "very nice, healthy specimen."[92] Even though the state game warden was "determined not to let [Baker] have the sheep," with the influence of Senator Teller and the lobbying of "Buffalo" Jones, who pleaded with the warden "to allow a thorough test to domesticate a few," a permit was finally and begrudgingly granted.[93] On June 24, 1901, the National Zoo received its first mountain sheep.[94]

Like most zoo mammals, however, the individual required companions. Since Colorado sheep seemed impossible to catch, and since the Smithsonian seemed to have exhausted its favors from the state government, Baker began to look elsewhere. In the 1901 summer, he forged a new, unofficial policy for catching bighorns. Baker articulated this to Miers Fisher of Denver, who was inquiring about a Smithsonian contract after reading about the adventures of Sharp and Moninger in local papers: "The National Zoological Park still desires to obtain four or five specimens, but it does not seem advisable to make formal contracts with persons to supply them as it has been found that even where the parties seemed to be most favorably situated there was no certainty that they could secure the animals . . . the Park will stand ready . . . to purchase specimens."[95]

While Baker stood ready, he pushed his mountain sheep network across the entirety of western North America. First, he looked to the state of Montana. He had formed relationships with three hunters from Kipp, Browning, and Main who agreed to pursue sheep. Montana, though, Baker discovered, only issued permits for mountain sheep serving educational purposes *within* the state.[96] They could not be sent elsewhere. Baker and Langley consequently appealed to the governor to grant the Smithsonian "official sanction" around Blackfoot.[97] The request, however, was denied.[98] Like Colorado, Montana had begun to take its wildlife regulations seriously. Baker corresponded with hunters from Idaho Falls, Hailey, Patterson, and Small, Idaho, and Langley with a collector from Aberdeen, South Dakota.[99] Baker looked to Flagstaff and Yuma, Arizona.[100] He looked to Eddy, New Mexico.[101] He looked to Salt Lake City, Utah.[102] He looked to Elizabeth Lake, California, for Sierra Nevada bighorns.[103] He looked to Parkman, Wyoming.[104] He even looked to Colorado again, for he had heard rumors that mountain sheep roamed upon the privately owned Glen Beulah Park, near De Beque in the western part of the state. Yet these sheep were still too wild to capture. Due to the park's canyons and cliffs, the manager claimed unsurprisingly that "it would be quite as difficult to capture one in the park as outside."[105]

Baker searched high and low throughout the continental United States, but he also looked north to Canada. Indeed, Hornaday, then director of the NYZP, believed that the white, horned sheep of northern

Canada and Alaska would survive better in captivity because they lived at lower altitudes.[106] Although these sheep would eventually be classified into different species (either *Ovis dalli* or *Ovis nivicola*, not *Ovis canadensis*), they could still serve as representatives of the disappearing mountain sheep. After communicating with Hornaday, Baker contacted a sheep-seeker in British Columbia.[107] He courted a "purser of a steamer" that ran between Seattle and Scagway, Alaska, who promised to catch a particular type of horned, white sheep found only along a specific Alaskan inlet (classified *Ovis fannis* by Hornaday).[108] Baker even pursued the possibility of capturing desert bighorns in the State of Chihuahua in northwestern Mexico.[109] These sheep were apparently in graver danger than their relatives across the border, for hunters from Paris and London frequently vacationed to the region to take advantage of low altitudes and lax game laws.[110]

No matter where in North America Frank Baker recruited hunters, trappers, collectors, and frontiersmen, mountain sheep resisted. They had long served as target practice for trigger-happy adventurers, but they refused to be captured alive. Bighorns evaded the imperial appropriation that tethered elephants, fenced antelope, and caged monkeys. The elusive cultures they established on cliffs, at the edge of human settlement, proved virtually impenetrable. Joseph Chandler admitted to Baker that "the goal is dangerous" because sheep were "only to be had on the peaks . . . seldom leaving the precipices."[111] James Fullerton believed it common knowledge that sheep were "the most difficult of all game to hunt."[112] W. H. Root joked, almost two years after the incident, that "Buffalo Jones had the consent of the Governor of Colorado but not of the sheep."[113] Winnie Harwood Phillips believed that the sheep must "know their safety lies in escaping to the tops of the waterless mountains."[114] Will Frakes declared Sierra Nevada bighorns "the most cautious and suspicious animal that I ever attempted to capture."[115] Moninger, after trying to catch damned sheep for over a year, scribbled angrily and almost incoherently to Baker, "they can go any place if any else can get them let them go ahead." Then, he raged that someone had poisoned his dogs![116] Sharp also continued to pursue sheep long after his contract expired. In fact, his commitment to catching sheep exploded into an obsession. Recognizing this, during the winter of 1902, Baker suggested subtly, "As you have now engaged in this undertaking for a year I think it will not be worth while for you to pursue the matter much longer."[117] Sharp ignored the advice. He was in pursuit of the wild.

Mountain Sheep Resistance and Western Knowledge

The campaign to catch mountain sheep was an arduous affair. The entire drama may have been the most sustained effort in history, by the US government, to procure a single species for the purpose of zoological display. By the spring of 1902, Frank Baker had grown "extremely anxious."[118] The citizens of Colorado had, at the same time, begun to demand that their state monitor its cherished wildlife more closely. C. W. Harris, Commissioner of Game & Fish, told Baker that "people in the state, are getting angry at this Department, on account of capture permits. . . . I just received a letter, stating where a man had a permit to capture one sheep and in trying to do so, he killed eleven, and captured none."[119] Wyoming also increased its surveillance. Only the governor could now grant permission to extract sheep, and DeForest Richards surely was not eager to rid his state of a symbolic animal that his constituents held dear.[120] It appeared that the Smithsonian would have to soon abandon its search for bighorns. Indeed, in a letter to Montana Senator William Clark, Secretary Langley explicitly blamed the states for making it difficult for the Smithsonian to achieve its mission.[121] Public resistance to catching sheep was mounting.

Then, on April 11, 1902, a second sheep arrived miraculously on the doorstep of the National Zoo.[122] Like the first, this sheep was owned previously by a resident of western Colorado, living in the town of Ouray.[123] Hearing word of Baker's desire for sheep, Mr. Dusenbery decided to sell his pet. Harris reluctantly granted a permit. And Baker realized that from now on, it would no longer be "advisable to ask further privileges" from Colorado.[124] This was it. At the end of 1902, four years after the chase commenced, two Rocky Mountain sheep (one from Eckert and one from Ouray) shared an exhibit in the nation's capital. Despite the adventures subsidized by the National Zoo, these particular sheep, fenced upon a rocky hill next to the aoudad enclosure, were unremarkably obtained through everyday Americans who had quietly tamed the animals that the federal government pursued around its supposedly "conquered" West.[125] Wild sheep tamed by "wild" Americans ironically provided the National Zoo with "pets" that told a story of a Nation "taming" a frontier. Because this lie lays hidden behind curation, these two sheep allowed the zoo, and the United States, to lay claim to the modernity of mastery pursued by nations the world over.

Indeed, during the first decade of the twentieth century, national power proved more chimerical than textbook narratives of the Gilded

Figure 5.3. This individual, probably "the second sheep" (formerly Mr. Dusenberry's pet), "skylining" in the National Zoo, circa 1905. Courtesy of the Smithsonian Institution Archives.

Age and Progressive Era may suggest. Despite the failed populist movement of the 1890s, despite American victory in the Spanish-American War, despite Teddy Roosevelt's big stick diplomacy, and despite the display of mastery exhibited by international expositions and artfully arranged zoo enclosures, the State remained far from omnipotent. Within the space of the zoo, power appeared to be neat and tidy. A zoogoer could observe North American mountain sheep in one enclosure, walk a few steps, and then watch the wild sheep from North Africa and southeastern Spain in another enclosure. A zoogoer could assume that the United States had conquered the *Ovis* ungulates of the world, or maybe even *the world* itself. The display of the zoo, however, as demonstrated above, masked the true story of hunters, locals, governors, permits, policies, public opinions, and failures. Undoubtedly, over the course of the nineteenth century, the state grew in both size and power. Over the same time span, an amateur tradition of natural history collecting gave way to professionalized trades in animals, plants, and other environmental artifacts that met the public's insatiable demand for natural history after the Civil War.[126] Nonetheless, these well-known transitions fail to tell the whole story. In mountain-sheep harvesting, many actors, diffused throughout the federal system and spread across American society, exercised a power (or a lack thereof) that shaped the national government and the business of animal collecting.

Since, by 1902, it was common knowledge that mountain sheep were "impossible to acclimate," no one expected the pair to live long—a fact plainly (and interestingly) stated in the guidebook that visitors carried from enclosure to enclosure.[127] The first sheep died in January 1903, killed by intestinal parasites that thrive at low altitudes and hot climates.[128] The second sheep apparently lived for several years; no record of their death can be found. The National Zoo did not cease pursuing bighorns after the first pair's procurement. The quest for mountain sheep was, in fact, only just beginning. Wild Westerners chased bighorns for the Smithsonian until the eve of the First World War. A few sheep arrived at the zoo. A few sheep died. Most absconded, retreating to the majestic and unreachable heights of the American West. As late as 1917, Richard Rathbun, acting secretary of the Smithsonian, told Texas Senator Morris Sheppard (who was trying to locate bighorns for San Antonio's zoo) that "the wild sheep is among the most difficult of animals to acclimatize."[129] That same year, Frank Baker purchased an individual from a young Mojave student of the Colorado River Indian School. Baker sent the boy ten dollars, 6 percent of the going rate for bighorns.[130]

Like much of the history of the American West, the intertwined stories of early zoos and mountain sheep twist around the theme of exploitation, familiar to both the historiography of the American West and contemporary criticisms of zoos. Gilded Age and Progressive Era zoo-building, as we have seen, exploited both humans and animals alike. Exploitation, of course, surrounded every zoo enclosure. From capture to transport, from containment to curation, from display to consumption, every life in the zoo—sheep included—embodied (literally) human dominion over the nonhuman, the natural, the exotic. Exploitation made the tableau of mastery, etched into the very purpose of the zoo movement, possible in the first place. Yet, seeing zoological parks, and its animals, through the lens of exploitation only gets at the surface of their complicated histories. Tracing the stories of a few individual mountain sheep alone shows us that countless zoo stories are awaiting the historian.

Two historical lessons cling, with mountain sheep, to the cliffs and canyons of North America. First, as the National Zoo chased sheep, it wove together a body of knowledge, a new "popular zoology," about these elusive animals. This knowledge emanated not from the new laboratories of scientists. This knowledge arrived not from the "lay networks" of amateur naturalists who regularly sent data to urban scientific elites.[131] This knowledge came from the heavens, assembled piecemeal by the North Americans who ventured to vertical frontiers. In many ways, these individuals appeared as the archetypal characters of Western myth—cowboys, hunters, adventurers, mountain men. However, as sheep-chasers, it was through their failures, their lack, that both zoological knowledge was built from the ground up and the North American West was colonized from the peaks down. Even though a few systematic zoologists had tried classifying the many types of North American mountain sheep, in 1922 a Milwaukee hunter still declared "the books . . . very confused and misleading."[132] Ned Hollister, a later superintendent of the National Zoo, admitted that "the sheep have never been thoroughly studied as yet, and the relationship between the forms are imperfectly known."[133] The Americans who lived among bighorns could care less about Aristotelian forms, but they knew the sheep best. And it was these individuals who provided, in exchange for payment, the raw material for science. Like the sheep, knowledge could never be captured. It had to be "salvaged," or extracted from, an assortment of unlikely actors—Wyoming hunters, New Mexico animal collectors, Alaskan sailors, Colorado governors, Montana professors, celebrity adventurers, game wardens, mountain dwellers, conservationists, Mojave children, and future presidents.[134] Knowledge, as animal historian Mieke Roscher puts it, is always "created historically through the taxonomization of animals."[135] It should come as no surprise that as animals were ordered, people were ordered too. And, of course, both ran away from their placements. Placing bighorns into order in the zoo facilitated the professionalization of zoology. Furthermore, historians and sociologists of science have been documenting for decades how sciences and knowledge systems have always been coproduced by nonspecialists and specialists alike.[136] And animal historians, working in a social history framework, have been demonstrating how the sciences and technologies surrounding a specific species—whether camels, orcas, parrots, or pigs—were first forged not by scientists but by those who lived and worked around that species.[137] The same holds true for mountain sheep.

Second, extinction and conservation cannot be viewed only dialectically. Early animal conservation efforts, as environmental historians have shown, mobilized to confront the specter of extinction.[138] Ironically, though, many of these campaigns, in the short term, often endangered the very species they sought to save. The forging of an American "conservationist empire," to borrow environmental law scholar Jedediah Purdy's phrase, was a costly endeavor.[139] Even though the National Zoo (and other zoos) sought to protect the mountain sheep's species, the process of catching individuals caused high casualties. In the long term, though, over the course of the twentieth century, zoo-sponsored conservation campaigns would reduce their recklessness.[140] The violences of capture in the wild lessened, opening space for more controlled,[141] more unseen,[142] institutionalized violences.

Sheep survived. Since the beginning of the nineteenth century, the population of *Ovis canadensis* has declined from more than 1.5 million to fewer than seventy thousand today. The Sierra Nevada and Peninsular subspecies each have fewer than six hundred remaining members. Nonetheless, due to conservation efforts and sheep resilience, mountain sheep numbers have been on the rise since the National Zoo first began its chase at the dawn of the twentieth century.[143] Mountain sheep still face challenges that threaten existence. A resurgence in mountain lion populations, "subsidized" by local cattle, has allowed predators to propagate more rapidly than prey, posing a serious threat to sheep, especially in the American southwest and California.[144] Americans are still pressing against the range and habitat of mountain sheep in myriad ways. Surface-mining, urbanization, hiking trails, military training sites, livestock grazing and its accompanying epizootics (like scabies, mange, and pneumonia), invasive plant species (like tamarisk), golf courses, the construction of "travel corridors," and the quickly changing drought- and fire-driven landscape alterations of anthropogenic climate change all lay claim to mountain sheep territory.[145] For zoos, mountain sheep continued to resist for most of the twentieth century—remaining difficult to locate, capture, transport, and keep in captivity. Occasionally, a mountain sheep would arrive at a zoo, either unpredictably or after prolonged adventures like the one narrated above. And frequently their lives were short.

In the 1970s, zoos began to breed sheep in captivity, sidestepping many of the logistical difficulties of procurement that confronted the National Zoo seven decades earlier. In 1980, according to the *International Zoo*

Yearbook, twenty-three zoos, scattered about the world, held 106 mountain sheep. Of these, eighty-one had been born in captivity. Yet, captive breeding has introduced new threats to captive herds of *Ovis canadensis*: reductions in viability, decreases in fertility, increases in mortality, and less genetic variability.[146] Zoos do not release captive-bred sheep into the wild, as they do with other species. Sheep born in zoos, stay in zoos, as do their offspring. Surely, these captive herds deserve a history of their own. They deserve a history that analyzes them—their bodies, their symbolisms, their cultures in captivity—as biopolitical products of the Anthropocene. At least 159 mountain sheep live in zoos today.[147] While none live in the National Zoo, mountain sheep can be found in cities like Calgary, Los Angeles, Edmonton, San Diego, Phoenix, and Denver. Tens of thousands of resilient others still stretch from northern Canada to northern Mexico. They still bind a continent together, living above the perception of most twenty-first-century Americans.

CHAPTER SIX

The *Zoo*notic Nature of Tuberculosis

Roosevelt's Gifts

Tuberculosis is an extremely contagious disease that has killed between one and three million people in every year of the twenty-first century and has contributed to the death of over one billion in the past two hundred years.[1] Scientists had long believed that the mycobacterium that causes tuberculosis first emerged around ten thousand years ago, as sedentary human and animal populations coalesced around agricultural villages. Specifically, it was thought that *Mycobacterium tuberculosis* first originated from *M. bovis*, spreading from cattle to humans.[2] Despite the strong cultural desire to pin a disease neatly on a nonhuman scape*goat*, contemporary genetic analysis has established a much deeper human history of tuberculosis, linking the disease to the most recent dispersals of anatomically modern humans out of Africa around forty thousand years ago and locating the bacterium's coevolution within human communities at as early as 73,000 BCE.[3] Based on this new understanding, humans were the vectors who first spread tuberculosis to each other and then to their (co)domesticates, though virtually all terrestrial animals have been susceptible to mycobacterial complexes for several million years. Tuberculosis, like all pandemic diseases, is a multispecies phenomenon produced by the circulation of microbes among entangled and densely populated human and animal communities.[4] By recovering historical animals erased by histories of medicine, this chapter will look at how individuals associated with the turn-of-the-century zoo began to reconceive tuberculosis, its zoonotic properties, and its airborne transmission.[5] It will also examine how fear, popular zoology, and public health discourses shaped one another.

On January 6, 1905, the *Evening Star* announced the deaths of two zoo celebrities, a lioness and a male baboon, both given as gifts by King Menelik II of Abyssinia to President Theodore Roosevelt. Even though the lioness, prior to her transport, lost an eye in an unfortunate accident, the postmortem examination showed that she succumbed to "inflammation of the kidneys and liver." The baboon, though, represented a "well-defined case of tuberculosis," displaying all the signature symptoms. Authorities believed that he had long suffered from the disease and that the "change of climate from Ethiopia to America" worsened the tuberculosis to its deadly "climax." In both cases, a pathologist employed by the Bureau of Animal Industry of the Department of Agriculture performed the autopsies. When completed, the bodies were preserved and enshrined in the National Museum.[6]

This baboon was far from the National Zoo's first tuberculosis victims. Just weeks before, a mule deer from New Mexico, two parrots from South America, a large elk, and a rattlesnake also likely died from tuberculosis. In all cases, their bodies were examined by physicians who concurred that changes in climate and environment exacerbated the symptoms of the disease. Despite these casualties, zoo officials felt "gratified to find comparatively such a small amount of tuberculosis among the animals, birds, and reptiles" when considering the high tuberculosis rates of other zoos. They believed that the tuberculosis found was probably contracted *before* the individuals were shipped to the United States, emanating from the "unsanitary [and subaltern] places where they were kept by dealers in animals."[7] Nonetheless, dead zoo animals in America offered researchers opportunities for study. For this reason, the National Zoo, and other zoos around the world, institutionalized the autopsy (literally, "seeing-for-oneself," in Latin) as an essential postmortem procedure for both explaining and preventing death.

The autopsy, however, always reached beyond its primary diagnostic function. In the wake of the lioness's and the baboon's deaths, one zoo representative spoke about a moose to a journalist with the *Evening Star*:

> The post-mortem work brings about interesting results [about] animal life. . . . For instance, it has been found that it is almost impossible to keep the moose in captivity. These animals usually die of chronic in-

flammation of the stomach and bowels from the lack of some element in their food. On their native heath, they live upon the branches of trees and other foliage [leaves, bark, shrubs, roots, water plants]. If the kind of food necessary cannot be supplied to them, their condition grows worse and death results.[8]

The National Zoo used the autopsy to create knowledge about animal life (zoology), relations with environment (ecology), disease (pathology), and preventive measures (public health). In addition, the autopsy served as connective tissue joining the zoo's institutional organs. Postmortem work brought surgeons, pathologists, zoologists, zookeepers, and veterinarians (the latter who mostly consisted of horse and livestock doctors) into a shared laboratory space, where they examined the deceased for a wide array of professional ends.[9] The autopsy not only brought together heterogeneous actors within the zoo, but it also, as always, connected the zoo to publics outside—scientists, medical professionals, animal dealers, and, most importantly, the zoogoing and newspaper-reading masses who consumed the elegies of their favorite animals.

The National Zoo and the zoo autopsy created a constructive space in and through which knowledge about disease was assembled. As historian of medicine Abigail Woods has argued forcefully, early zoos in Europe had long dealt with disease through "three modes of medicine that were typically applied to humans: public health, bedside medicine and hospital medicine." The practice of both surgery and prosection tied zoos to this medical milieu.[10] Disease and death were certainly a familiar part of zoo life. Zoo animals died frequently from pneumonia, aspergillosis, enteritis, gastroenteritis, intestinal coccidiosis, nephritis, and quail disease. Less frequently, but still often enough, zoo animals succumbed to rabies, abscesses, cancers, impactions, nematode worms, duodenal obstructions, and emotional "shocks" (when external stimuli produced an abrupt life-ending event). The list of conditions plaguing zoo inhabitants grew longer as years passed. The professionalization of human and veterinary medicine in society at-large provided an ever-expanding medicalizing lexicon for describing bodies and diagnosing illnesses, and the technological developments that accompanied this professionalization propelled medical knowledge along a path of increasing complexity. Each of the conditions listed above possess their own complex histories. Rather than generalize about all health conditions, this chapter explores the single disease of tuberculosis. Woods acknowledges that "as the zoo became medical, so medicine became zoological."[11] Echoing the importance of her contention, this chapter also demonstrates that zoo archives offer untapped storehouses for historians of medicine to integrate into histories of disease, public health, and medical knowledge.[12]

The story below laces together several moments in the history of tuberculosis as seen through this site study of the zoo and its entanglements. The first section deals with federal policies, discovered in the archives of the National Zoo, designed to subdue the "Great White Plague" within federal institutions of the District of Columbia. The second describes the "discovery" of tuberculosis within the National Zoo and outlines the prevalence of the disease among its inhabitants. The chapter then turns to the year of 1912, when the public began to worry about contracting the disease from zoo animals. The fourth section examines the correspondence of zoo officials, highlighting both the contestedness of tuberculosis and the interconnectivity of zoo medical networks, as Baker communicated with zoos in Philadelphia, New York, Cincinnati, and London. The fifth section zooms in on processes established for treating tuberculosis with iodine. And the chapter concludes by showing how the zoo sat at the center of knowledge-production as it paved the way for understanding tuberculosis as primarily an airborne pathogen. This final section examines the surprising career of William Evans, pathologist of the Lincoln Park Zoo, who reconfigured national understandings of tuberculosis and public health with lessons learned in the primate house. These lessons astounded Frank Baker and influenced American zoos, as well as other modern institutions that held captive populations.

In 1967, Herbert Ratcliffe, director of the Penrose Research Laboratory of the Zoological Society of Philadelphia (located in the Philadelphia Zoo) and professor of comparative zoology at the University of Pennsylvania, presented a paper before the American Philosophical Society titled "Contribution of a Zoo to an Ecology of Disease." Ratcliffe operated from a Cartesian assumption (one that also may prefigure the language of animal behaviorist Jacob von Uexküll) that "organisms are self-regulating, self-perpetuating physicochemical entities, which in health maintain dynamic equilibrium with environment through neural and neuromuscular responses which, in sum, constitute behavior." He argued that zoos, as "experiments by which mammals, birds, and other vertebrates are assembled into . . . densely populated urban areas," unavoidably caused "disturbed behavior." Ratcliffe contended that even though the improved diets of zoo animals after 1935 lessened malnutrition in zoos,[13] lowering disease rates generally, zoo environments still caused "disturbed behavior" that made animals susceptible to diseases. For example, Ratcliffe explained that grouping white-tail deer into a single exhibit, ignoring their typically "nongregarious" personality in the wild, caused them to experience "wasting, dwarfing, diarrhea, and deformed antlers." Most significant, Ratcliffe's paper demonstrated that zoo records can be used to "develop methods of disease control," for "imaginative use of the zoo may also aid man as he attempts adjustment to his self-induced dilemmas, both of survival as a member of mass society and in developing explanations for disease; that is, in developing an ecology of disease."[14] This chapter continues down the methodological path forged by this 1967 zoologist, exploring zoo records in order to deepen our understanding of tuberculosis. This chapter also shows that Ratcliffe was not the first to believe that the ailments of zoo animals can teach us about the ailments of human animals.

"Prevention of Tuberculosis Among Government Employees"

For most of the nineteenth century, tuberculosis was known as "consumption" and was diagnosed by observable changes in a cough. They would begin as "frequent and harassing," continue as "hollow rattles," and conclude as "graveyard coughs." Paleness, weakness, fatigue, and mucous accompanied the cough, and as conditions worsened along its spectrum, mucous shifted from green to bloody. "Consumption" was believed to be "hereditary and noncontagious," but it affected a large proportion of the American population. In 1882, after German microbiologist Robert Koch saw, beneath a microscope, the bacillus that caused the disease, "tuberculosis" gradually replaced "consumption," and a pervasive yet mysterious ailment transformed over the following decades into an identifiable pandemic with a specific bacteriological cause. Fear of tuberculosis gripped the masses, and this fear sparked campaigns to educate the public in how to contain the tubercle bacillus.[15]

Almost one year after the deaths of his lioness and baboon, Roosevelt issued an executive order that established a committee of three surgeons general to design measures for preventing tuberculosis in government buildings. Soon after, another order required that each Executive Department adhere to the established guidelines. Every department needed "to ascertain the names of any person . . . with tuberculosis, and to present to them the printed rules prescribed." Furthermore, this second order required a "thorough sanitary inspection of all public buildings," resulting in a list of all unsanitary conditions "immediately remediable" and all unsanitary conditions "requiring structural changes."[16] As federal buildings, the animal houses of the National Zoo would be inspected.

The report that the executive orders culminated in, "Prevention of Tuberculosis Among Government Employees," declared phthisis (the ancient Greek term for the disease) "the most widely spread and deadly disease that affected humanity."[17] It described the cause of the disease as a "living germ, a minute plant called the *bacillus* of *tuberculosis*, which enters the body in various ways and there causes . . . destruction of tissue and a general bodily infection, with wasting away and ultimate death if not checked at the beginning." Sometimes these "minute plants" would remain localized in specific parts of the body, or when ingested they could produce an intestinal form of tuberculosis. Most commonly, though, humans inhaled these "minute plants" into their lungs and "consumption, phthisis, pulmonary tuberculosis, or tuberculosis of the lungs" would develop. This type of tuberculosis was the deadliest and most contagious, responsible for (the report stated) one of every four deaths of individuals between the ages of twenty and fifty.[18]

"Prevention of Tuberculosis" emphasized the dangers of expectoration and sputum, stressing that "the disgusting habit of spitting upon the sidewalk . . . distributes millions of disease germs." After drying, the dribble of the tuberculated released germs into the air, where the healthy breathed them in and became sick. "Moist sputum," especially "virulent," abounded with germs, and the report emphasized that everyone should avoid contact with everyday vectors that could have touched the mouths of others—drinking glasses, towels, handkerchiefs, and tableware. The report (which sat at the historical juncture of contagion theory, miasmatic theory, and germ theory) underscored the threat of phlegm releasing germs into the air over the dangers of coughing directly *into* the air. However, in 1905, medicine had not yet fully grasped the central role of "the cough" in spreading pathogens, nor had it fully theorized "the air" as a legitimate medium in the ecology of disease.[19] Sputum, then, played the role of tying older understandings of disease with newer ones—Americans still needed a drying yet moist tangible object to ground a paradigm shift.[20] It would not be until the 1930s and 1940s, with the research of biologist William Firth Wells, and the 1950s and 1960s, with the work of medical researcher Richard Riley, that tuberculosis would become understood primarily as an airborne pathogen, spread through minuscule water droplets hacked directly into the air. These droplets encased tubercle bacilli, and the bacteria were so light that they could float in the air for hours, even after its water encasement evaporated. With this understanding, by the mid-twentieth century, coughing would eventually trump spitting in importance (though spitting would remain socially stigmatized).[21] In 1905, most urban Americans imagined contagious sputum caking the walkways of growing cities. And surely those who felt inclined to spit would not hesitate to do so upon zoo walkways or into the enclosures. The zoo, like the city, must be literally covered in tuberculosis.

Typically, the "people most liable to be infected are those who live unhygienic lives, or who are compelled . . . to work amidst unhealthful surroundings." The report depicted a disease linked to both work and class, for "overcrowded, unventilated dwellings," "sedentary occupation," "trades causing dust," "poor food," "insufficient clothing," and "uncleanliness" were all factors predisposing persons to consumption. The report clarified that tuberculosis did not simply "spontaneously generate" among the "uncivilized" lower orders, a notion that would have been commonly accepted by medical experts in the early nineteenth century, nor was tuberculosis merely a miasma that emanated from polluted environments. Instead, in the wake of medical practitioners like John Snow, Louis Pasteur, Joseph Lister, and Robert Koch, as germ theory (or the pathogenic theory of medicine) became commonly accepted, the report made sure to specify that "nothing can actually *cause* consumption except the entrance of the germ into the body." While the working classes may have been more likely to contract tuberculosis due to their own exploitation (or moral failings, depending on who was doing the narrating), and while some individuals may inherit a predisposition to the disease, the bacillus itself was a democratic "plant." Indeed, in 1905, the only way to prevent its spread was to "*destroy the sputum of every consumptive*, for it is almost always by means of the sputum that a consumptive infects another person." No matter where it lay—"floors, sidewalks, on clothing or utensils"—sputum was the endgame.[22]

"Prevention of Tuberculosis" concluded optimistically, emphasizing that Americans need not fear and tremble. Tuberculosis could be cured "when intelligent treatment is undertaken early in the disease." Many who could afford to do so traveled to the warmer and drier climates of southern California and Colorado, places defined by "sunshine, equable temperature, and absolutely pure air."[23] However, consumptives could convalesce anywhere if they found respite in local sanatoriums. "It is to be hoped," the report pleaded, "that before long all communities will understand that, to provide for indigent consumptives in their midst, sanatorium treatment at the public expense is not only a humane measure of the first order but an economical measure in which every citizen has a vital interest." In the meantime, until the nation subsidized the "public good," Americans needed to individually practice good hygiene.[24]

The following preventative plan could "be understood by anyone with ordinary intelligence," and taking these steps would, it was believed, halt tuberculosis in its tracks.

1. Sleep alone.
2. Use no hangings, upholstered furniture, or useless floor coverings.
3. Painted walls are preferable to those covered with wallpaper.
4. Expose the bedroom freely to outside air. . . . Sleep with windows open.
5. Do not be afraid of cold weather as long as the body is protected. . . . Keep feet dry.
6. Keep the body warmly clad.
7. Take plenty of nourishing food. Consumptives often need more nutriment. . . . Milk, eggs, and fatty foods are especially valuable.
8. No known medicine can *cure* consumption. Medicines for the relief of cough and other symptoms should be taken only on the advice of a physician.
9. Lead a temperate life in all things.
10. Be scrupulously careful not to infect other members of your family by distributing the germs contained in your sputum. Refrain from coughing . . . but when it is necessary turn your head and hold a handkerchief over your mouth.
11. Use a destructible portable spit receiver, which can be bought for a few cents; use one or more a day and destroy them by burning.
12. Never swallow material brought up from the lungs; it may cause infection of the digestive tract.
13. It is not best to use handkerchiefs to receive sputum. Japanese paper napkins or squares of old linen, to be burnt when soiled, may be used; but these are not as cleanly as portable spittoons.

14. Avoid dust, disorder, dampness, and darkness.
15. Be hopeful and expect a cure.[25]

The above regimen represents, in 1905, the most up-to-date understandings about how to treat tuberculosis. Yet it also represents the imprecision of this knowledge. A government employee with tuberculosis might have skeptically read this list as merely a placebo. Rule #9 vaguely instructed the sick in how to lead a "good life" in the Progressive Era, but it failed to offer any direction for getting better. Rule #13 directly contradicted Rule #10. Rule #14 failed to define what constitutes "disorder" and neglected to explain how one could adhere to Rule #1 while simultaneously avoiding "darkness." And Rule #8, hidden in the middle of the list for the careful reader, encouraged the ill to grasp at Rule #15, knowing that the cure hoped for may remain elusive.

The report also established a plan of action for government properties like the National Zoo.

1. All persons in Government employ are forbidden to spit upon floors.
2. Rooms, hallways, and lavatories shall be freely aired and cleaned at least once a day.
3. Spittoons shall receive a daily cleansing with very hot water.
4. Dust must be removed as completely as possible by dampened cloths or mops. It should never be needlessly stirred up by broom or duster, as this practice only spreads germs.
5. Floors of tiling, brick, or stone must be frequently scoured with soap and water.
6. Senior clerks in charge of workrooms will secure the admission of as much fresh air and sunshine as conditions will permit.
7. The use of individual drinking glasses is recommended.
8. Persons in Government employ who suffer from pulmonary tuberculosis shall be separated when possible from others while at work.
9. Such persons will not be permitted to use the public spittoons, but must provide themselves with individual sputum receivers, preferably of easily destructible material. . . . They will be held strictly responsible for the disposal of their own sputum.
10. Such persons must provide their own drinking glasses, soap, and towels and shall not use those provided for the general use.

11. Plainly printed notices: "Do not spit on the floor; to do so may spread disease," shall be prominently posted in rooms, hallways, and lavatories.[26]

These regulations imposed straightforward guidelines that not only sought to cleanse the federal government of germs but also provided a protocol of behavior for sick employees. If they failed to comply with Rules #1, #9, and #10, they could, "in the discretion of the head of their Department," be dismissed from the civil service.[27]

The National Zoo, like the entire federal government, received the above report. Shortly after, circulars about preventing tuberculosis were disbursed to all zoo employees (about eighty-two people) and anti-spitting signs were placed throughout the park.[28] All laborers, groundskeepers, mechanics, and zookeepers were expected to comply with the White House's regulations. Any violation whatsoever, as communicated directly by Frank Baker, needed to be reported immediately.[29] However, in comparison to the vile anti-public-health backlash of 2020 and 2021 fostered by a delusional conservatism addicted to social media and bizarre applications of "liberty," it seems as though these circulars were well-received by zoo workers, who were personally acquainted with the seriousness of tuberculosis. William Blackburne responded dutifully to Baker: "I will instruct the men under my charge to comply + do my utmost to see [the rules] are carried out accordingly."[30] Nonetheless, filled with thousands of diverse animals and humans, the National Zoo was an institution far different from others in Washington. The zoo would need more than the latest spittoon technology to contain tuberculosis.

Tuberculosis in the Zoo

Tuberculosis had gained the attention of the zoo before Roosevelt's report. By the first years of the twentieth century, autopsies on zoo animals were performed regularly by the Bureau of Animal Industry. Tuberculosis appeared increasingly in reports. On March 19, 1903, the zoo shipped the body of a young agouti to the bureau, and after the discovery of small lesions through microscopical examination, the surgeon diagnosed the rodent with "miliary tuberculosis [characterized by lesions that resemble millet seeds] in the lungs and mediastinal lymph glands." Corroborating this diagnosis was a four-inch tumor, embedded in the omentum, containing a "yellow pus that also proved to be tubercular," that is, containing the tubercle bacillus. Due to this, the agouti experienced extreme swelling of the "sublumbar and superficial inguinal glands," located in the pelvic region of mammals. This description was typical of zoo animals who died from "generalized tuberculosis." Yet Daniel Elmer Salmon, chief of the bureau and the first doctor of veterinary medicine in the United States (a degree earned at Cornell in 1876), wanted to know more about this case and asked Baker, "Will you kindly inform me the origin of this animal, the length of the period of its captivity, and how long the animal had been sick before death?" In this same report, Salmon also presented the postmortem diagnoses of a monkey who died from "broncho-pneumonia" and a lynx who succumbed to "gastro-intestinal catarrh," but he did not request more information for these deaths. Autopsy reports usually provided information rather than raising questions. An agouti with tuberculosis sparked curiosity in Salmon.[31]

In May of that year, Salmon concluded that "miliary tuberculosis of the lungs" killed a peccary. Unlike the agouti, the telling lesions were not visible. Instead, the animal's "intensely congested" lungs signaled that acute bronchitis was the culprit. Closer examination under a microscope, though, "revealed the presence of the tubercle bacillus in great abundance."[32] Diagnosing a dead zoo animal with tuberculosis usually required the microscopic confirmation of the bacillus. Sometimes the millet-like lesions that often accompanied tuberculosis would appear under the microscope, but other times, they would not. Sometimes yellowish pus would be found in tissues contaminated with the bacillus, but other times this indicative purulence with its signature color would be absent. And often tuberculosis would cause the visible swelling of various glands of the lymphatic system; however, swelling could also be produced by other diseases that put systemic stress on the body. Therefore, knowledge about tuberculosis in zoo animals relied heavily on the microscopes of the Bureau of Animal Industry. And with an increase in the number of microscopes, the number of animals diagnosed with tuberculosis rose steadily.

In 1903, a Philippine deer (with masses of "soft yellowish white purulent material" dotted along her lungs) tested positive for the bacilli and was diagnosed with "pulmonary tuberculosis."[33] In 1904, a South American female raccoon, with yellowish material in her thoracic cavity, died from "generalized tuberculosis" of the diaphragm, lung, and liver.[34] A baboon died from tubercular pulmonary nodules and various tubercular glands in the folds of the mesentery; a male monkey died from tuberculosis in his "lungs, spleen and the lymphatics of practically all of the viscera"; a diamond rattlesnake died from tuberculosis, the lethal baccili housed in her lungs and lymph nodes; and the death toll continued like this, ever increasing.[35] For the remainder of 1904, the National Zoo lost a raccoon, a macaque, an elk, a kinkajou, and five various types of deer.[36] In 1905, casualties included a sooty mangabey, a bison, a European deer, a Sambar deer, a coatimundi, a baboon, and two wallabies, in addition to the victims listed above.[37] In 1906, as Baker dispersed the public health circulars, the death toll expanded, claiming the lives of a kangaroo rat, a monkey, a fox, a camel, a macaque, a beaver, a rat kangaroo, a snowy owl, a hutia, a great gray kangaroo, an elk, two pigtailed monkeys, three rhesus monkeys, three rat kangaroos, and eight baboons.[38] And in 1907, tuberculosis killed a number of newly purchased primates—a rhesus monkey, two baboons, a pigtail monkey, and two other monkeys of unknown species. That same year, three Hungarian quail, three crested pigeons, and a giant kingfisher, an Esquimau dog, a snake, a nilgai, a kinkajou, and a beaver succumbed to tuberculosis.[39] You should be getting the idea.

Usually, zoo animals were diagnosed with tuberculosis after death, *after* autopsies could be performed by the bureau. The exact number of animals *living* with tuberculosis, then, remains unknown. The specific pathogeneses of tuberculosis depended greatly on the "resistance and susceptibility" of an individual host. Some of the inflicted could live full lives with the latent form of the disease, while others would die within weeks of contraction. So it can be assumed that at any given point during these years the number of animals living with tuberculosis far surpassed the number of tuberculated carcasses.[40] Occasionally, living animals were diagnosed. For example, in 1907, the zoo sent a living passenger pigeon (by then, extinct in the wild) with "nodular formations under the eye, on unfeathered parts of the wing, and on the feet." The chief of the Pathological Division of the Bureau confirmed that the nodes indeed housed the bacillus.[41] Animals living with systemic tuberculosis, though, rarely received diagnoses, since the bacillus lay undetectable within the viscera. However, even when external nodules or sputum (especially in the case of primates with pulmonary tuberculosis) could be tested, the costs of testing made these living diagnoses rare.

Zoo animals died frequently from infections, diseases, and ailments of all sorts, for diagnoses like pneumonia, enteritis, gastritis, nephritis,

Figure 6.1. National Zoo death toll by confirmed tuberculosis in formal autopsy reports, 1903–1916.

gastric catarrh, and liver cirrhosis appeared often in autopsy reports. Zoo animals also died from parasites and injuries sustained either in shipment or by altercations with other animals. Nonetheless, as tuberculosis spread through the human society of Washington, it also reached the animal community of the zoo. The National Zoo needed to learn how to deal with this contagion.[42]

1912: Dead Bodies and the Smithsonian

Tuberculosis continued to take lives. Some even worried that zoo animals could be spreading the disease to zoogoers, since there had been talk since the 1870s of exchange between species.[43] In a letter to Baker, Frederick William True, the first head curator of biology in the National Museum and a specialist on whales and dolphins, raised a concern about transporting carcasses. Alarmed, True typed:

> The enclosed letter of Mr. Geo. B. Turner [chief taxidermist] . . . relates to the health of Museum employees who may be involved in handling animals that have died in the Park from tuberculosis. In so far as his remarks relate to the practice (if it be one) of transporting the carcasses of animals which have died from infectious diseases in the same wagon that carries supplies to the living animals, I think

> his point is well taken. The practice, if it exists, should be stopped at once, and other means found for carrying dead animals. I am certain that you are fully alive to the importance of preventing the spread of disease . . . from the animals to the employees, if that be possible.[44]

True urged Baker to "prepare a full report on the subject." And he added that the practice of sending dead animals to the Museum should be abolished for the sake of public health, except for cases concerning specimens of "considerable scientific value." He concluded by leveling a terse critique against Baker, claiming, "I have understood you to say that no steps are taken at present to disinfect the animal cages in the Park, and that it would be impracticable to do so." Although True admitted that he might not have the "right impression," he ended his diatribe by emphasizing that some Washington parents refused to allow their children through the zoo's gates "for fear of contagion."[45]

Baker responded by describing the process of disposal. When an animal died, the body was transported from the Rock Creek to the Museum in a "covered wagon with tight bottom" that presented "no danger whatever of scattering contagious material on the streets." To prove its cleanliness, Baker emphasized/admitted (curiously) that "food supplies were transported afterwards in the wagon, placed in boxes, barrels, cans or other receptacles so there is little or no danger of contamination." To strengthen this tenuous reasoning, Baker added that the wagon was washed out frequently and stressed that he knew "of no case of disease traceable to this source."[46]

Once he dismissed the possibility that carting tuberculated bodies through the streets could spread disease, Baker then informed True about tuberculosis specifically. First, he maintained that in 1911, the previous year, tuberculosis was responsible for only 8 percent of animal deaths and promised True that he would "at once endeavor to obtain," for the sake of comparison, similar statistics from other zoos. Second, he explained that since the zoo's founding, only two employees had died from tuberculosis, "one a laborer who never had anything to do with the animals, the other a keeper who was probably infected before employment, as his father died of the disease," drawing on the long-standing belief that consumption could be passed down by birth. Many employees had been working there for years, demonstrating to Baker that the park must be a sanitary place, otherwise these workers would have fled long ago. However, Baker, through his third point, seemed unsure about whether tuberculosis *could* pass from one species to another:

> Koch [Robert Koch, the first to pinpoint the bacillus under a microscope], only recently dead, and who when living was the highest authority of his time on this subject, declared after repeated experiments that the transmission of tuberculosis from animals to man did not take place. Many American authorities dissent from this opinion, and the general consensus seems to be that infection occurs only by the mouth.

Baker avoided taking a decisive stand on this debate, one that roared for a decade in Britain, as researchers studied bovine tuberculosis and its contamination of the meat supply. His remembrance, though, of Koch as the "highest [tuberculosis] authority of his time" suggests that Baker remained skeptical about the disease's zoonotic nature. It also shows that Baker had not followed the debate closely, for in 1911 the Royal Commission on Tuberculosis "announced unambiguously" that tuberculosis could cross the species line.[47] Baker was correct, though, in that the emphasis of the debate was on ingestion, not inhalation. Finally, Baker concluded his response to True by recommending that all dead animals be first sent to the Bureau of Animal Industry, where their bodies could be tested for disease and properly destroyed if contaminated. While the zoo had been sending potentially diseased flesh to the bureau for quite some time, Baker agreed to routinize this practice, varying away from the convention (forged by Hornaday in 1887) of shipping the dead directly to the museum for taxidermic preservation.[48]

After addressing tuberculosis, Baker defended himself by adding more suggestions about public health in the zoo. For example, he hoped for a hospital and quarantine station, a place "where sick animals could be isolated and new accessions kept until it could be ascertained that they have no infectious diseases." Typically, diseases were, Baker continued, "brought into the Park, not bred here," the common assumption of nearly all zoos and an extension of broader nationalist/xenophobic stances that framed immigrants as vectors. In addition, Baker expressed desire for a pathological laboratory on site so that animals would not have to be wheeled around Washington in the first place. The superintendent closed his letter by reemphasizing the zoo's cleanliness (remember, from chapter 3, Baker had patented his own sterilizing solution). The zoo's cages and enclosures were washed routinely and disinfected regularly, making the "buildings compare favorably with many hospitals."[49] (Due to financial constraints, Baker's dream for an actual hospital within the park would not come to fruition until 1915.)[50]

Whether True was correct believing that Baker had failed to prioritize public health, take tuberculosis seriously, and uphold a high standard of cleanliness can never be known. Certainly, the zoo frequently ignored animal welfare. And zoos in general were often filthy places. Nonetheless, in 1905, almost seven years earlier, Baker did try to persuade Congress to allocate funding for the buildings he mentioned—a hospital, quarantine building, and laboratory. In fact, in support of this project, he proclaimed (like Ratcliffe in 1967) that "very little is known concerning the diseases that affect wild animals and the parasites that associate themselves with them. An extension of our knowledge in this direction would undoubtedly be of benefit to those who are studying the diseases of man."[51]

True was not alone in wanting answers about tuberculosis risks. Richard Rathbun, biologist and assistant secretary of the Smithsonian, first received the polite letter from George Turner that prompted True's criticisms:

> There has never been, in the history of man, a time when science has made such a determined effort toward public health . . . we hear it on all sides, fine foods, isolation of contagious diseases, precautions against tuberculosis, etc. Now the question arises in my mind are we, as an institution, doing all we can toward betterment of public health? Are we doing things we ought not to do? Let us look + see + form a definite conclusion as to whether we are right or wrong.[52]

Turner continued by informing Rathbun that "a doctor said to me recently that about 90 percent of the deaths at the zoo were due to tuberculosis." Though Turner assumed that this estimate was most likely too high, he did suspect that "a very large percentage of deaths" were probably due to the disease. Most likely, tuberculosis did play a role in far more than 8 percent of zoo deaths, and tuberculosis absolutely resided undiagnosed in many living zoo animals.[53]

A Contested Contagion: Tuberculosis, Its Zoonotic Nature, and the Zoo Movement

Baker believed that Turner's and True's concerns were misguided. Still, he kept his promise. On October 19, 1912, Baker typed a letter hastily to his zoo-leader peers, asking for data about tuberculosis in their zoos. The letter stated:

> A question has been raised here as to the extent to which tuberculosis exists among the animals in the collection of the National Zoological Park and the possible danger of infection of employees of the Park from such animals. . . . I have been asked to ascertain, so far as such information is available, what the percentage of mortality from tuberculosis is in other zoological collections as compared with the total mortality, also the percentage of the employees of zoological gardens who have been affected or have died from tuberculosis as compared with the same figures for the city as a whole. It is also desired for this purpose to consider separately employees who are directly connected with the care of the animals and those employed on other work, such as grounds, construction, etc. It is further desired to know what special precautions are taken to prevent the communication of tuberculosis from infected animals to others, and, also, to employees.
>
> It is perhaps unnecessary to say that there seems to be on the part of those not acquainted with the actual conditions in zoological gardens a greatly exaggerated opinion as to the extent to which tuberculosis prevails among the animals that are kept there. The percentage of death here has varied in recent years from 8% of the total mortality to considerably less than that proportion.
>
> Any information or statistics that you can give along this line will be cordially appreciated.

Baker mailed this letter to the most prominent zoo men in the English-speaking world: Dr. P. Chalmers Mitchell, London; Dr. W. B. Cadwalader, Philadelphia; Dr. W. T. Hornaday, New York; and Mr. S. A. Stephan, Cincinnati.[54]

Figure 6.2. Cincinnati Zoo director Sol Stephan feeding a baby elephant. Courtesy of the Cincinnati Zoo & Botanical Garden.

These zoo superintendents largely affirmed Baker's opinions. Sol Stephan, general manager of the Cincinnati Zoological Company, informed Baker that Cincinnati's zoo was "very fortunate in not having many of the animals affected with tuberculosis," possessing death rates, in Stephan's estimation, of easily less than 6 percent. When animals did contract tuberculosis, he explained that they would be quarantined in an isolated building. If the cases were severe, the animal would be "destroyed" and cremated to ensure the elimination of the bacillus. However, these cases were rare. Drawing off common nineteenth-century notions of miasmata, which emphasized the role of environment in spreading disease, Stephan explained, "Our grounds are high and rolling and drain very quickly. Our enclosures are always dry a couple of hours after a heavy rain. Animals do better on high rolling ground than on flat ground." As far as humans were concerned, Stephan, in the thirty-seven years he had devoted to the zoo, did "not remember . . . an instance where any of the employees or keepers were affected with tuberculosis." Yet he did

Figure 6.3. William Hornaday reading from his office at the NYZP (Bronx Zoo) in 1910. © Wildlife Conservation Society. Reproduced by permission of the WCS Archives.

mention, as an aside, "of course you are well aware that there are more cases of tuberculosis found among the monkeys."[55]

Hornaday, at the NYZP, echoed Stephan, estimating that the tuberculosis death rate, for 1905 to 1909, was almost exactly 5 percent. And "*only one employee*," which Hornaday underlined for emphasis, had died from tuberculosis. This individual, Frits Dahl, had been hired directly from the Spanish-American War with deeply infected lungs and placed in the Reptile House as a keeper with the hope that the arid environment might save his life. Unfortunately, he died shortly after his hire. Yet Hornaday asserted confidently that Dahl neither contracted tuberculosis from, nor gave tuberculosis to, any zoo animals. In fact, tuberculosis concerned Hornaday so little that he did not "regard it necessary to take any special precautions to prevent keepers from contracting tuberculosis from our captive animals." They did not, Hornaday insisted, "associate with their charges sufficiently close to render infection a possibility." In a sarcastic tone, he informed Baker that "if there is any proof on record that a human being acquired Tuberculosis from an animal in captivity, I would like to see it, and be furnished with all the facts in the case." While Hornaday dismissed tuberculosis, he admitted that some of the primates had acquired the disease (and two had died in 1907), likely obtained from visitors coughing "directly into their faces." Hornaday suggested that the animals were the ones in danger of infection, not the keepers.[56] (If he knew the graveness of that danger, his response to Baker surely would have been different. Less than fifteen months later, Hornaday's world-renowned great ape collection, all five orangutans and four chimpanzees—who had each, on average, lived in the collection for more than four years—would die quickly and mercilessly in a three-week lymphatic tuberculosis epidemic. The die-off made international news and crushed Hornaday's spirits, especially because he admitted that he did not have the "faintest clue" about how tuberculosis spread. All he could do was blame it on the rats.)[57]

Cadwalader, of the Zoological Society of Philadelphia, did not provide Baker with statistics, but he did state definitively that tuberculosis had not killed a zoo employee since 1874, the year the zoo opened. His three-sentence response answered few of Baker's questions; however, the letter's brevity was itself indicative of Cadwalader's worries.[58] (This is particularly troubling since the 1939 report of the zoo's Penrose Research Laboratory stated unequivocally that a large number of primates died from tuberculosis in 1904, 1905, 1906, 1907, 1912, and 1913. In 1906, the disease in question killed more than half of the collection.[59] *Was the zoo's superintendent ignorant, lying, or in denial? Or was tuberculosis simply still a mysterious disease?*)

Three of the four zoo men with whom Baker corresponded sent responses that could certainly defend him against the critiques of Turner, True, and worried District citizens. Stephan, Hornaday, and Cadwalader did not view the zoo as a threat to public health. They all dismissed the possibility of employees contracting tuberculosis from the animals. Furthermore, they contended that tuberculosis caused fewer than the 8 percent of deaths estimated by Baker. Tuberculosis was certainly not an epidemic that warranted panic, or even immediate reprioritization, especially for an institution so used to large numbers of dying animals. Stephan and Hornaday both acknowledged a mild concern for primates, but neither seemed worried. Peter Chalmers Mitchell, secretary of the Zoological Society of London, however, told a different story from that of his American counterparts.[60]

Instead of providing precise death rates, Mitchell (who himself had made early forays with tuberculin tests in his zoo[61]) offered Baker total casualty numbers, showing that the "mortality from tuberculosis in our Gardens has been going down in the last few years." These numbers proved astounding, especially when contrasted with the scant attention given to the matter by the other directors. In 1908, Regent's Park lost 164 animals to tuberculosis, and in 1909, 1910, and 1911, it lost 94, 190, and 147 animals, respectively. Mitchell, like Stephan, Hornaday, and Cadwalader, expressed little anguish over the risk posed to zoo employees; however, the data Mitchell gathered on this topic differed:

> With regard to . . . keepers, gardeners, and labourers . . . the only cases of tuberculosis of which I know in the last ten years, were one keeper, who died; one indoor clerk, who as a matter of fact had nothing to do with the animals, who died; and one gardener, who was sent to a sanatorium, and has been released, and is back at work apparently practically cured. Even if there is a greater chance of infection from contact with the animals, which I doubt, I think this is more than compensated for by the large amount of outdoor work which the Gardens staffs have to do. Certainly, tuberculosis is not one of our troubles with the staff.

While optimistic that tuberculosis was lessening among the animals and skeptical about the likelihood of zoo animals infecting humans, Mitchell's letter demonstrated that tuberculosis could indeed claim a great number of animal lives. Also, his letter at least opened the possibility that, in the case of the dead zoo keeper, tuberculosis *could* be zoonotic.[62]

Mitchell concluded his letter by describing the battle he led against tuberculosis, which focused on quarantine, cleanliness, and disinfection, all the precautions that True advocated for:

> New arrivals are observed in the quarantine house . . . and if we have any reason to believe that an animal . . . is suffering from tuberculosis, it is removed to the sanatorium. Every time an animal dies, the cage or enclosure in which it has been living is cleaned and disinfected. It is part of the routine of the pathological department to give information every morning if there is any [death] which seems to require special disinfection. The house in which the smaller monkeys are kept, every

> now and again, we have emptied, the walls scraped and washed, and then disinfected, by burning formalin cartridges, and leaving it shut up for twelve hours.[63]

While Mitchell may have been optimistic, he would not have deliberated about sanatoriums and formaldehyde unless he acknowledged tuberculosis as a real threat. Also, unlike the three Americans, Mitchell cited large numbers of tuberculosis casualties. And he was the only zoo director to not *completely* rule out the possibility of humans (in this case, a keeper) contracting tuberculosis from captive animals.

Baker, Cadwalader, Hornaday, and Stephan may have been reluctant to publicize data about tuberculosis within their zoos out of fear that acknowledging its presence could be bad for business. Clearly, tuberculosis was killing animals. These zoo men could have been, either intentionally or subconsciously, denying tuberculosis its agency in order to preserve their own. While this interpretation may make sense when discussing tuberculosis to the public, it does not seem likely that these leaders would downplay tuberculosis to their zoo-leader colleagues if the disease legitimately concerned them. Their different responses were not motivated by "saving face" but rather were symptoms of "uncertainty" about the disease itself.[64] In 1912, knowledge about tuberculosis still remained contested, especially in understanding transmission.

By the end of the nineteenth century, publics living adjacent to large cattle, sheep, and swine populations were aware generally of the zoonotic potential and nonhuman origins of many diseases. As historian of medicine Lise Wilkinson has shown, thorough examinations of animals as they relate to disease began with the English cattle plagues of the early eighteenth century, continued through the rise of the veterinary discipline, and intensified with controversies surrounding rabies, glanders, and anthrax through the mid-nineteenth century.[65] Tuberculosis research appeared at the end of this timeline, and its zoonotic nature was still being debated by researchers during the rise of the international zoo movement.

Since the 1870s, public health officials in Britain had worried about the contamination of both meat and milk by bovine tuberculosis, and they took systematic efforts to eliminate this threat. They increased, for example, the regulation and inspection of dairies, abattoirs, and farms in order to ensure higher hygienic standards. While the zoonotic nature of tuberculosis had been formally recognized since 1876, medical authorities generally agreed that less than 1 percent of human deaths caused by tuberculosis could be traced to the bovine strand of the disease, though other estimates were higher, especially for Great Britain.[66] No matter the percentage, bovine tuberculosis still killed hundreds or thousands of humans each year, depending on the country. And usually, the tubercular lesions would be found in the stomach or intestines (consumption of the bowels).[67] J. M'Fadyean and G. S. Woodhead, coauthors of "On the Communicability of Tuberculosis from Animals to Man," published in 1891, still claimed that cows, beef, and milk were certainly not propelling the great white plague.[68] And the increasing pasteurization of milk from 1898 through the 1930s also certainly decreased any of this existent spread from cows to humans.[69] Yet milk did set a precedent for the zoonotic fears surrounding the zoo. The public knew that tuberculosis had previously spread from animals to humans; therefore, imagining animals transferring tuberculosis to humans in other ways, even simply by proximity, *felt* possible to some. Frank Baker nonetheless expressed confidence that tuberculosis presented no threat. In fact, he remained skeptical of the very notion that the bacillus could cross species lines at all. For others, after all, saliva was seen as the pathogen's identifiable vehicle of transport, and it was hard to envision zoonotic infection since zoo workers and zoogoers rarely shared utensils and spittoons with the encaged, nor did they milk the elephants.

Later in 1912, an anonymous person within the Smithsonian community scribbled a note to Baker about the incidence of tuberculosis for Smithsonian taxidermists. If tuberculosis was truly a zoonotic threat, who would be more at-risk than individuals whose hands were buried regularly in dead bodies? The missive was written in a scrawl of hurried script, full of cross-outs, editing arrows, and marginalia:

> I have taken the trouble personally to look into the record of taxidermists and other persons whose duties have involved handling the dead animals from that establishment [the National Zoo]. . . . A few persons may have been overlooked, and the records of a small number are either inaccessible or obscure. It may safely be asserted, however, that the persons included in the following list comprise practically the entire number, and that the records of the few that are lacking would not . . . affect the general conclusion.

Among the list of twenty-seven names, only two taxidermists were linked to tuberculosis—Jenness Richardson and G. B. Turner (the chief taxidermist who had raised the concern about tuberculosis in the first place). Taxidermist Richardson "died of tuberculosis," but, according to the researcher, "he was a frail-looking man when I first saw him and I have been informed (by Dr. Benedict I think) that he contracted tuberculosis before he came to Washington." The author then described Turner as "rather thin and pale, but seems very full of energy, and has not been sick very often, so far as I am aware." After emphasizing Turner's vitality, the researcher stated that "whether he [Turner] has tuberculosis is uncertain. He seems to have the disease very much on his mind and this may be due to his fear of it or to symptoms which he himself has noticed." Although this writer hesitated over Turner's conditions, he confidently diagnosed the chief taxidermist with a case of hypochondriasis.[70]

It is evident from tone that whoever produced this report did so for the purpose of corroborating Baker's stance that tuberculosis was not a zoonotic threat for the zoo. Taxidermist George Marshall, who "has done all kinds of taxidermic work for the last 25 years and has probably been exposed as much as anyone," for example, was described as healthy because "there was no reason to suppose that he contracted it." The researcher asserted similar reasoning regarding D. B. Mackie, assistant in the Division of Mammals, despite mentioning his "fits," which supposedly arose after getting hit "on the head by a cannonball when young." The researcher failed to describe what he meant by "fits," a word that was used to describe both a consumptive's coughing episodes as well as seizures. Taxidermist Henry Marshall "never showed any signs of tuberculosis so far as [the researcher] was aware." Taxidermist A. H. Forney was considered nontubercular because he died from gastric enteritis, W. L. Halen because he died of drowning, and osteological preparator J. W. Scollick because he "suffered . . . from bad eyesight," as if a taxidermist could only be inflicted with one ailment at a time. Indeed, taxidermist A. H. Cloves was "frail looking," but the writer never entertained the possibility that tuberculosis caused this appearance even though Richardson, the only person the author admitted died from the disease, possessed a similar "frail" condition. A. B. Thorne, a zoo laborer who had transported many dead bodies from the park, looked thin, yet was "not known to have tuberculosis." The researcher may not have been wrong, but Baker's ally certainly did not look for the disease among taxidermists as thoroughly as the surgeons of the Bureau of Animal Industry looked for tuberculosis among deceased animals. Instead, this author presented these impressions, drawn from nineteenth-century colloquialisms about disease, to assuage fear and praise both the zoo and its superintendent.[71]

This crude survey was certainly constrained by evolving discourses of disease, diagnoses, and a nascent technics of measurement and inspectability. Tuberculosis manifested itself differently in different people. It could remain undetected for decades, masked by other ailments, or it could produce symptoms of varying severity in different periods of a victim's life, creating an inconsistency that made straightforward diagnoses difficult. Tuberculosis was also inherently undetectable (except in cases of severe symptoms that signaled a classic example of consumption) without evidence of the tubercle bacillus, seen under the microscope. In 1912, such evidence could simply be harvested more easily from dead animal bodies than from living human bodies, and the tuberculin skin tests (created by Austrian physician Clemens Freiherr Baron von Pirquet in 1907) used to detect the disease would not become commonplace until the 1930s.[72] Furthermore, sputum-centeredness surely encouraged the researcher to downplay zoonotic potentials.

From a twenty-first-century vantage, the unknown author, from the evidence provided, could not claim that zoo animals did not spread tuberculosis to humans. At best, the author's data was vaguely suggestive, showing that the twenty-seven taxidermists did not display the violent hallmarks of consumption. At worst, the results shed doubt on the very argument he was advancing, for the fingerprints, however "frail," of tuberculosis were scattered throughout the sparse information he provided. Nonetheless, from either perspective, the research successfully demonstrated two truths about tuberculosis in 1912. First, death was central to understanding tuberculosis. If someone, human or animal, did not die, their experience with the disease was often overlooked. Death made tuberculosis visible. Second, even though tuberculosis's zoonotic nature remained contested, conversations about the zoonotic potentials of the disease echoed around the zoo. Some citizens of the District even worried about other zoonoses. Two years later, pet store owner Edward Schmid (who sold koi to the zoo) wrote a letter to Baker about the popular misconception that diphtheria was a zoonotic disease. In fact, when Schmid's own daughter died from diphtheria, a faction arose against him claiming that the rabbits of his store killed her![73] The zoo was the most fitting place to deliberate about any and all zoonotic possibilities, from mange to rabies to brucellosis to equine influenza.

Newspapers had been printing articles on this topic since the 1890s, demonstrating that everyday Americans were aware of both the prevalence of tuberculosis in zoos and its zoonotic potential. Anxiety had long transcended the politicized issue of meat and milk. On July 16, 1898, the *New Haven Evening Register* published "Illness at the Zoo," which discussed the ailments of the London Zoo's animals, including lion and bear cubs with mumps and camels with kidney complications. The article stressed that the "monkey tribe is the hardest to keep in sound health [because] . . . they are liable to catch pneumonia or develop tuberculosis in a single night."[74] On May 28, 1893, the *Idaho Statesman* published "Tuberculosis in Monkeys," which described an outbreak among the primates of the Central Park Zoo (two thousand miles away), blamed, like so many diseases, on rats and mice.[75] The *Philadelphia Inquirer* informed the public about a famous hippopotamus of the Philadelphia Zoo, Princess Agwan, who succumbed to tuberculosis and was discovered by her keeper "cuddled up in a corner of the cage dead."[76] The *Daily Express* of San Antonio featured two columns about how the "unnatural life led by the snakes in the park [referring to the NYZP] is the cause of many cases of illness." Most of this article explored the challenges of caring for the king cobra; however, it ended mourning the death of a tubercular boa constrictor held in Chicago's Lincoln Park Zoo.[77] The *Kansas City Star* reported the deaths of hundreds of birds with tuberculosis, also caged in Lincoln Park, who supposedly communicated the disease to human zoogoers.[78] The *Evening Herald* of Bellingham, Washington, similarly reported an avian tuberculosis outbreak in New York's zoo, also emphasizing the spread of the disease from avians to humans.[79] And Duluth citizens read about the deaths of Roosevelt's gifts from King Menelik a few days after Baker had their corpses shipped away.[80] News about tuberculosis outbreaks was contagious, spreading through the closely knit journalistic networks established after the Civil War. When an animal perished from tuberculosis in New York, Philadelphia, Washington, DC, or Chicago, the whole nation knew about it.

The Zoo as a Medical Laboratory: Treating Tuberculosis

Pearle Rea Hutson was a secretary of an Illinois branch of the American Red Cross. In 1913, while researching the relationship between diet and pulmonary tuberculosis, she sent a letter to Frank Baker asking for "statistics concerning this disease among the Quadrumana" (a now-obsolete taxonomical category intended to force a line between *four-handed* primates and *two-handed* humans, or *Bimana*). Hutson asked three questions. First, "How many monkeys, apes, gorillas or other Quadrumana have you in the zoo?" Second, "What is their average length of life in captivity?" Third, "What percentage dies of pulmonary tuberculosis?"[81] To the first question, Baker replied that the zoo usually possessed between forty and fifty monkeys and lemurs. To the second, Baker pleaded ignorance, claiming, "to calculate the average longevity, so that the figures obtained will represent the real viability of each species in captivity, requires very careful work, and we have not, as yet, prepared such figures." To the third question, Baker informed Hutson that in 1911, out of eighteen monkeys to die, five had tuberculosis. In 1912, of twelve monkeys to die, four had tuberculosis.[82] Hutson thanked Baker for the information and promised to send her conclusions as soon as she completed her research.[83]

The National Zoo was not only a public institution entangled with popular discourses about tuberculosis, but it also became important to cutting-edge research. In summer of 1913, Bertram Waters, one of the world's leading tuberculosis experts, communicated with Baker, through the War Department's Office of the Surgeon General, and offered to mail his recently published paper about treating tuberculosis with a pentane-ozonide solution.[84] Waters worked as the medical director of the Ferry Boat Middletown Day Camp sanatorium (serving both Manhattan and the Bronx) and, between 1904 and 1910, he acted as the physician-in-charge at three New York dispensaries and served as director of Middletown's Tuberculosis Camp and Open Air School for children.[85] Waters published several key papers—most important, "Destruction and Repair in Pulmonary Tuberculosis." This paper explained how tuberculosis lesion sites could heal and be reactivated, demonstrating, for the first time, why tuberculosis frequently ebbed and flowed in severity throughout an individual's lifetime.[86]

Tuberculosis specialists frequently contacted Baker about the zoo's primates. By 1914, autopsy reports on dead monkeys were routinized and thorough. When "Dusty," a mangabey, became extremely ill, he was sent for diagnosis and surgery. And he died upon the operation table from a lethal dose of chloroform, for the surgeon, a Dr. Harrington, "used a great deal in order to obtain the physiological effects sufficient" for examination. In hindsight, he regretted the decision, but in the moment it did not seem "practical to use cocaine and induce local anaesthesia . . . as is my intention . . . with human subjects" because mangabeys will "bite and claw."[87] Upon opening Dusty's body, Harrington found "that about all the organs . . . show evidence of general miliary tuberculosis." With the report, the surgeon sent Baker an X-ray photograph that showed a missing left lung, which was "largely, if not wholly, obliterated . . . due to its destruction by the pulmonary form" of tuberculosis. Dusty's right lung "had not suffered to any appreciable extent, the upper lobe showing only superficial infection or small necrotic areas involved here and there." On the image, the surgeon marked an "X" on a location "between the first and second ribs" where "the slowly soluble iodine cartridge, or mass, was inserted by the hypodermic needle at a point directly into the apex of the lung." For photographic purposes, argyrol (a solution containing silver and usually used to treat gonorrhea) was employed to pigment the iodine solution so that Baker could clearly see the effects of tuberculosis on Dusty. Harrington concluded the report: "I am taking this opportunity of thanking you . . . for the scientific interest you have shown in the *proposed method* [emphasis added] of treating pulmonary tuberculosis." Then, he added, "additionally, my thanks are due to Chief Keeper Blackburne, for his skillfulness in handling the animals, and in aiding me during the experiment. He has indeed proven himself to be 'a graduate of the Great University of Experience' in this regard."[88] Unsurprising, considering the history of popular zoology, the zoo (and "Experience") played an important role in shaping Harrington's understanding of tuberculosis.

While it remains unclear exactly what "proposed method" Baker offered or discussed with Harrington, he must have passed along Bertram Waters's method of treating tuberculosis with pentane ozonide, an ozone compound that liberates iodine from potassium iodide in order to create a pure iodine cartridge that could be injected directly into the lungs. This method marked a pioneering approach in therapeutic medicine; however, the agent used was not new. Iodine (alongside mercury and creosote) had long been inhaled into consumptive lungs. Some practitioners, after the Listerian turn of the 1870s and 1880s, believed that iodine possessed an antiseptic effect.[89] Others, during the hematological turn around 1914, believed that iodine had a "stimulating effect" on a body's leukocytes. And still other practitioners viewed this treatment as "pseudo-scientific."[90] Iodine certainly circulated throughout the medical marketplace of the nineteenth century. Regardless of the historical-medical debates surrounding this mineral (which deserves a study of its own), Baker, Blackburne, and their zoo functioned as a medium through which medical knowledge about tuberculosis was transmitted from one doctor to another. Baker may have also offered Harrington his own insights on primate anatomy. Yet whether Baker added his own opinions or whether the "method" Baker "proposed" was simply making Harrington aware of Waters's recent research, this process of knowledge-translation was essential to knowledge-production. Stories like these are all-too-frequently lost beneath progressive narratives of medicine. Baker and Blackburne showed Harrington how to apply Waters's treatment to primates, which helped him later treat humans.

The Zoo as a Medical Laboratory: Understanding Tuberculosis as an Airborne Pathogen

The National Zoo, as Frank Baker knew, was not the only American zoo situated at the center of tuberculosis research. In 1910, Burton Hendrick, muckraker journalist for *McClure's Magazine* and future Pulitzer Prize winner, published an article titled "Oxygenizing a City: An Attempt to Make Two and a Quarter Million People Work and Play Under Sanitary Conditions" that emphasized the importance of zoo primates in advancing tuberculosis research and rejuvenating Chicago's indoor air. This article was read by all zoo directors around the United States, including Frank Baker. Hendrick opened this article by remembering that in the Lincoln Park Zoo, "a few years ago, tuberculosis was a more destructive disease than it is to-day in the East Side tenement sections of New York," where Bertram Waters faced outbreaks.[91] He emphasized that "virtually no animal seemed to be immune. Lions, tigers, mountain-sheep, reindeer, buffalo, ostriches, pythons—all frequently fell victim to the disease." The monkeys were "especially virulent," and Hendrick explained that next to fighting ("monkeys are persistent fighters") tuberculosis was responsible for almost all of the deaths in the monkey house.[92]

Environmental historian Sean Kheraj, drawing off the important work of Martin Melosi and in regard to the equine influenza pandemic of 1872, has demonstrated that "North American cities were networked ecosystems with . . . the capacity to behave as unified disease pools."[93] Zoos, as networked institutions within these already-networked ecosystems, were, of course, no exception. It is not surprising that as tuberculosis spread throughout the National Zoo, it also spread throughout the Lincoln Park Zoo. In August 1901, the *Chicago American* reported on the first African boa constrictor in the United States to be killed by consumption, noting that surgeons from the Academy of Sciences found "perfect specimens" of the bacillus in his lungs, bronchial tubes, and liver that would certainly "add so much to scientific knowledge." Also, as a telling aside and keeping with contemporary understandings of the times, the article attributed the boa's tuberculosis to the *sputum* of zoogoers.[94] Later that same year, an ostrich, after suffering for months, also died from tuberculosis, and her body was restored in the Academy of Science.[95] From there, tuberculosis continued to pass through the zoo, from enclosure to enclosure, from Tanner (the Yellowstone black bear captured by a former Illinois governor) to countless others.[96] In April 1904, Chicago mourned the death of Leo, their famed lion, who had struggled with tuberculosis and was put down to prevent spread.[97] In fact, as early as November 1903, Chicago activist Mrs. H. J. Hall (who should be seen alongside other anti-zoo feminists[98] like Miss M. Gunderson of chapter 4) warned the public to stay away from Lincoln Park so as not to "communicate" tuberculosis from the animals.[99]

Just like the National Zoo, the Lincoln Park Zoo, through zoo pathologist William Evans, performed autopsies on dead animals. According to his records, as reported by Hendrick, "there were from fifty to sixty deaths every year, and . . . 80% of these were caused by tuberculosis," a rate far higher than the ones reported by Baker, Hornaday, Stephan, Cadwalader, and Mitchell. "Cy" DeVry, who "enjoyed a national reputation as an intelligent animal-keeper," stated that he did all he could to reproduce "the same physical conditions under which animals lived in their native lands. Those that had come from a cold climate were kept in a cold atmosphere; those accustomed to a moderate temperature were supplied with moderately warm air; and those that lived naturally in the tropical zone were conscientiously provided with tropical heat." As mentioned above, susceptibility to tuberculosis had been often linked to climate, and, therefore, given DeVry's thoughtful effort in manufacturing the appropriate temperatures for each animal, the high death rate proved all the more shocking to the public. Was this mortality "the result of ignorant training, of malnutrition, or of inconsiderate exposure to the disease?" Hendrick queried.[100]

Figure 6.4. Lincoln Park Zoo superintendent Cy DeVry posing with two peccaries, held against his bosom, circa 1903. Courtesy of the Chicago History Museum.

These questions were similar to the ones that Hutson asked Baker. Yet the Lincoln Park Zoo, unlike that National Zoo, employed someone who would become uniquely capable of answering these questions. Zoo pathologist William Augustus Evans was born in Marion, Alabama, in 1865, the son of a mother who was a descendant of an original Jamestown colonist and colonial Virginian governor.[101] Two years after graduating from the Agricultural College of Mississippi, in Starkville, in 1883, he earned an MD from Tulane University. After practicing medicine with his father from 1885 to 1891, Evans then moved north to work as a professor of pathology in the College of Medicine at the University of Illinois.[102] In Champaign, he became increasingly connected in Chicago's medical community (especially as he helped found the Columbus Laboratory, the first in Chicago, in 1893). Like the National Zoo, the Lincoln Park Zoo offered unique and high-profile opportunities—for example, the chance to perform Chicago's first orangutan autopsy on the famous Miss Dooley, who died from acute digestion.[103] While it is unclear exactly when Evans began his employment in the Lincoln Park Zoo, we know by his newsprint presence that he was indeed working there by 1901 while he simultaneously held his faculty appointment.

The monkeys demanded the greatest concern at the zoo, and Evans would pay them the utmost attention. Hendrick relayed a vivid scene from his interview with DeVry:

> "Just take my monkeys, for example," said Mr. DeVry. "The average temperature of the places from which they come is eighty-five degrees Fahrenheit. Now, the thing to do is to keep them always in eighty-five-degree air. In the summer-time they get this easily in the open. When winter comes, I take them all inside, turn on the steam, and keep their quarters at about eighty-five degrees. They like it, too. Look here," and he led the way into one of the modern steam-heated monkey apartments. Here twenty or thirty shivering creatures were making heroic efforts to keep warm. In one corner, a larger radiator was sending forth enormous gusts of hot air, and a wooden shelf on top of this radiator was the most popular quarter of the cage. Every inch was filled with huddling monkeys, rapturously basking in the heat.

William Evans then commented to DeVry, in this remembered dialogue, that the "scheme seems logical enough. . . . You are trying to do for your monkeys what nature does for them in their own homes." Yet they still contracted tuberculosis. "I think we had better try something else," Evans reasoned.[104]

The next fall when a new shipment of monkeys "fresh from the tropics" arrived in Chicago, Evans—who had just delivered a lecture in the Matthew Laflin Memorial Hall with the title "Overcrowding and Fresh Air"—proposed an experiment.[105] He suggested that twenty healthy monkeys be taken inside the monkey house while five sick monkeys be quarantined outside. "It will be interesting to see what will happen," Evans exclaimed. As winter approached, the sick monkeys were "kept in a place where they were constantly exposed to . . . chilling drafts." The zoo provided these ill primates with a thatched-roof shelter, lacking artificial heat but still offering protection from the wind. Surprisingly, "the invalids seemed to care little for this shelter," except for at night when they slept under it. The sick monkeys "showed as natural an inclination for the cold open air as their healthy brothers did for the hot drafts inside." In fact, the outdoor monkeys grew "thick brown furry coats" and these previously "sluggish creatures" suddenly "started into life," resurrected from sickness to health. The monkeys' muscles "had grown large and strong; they ate eagerly and manifested an increased desire for the favorite simian pastime—fighting." (As a growing ethological literature has shown, all primates use rough-and-tumble grappling, chasing, wrestling, and play fighting to develop social, cultural, and language skills, though the specific content of play varies greatly among species.)[106] Soon, the outdoor monkeys became "the most popular curiosities of the Zoo." Zoogoers in Chicago delightedly watched their tropical ancestors sitting atop snowbanks feasting on bananas.[107]

In contrast, the twenty healthy monkeys placed in the heated house died from tuberculosis. Every single one! The artificial environment "had killed them, as it had killed hundreds of their predecessors." None of the outdoor monkeys, in comparison, had contracted tuberculosis. Even five years later, two of these five outdoor monkeys were still alive and in good health, enjoying a longevity rare for zoo primates in 1910. The other three who died within five years following Evans's experiment succumbed to causes that were not disease-related. One broke his back, another died giving birth, and the last became unexpectedly paralyzed. The monkeys forced to endure the fierce winter of the Windy City escaped the tuberculosis that plagued their indoor peers.[108] Word, like bacteria, spread through zoo networks, inspiring zoo directors like Sol Stephan of the Cincinnati Zoo to follow in Evans's footsteps and move the monkeys outdoors for winter.[109] Frank Baker, having read Hendrick's article in Washington, was simply aghast and immediately wrote letters to both Evans and DeVry: "I have kept out some kinds of monkeys such as those that naturally inhabit somewhat temperate climes such as the *Macacus fuscatus* of Japan [Japanese macaque], but I should certainly hesitate to expose animals from the valley of the Amazon or similar tropical regions."

The director of the National Zoo could not believe what Evans was trying in Chicago![110]

This experiment not only demonstrated the contagiousness of tuberculosis but also revealed that the tubercle bacillus could be passed through the air, a basic fact that would not be accepted and understood scientifically for several decades. In the classic experiment that reconceived tuberculosis as *primarily* an airborne pathogen, Richard R. Riley, William Firth Wells, and fellow researchers in Baltimore pumped the air from a six-room tuberculosis ward into an adjacent room containing 156 guinea pigs. Seventy-one of them contracted tuberculosis.[111] Riley and Wells's experiment led to the landmark 1959 essay "Aerial Dissemination of Pulmonary Tuberculosis: A Two-Year Study of Contagion in a Tuberculosis Ward." Amazingly, Evans and DeVry's zoo experiment demonstrated similar conclusions more than fifty years earlier.[112] Their investigation showed that monkeys held in artificially heated buildings contracted tuberculosis at accelerated rates. They could not yet prove the ultimate conclusion that the heating system of the monkey house recirculated bacteria, captured and encased in water droplets exhaled from primate lungs, simply because they had neither the microscope technology nor the bacteriological expertise and pathological lexicon to make such claims. All Evans, DeVry, *McClure's* subscribers (via the journalism of Hendrick), and other zoo leaders active in zoo networks could know for certain was that there was something malignant about the structure of the monkey house.

While a pathologist, zoo superintendent, and journalist could not completely redefine tuberculosis as an airborne pathogen, they could rethink zoo planning and exhibit layout in response to the disease, despite its elusive bacteriological properties. Cy DeVry, immediately following the experiment, "revised his philosophy of animal hygiene; he recognized that the point was not to make the climate adaptable to your animal, but to make your animal adapt to the climate." After the epidemic in the monkey house, the Lincoln Park Zoo became "an open-air, cold-air zoo." The heating systems were removed from all the buildings, and the windows opened. Now, zoo animals in Chicago spent their days outside, in the fresh air. Zebu and antelope from India, wild pigs from Mexico, kangaroos from Australia, and ostriches from Africa all adapted to snowy Chicago.[113]

Five years after Evans's experiment, zoo animals still died at the claws of disease, but tuberculosis was mostly eradicated from the list of killers. Bison calves became a common sight, and the Lincoln Park Zoo (alongside the Bronx Zoo) emerged as a leading bison breeder and exporter, supplying the world's zoos with the very species Hornaday had "saved" two decades earlier in Washington.[114] While the animals adapted to living outdoors, the Lincoln Park Zoo redesigned its buildings one at a time, as it could afford, to accommodate new heating systems that would more effectively draw in air from the outside, heat and circulate it quickly, and then eject it efficiently through an exhaust. This sort of tempered air ventilation system technology was installed, for example, in the new Monkey House built in 1923.[115] Evans's findings, though, had an immediate impact. The very first building constructed after his outdoor experiment—the Lion House in 1912—was designed around tempered air ventilation. To demonstrate how Evans's insights became translated into architectural language, let us take a look at the heating specifications from the architectural firm that built the Lion House: "Steam will be supplied from an *outside source*, as marked on plan. The contractor will connect to the outside steam and return lines at the wall of this building. A two-pipe return line vacuum system will be used. Direct radiation will be installed in all important rooms, and the building will be ventilated by electrically driven fans." The lions would no longer huddle around radiators placed in the corner. No longer would heated air *be* recycled air. This new system, known as "blast radiation," prioritized both "rapid heating" and "thorough venting," incorporating "the outdoors" and "fresh air" into its circulatory system, thus increasing the likelihood of dispersing the tubercle bacillus.[116]

The elimination of tuberculosis within the Lincoln Park Zoo catapulted William Evans from the zoo laboratory into the public sphere. In fact, Evans became one of the leading tuberculosis experts in the United States. In 1907, during a scarlet fever outbreak, Chicago mayor Fred Busse appointed Evans as health commissioner. In this position, in a decade when public health departments were still emerging in American cities, Evans worked to move the city of Chicago from sputum-centered to airborne-centered understandings of tuberculosis. In his first report as commissioner, published in 1911 for the years 1907 through 1910, Evans outlined, in detail, the various mediums through which a bacterium could spread—liquid, air, soil, and surfaces. However, he emphasized the importance of air to the spread of the tubercle bacillus.[117] This paradigm shift did not, of course, happen overnight. It took Evans several years to convince policymakers of the lessons that were so nakedly on display in the beastly zoo. Put another way, it took Evans time to translate medical facts across the ontological divide that divorced "the animal" from "the human," even if the zoonotic bacterium themselves had never acknowledged the sanctity of this boundary. In the city council proceedings from 1907 and 1908, when the Department of Health enumerated "rules" concerning contagious diseases, sputum-centeredness still clearly reigned: Rule #38, dealing with tuberculosis, legally required all employers of indoor labor to post official "No Spitting" signs in their workplaces, just as in the National Zoo. Though the department, during Evans's first year as commissioner, established eleven rules for preventing the spread of tuberculosis, not a single one acknowledged that the disease spread through the air.[118] Only one year later, though, Evans began to influence wider municipal audiences when, through a department bulletin, he publicized the connection between tuberculosis mortality and a lack of ventilation, popularizing the lessons he learned in the monkey house.[119] This feat certainly did not go unnoticed, since Evans, in November 1908, accepted a silver medal from the International Congress on Tuberculosis for proper ordinances.[120]

Critiques of the metropolis had long relied upon the suffocating imagery of air pollution to bemoan the dark side of urbanization and industrialization.[121] However, Evans devoted his influential civic position to publicizing ventilation not for the purpose of eradicating a Victorian-monolithic, miasmal specter that loomed over the city but in order to evacuate the airborne pathogens, floating *in* the air, from indoor spaces, thus contributing to shifting understandings of air, as well as of tuberculosis. As commissioner, Evans published widely about ventilation. In his pamphlet "Hygienic Principles of Ventilation" (1910), Evans, for the first time since his zoo experiment, laid out his ventilation philosophy in full. At the end of the treatise, he elucidated his "eleven principles of ventilation." These speak amazingly for themselves in how they were inspired by monkeys. They also contrast starkly with the principles of the National Zoo's "Prevention of Tuberculosis Among Government Employees," underscoring how dramatically bacteriological thought changed between 1905 and 1910.

1. Attention must be concentrated on the *breathing* zones of the habitually occupied portions of the room . . . from 4 feet above the floor to 6 ½ feet above it.
2. In this zone there must be the greatest possible effort to prevent *(a)* stagnation, and, *(b)* dilution.
3. Dead zones and corner eddies must be encouraged as settling places for bacteria. In them there must be the smallest possible air movement. Such spaces must be cleaned daily with moist methods. Therefore, they must be readily accessible for cleaning purposes. . . . If such nooks are occupied, *e.g.* a working desk is placed therein, conducting ducts with air inlets must be made to open into them. Ventilation of such crannies by dilution is a myth.
4. The breathing zone must be occupied by an even up-current of air, moving at the rate of at least 60 feet a minute from multiple floor inlets to multiple ceiling outlets. If the natural upward flow is met by a down flow of air, the expired air is buttressed right in the head zone and does its maximum harm.
5. The ventilating air in mechanical systems should never enter the room at more than 40 degrees below body temperature; 30 degrees below is better still. Where men are working hard, a high temperature, say 70°, is required on account of the chilling of perspiration. Where men are very still and sluggish, a room temperature of 68° to 70° is required. Where men are mildly active, 60° to 65° is better.
6. The air must be protected from the introduction of dust from the outside and the dissemination of dust produced inside.
7. It is of great importance that the relative humidity of the inside air should not be materially greater than that of the outside air. It is of some importance that it should not be more than 10 degrees less than the outside air.
8. The volume of air per person per hour should be around 500 cubic feet, varying with the directness of removal of the polluted air.
9. The arrangement of the ventilation must be such that the windows can be opened and the room blown out at least once in each 24 hours.
10. *Recirculation of air is not to be tolerated.* [emphasis added]
11. The air must be below 6 parts per 10,000 in CO_2, and low in dust and bacterial content.[122]

In creating these eleven principles, Evans, after five years, learned how to package the lessons he learned from the monkey house into a technologically precise language that architects and planners would find useful in the installation of new ventilation systems like the blast radiation installed in the Lion House. Though "Hygienic Principles of Ventilation" was inspired in the zoo, there was no mention of either the zoo or animals anywhere in its twelve-page text.

Evans also carried ventilation to public schools, where he additionally formulated the idea of a humidifier. By applying his monkey-house vantage to the captivity of the schoolroom, Evans discovered that the air circulated through Chicago's schools was far less humid than the air outside. The lack of humidity created indoor atmospheres that extracted moisture from the throats and lungs of children, producing coughs, sneezes, and soreness while simultaneously depositing bacteria. To remedy this, Evans not only called for schools to be heated in the same manner as the street cars that first integrated the principles of ventilation (with air intakes at the floor that pass through radiators and then exit the car through ceiling ducts, just like the Lion House) but also introduced a humidification process that directed heated air "through jets of water or steam" before entering the classroom.[123] Many schools, though, embraced an even more direct lesson from the zoo and became open-air schools, filled with teachers and students wearing thicker coats, just like their primate cousins who played atop mountains of snow.[124] Of course, ventilation was only a part of the regime of medicalization that Evans forged in order to contain tuberculosis. He also improved data-keeping protocol to track cases, pushed increased pasteurization, and established a network of sanatoriums and tuberculosis centers throughout Chicago.[125] All of this suggests that what is often analyzed by historians of medicine as the emergence of state medicine between 1870 and 1914 has origins in local municipal governments, including their entanglements with zoos, as much as in nation-building governments of the federal state.[126]

Evans's influence reached beyond the city of Chicago. After resigning as health commissioner in 1911, he served as a professor of hygiene and sanitary science (called "public health," a field that Evans himself helped define, by 1921) at Northwestern University's medical school.[127] In this capacity, Evans used the classroom to change popular and scholarly conceptions of tuberculosis. In his notes for the fourth-year lecture course in public health, he depicted tuberculosis as a respiratory disease.

Anticipating Riley and Wells, he taught his undergraduates that airborne diseases spread because "organisms coughed up float in the air and may be quickly inhaled." He even, amazingly, hypothesized that the bacillus became airborne by either sitting inside a small water droplet or by resting upon a particle of dust. As he did in the Lincoln Park Zoo and in the Department of Health, Evans preached ventilation, ventilation, ventilation, even etching the word into his survey course description.[128] He taught at Northwestern until 1927, and by then tuberculosis mortality rates had declined significantly, moving from the deadliest disease in America around 1900 to, Evans estimated, "sixth or seventh in the list of causes of death." Evans also illuminated for his students the link between tuberculosis mortality and poverty.[129] In so doing, he anticipated the concerns about structural inequality, social justice, and tuberculosis contraction that would emerge in the postwar world and continue increasingly until our present day, where one million people (mostly in the Global South, and many with HIV) die from tuberculosis every year.[130] During his tenure at Northwestern, not only did Evans influence thousands of medical students, but he also served as president of the American Public Health Association, as an editor for the health department of the *Chicago Tribune*, and as a frequent author of nationally syndicated columns on public health and hygiene.[131] According to the *Mississippi Doctor* in 1934, Evans possessed "the ability to take the most intricate medical fact and digest it down, condense, simplify, sum it all up and put it in a 'nut's shell,'" which might explain why he received 1,087,447 letters from his readers.[132] This cultural adroitness, though, should not be too surprising to us, for if we replace "medical" with "scientific," this praise of Evans could just as easily be a zoo booster's trumpeting of any zoo's educational mission statement. How fitting for a man who made his first breakthrough with tuberculosis in the Lincoln Park Zoo.

Though surely a few in-the-know Chicagoans remembered Evans's accolades as zoo pathologist, once he became health commissioner, this zoo background vanished from *his* biographical and *our* historical record. Other than the 1910 article in *McClure's*, not a single, lowly reference to the Lincoln Park Zoo, to the monkeys in the snow (or for that matter, to *any* animals with tuberculosis), or even to the retrofitted primate house equipped with cutting-edge ventilation technology, exists in either the biographical salmagundi of Evans's life or in the canon of his medical, scientific, or popular writings. Could he have been embarrassed by his gig in the zoo, a place so often seen as an amusement for the masses? Could he have been afraid to acknowledge that a zoo pointed him toward the airbornedness of tuberculosis, that knowledge essential to public health was forged in a mere monkey house? Or was the Lincoln Park Zoo simply not worth mentioning, remembering, within the impressive career of a doctor of national renown? Probably, all of these. Yet, could it be with Evans, indeed could it be with the history of tuberculosis in general—the zoo in general—science in general—that the absence of animals is due to an expulsion?[133] Could it be that even *successful* experiments (like chapter 3's *failed* ones), when put under the microscope of scrutiny, reveal hidden animalities, clandestine animals, concealed within the structure of an experimental apparatus, veiled within narratives of human progress?

Quite possibly. William Evans, a renaissance man, was authored, as protagonist, into these narratives. And it is indeed remarkable that he could turn toward the margins of his page to find new ways of conceptualizing tuberculosis. Like William Garner and Samuel Langley, Evans—alongside the other individuals of this chapter, from Frank Baker to Pearle Rea Hutson to Bertram Waters—looked to animals to become "modern," to conquer disease, or assert that it had been conquered. Animals served as the prerequisites for human self-reflection, as tools for expressing human fears, as the stage for tales of human triumphalism. The stories of this chapter provide us with several important lessons about the history of tuberculosis. First, zoos functioned quite literally as laboratories where experiments were conducted, operations performed, and diseases diagnosed and where medical knowledge was formed, contested, and transmitted. Second, no simple intellectual history of tuberculosis exists. Knowledge about the disease was not assembled along the neat trajectory of a timeline but was instead fashioned across multiple timelines, layered upon each other. In 1912, as Baker, Cadwalader, Hornaday, Stephan, and Mitchell deliberated about the prevalence of tuberculosis in their zoos, the proper methods of disinfecting zoo exhibits, the death rates caused by the disease, and the threat of zoonotic spread, the Lincoln Park Zoo had already redesigned itself to prevent the transmission of tuberculosis through the air, even though an understanding of the bacillus as primarily an airborne pathogen would not become commonly accepted until well after World War II. Knowledge about tuberculosis formed piecemeal, and information moved, in fits and starts, throughout zoo networks. Third, the zoo as a public institution sat at the center of popular discourse, public policy, technological development, and scientific experimentation. Policies formed inside zoos shaped policies beyond its fences, and vice versa. Last, as zoo animals lived and died at the mercy of contagion, they taught zoo officials to think about tuberculosis in new and innovative ways. I realize that the mask I wore today while teaching my animal history course has an animal history itself, a history of entangled ecologies, spaces of captivities, radiators, air bubbles, lesioned lungs, and resilient and adaptable monkeys who always ran away from what they were supposed to be, yet, because of the zoo, could never fully become who they actually were, or would have been.

CONCLUSION

The National Zoo Movement

From its beginnings, the National Zoo attracted hunters, conservationists, and peddlers of animals. The zoo snared legislators sitting atop Capitol Hill, enamoring some and enraging others. The zoo excited the populous as it released wild animals onto the boulevards of the public imagination. The zoo intrigued scientific Americans—professional scientists, showmen of science, and a growing body of citizens interested in zoology. The zoo seized the attention of animal activists and pushed the public to think (or not) about the welfare of animals. The zoo connected a diverse group of Americans associated with bighorn sheep and in so doing forged zoology from the ground up and created a complicated conservationism from the state down. And the zoo encouraged zoo leaders, surgeons, doctors, and public health experts to rethink contagion. The National Zoo captured a nation.

Frank Baker helped steer the zoo into national significance. Unlike business moguls of the Gilded Age and Progressive Era, who kept information to themselves to ensure a competitive advantage, zoo directors shared knowledge. Zoos rarely competed with one another since a single city could not support multiple zoos. Therefore, in the zoo marketplace, exchanging information was usually beneficial. When Baker wrote to the Lincoln Park Zoo regarding their tuberculosis experiments, he knew Cy DeVry would help how he could, for they had long shared data about all sorts of topics—zoo keeper hiring practices; pay scales for laborers; the availability and costs of dingoes, yak, gray wolves, and Cuban tree rats; and about the best brand of barbed wire.[1] Zoo directors viewed each other as friends, as cordial members of a gentlemen's club built upon shared interests. And Baker consistently engaged his peers to improve his own zoo and create a zoo-networked nation.

In the summer of 1901, G. A. Parker, chairman of the American Park and Outdoor Art Association (APOAA) and superintendent of Keney Park in Hartford, Connecticut, sent Baker a letter regarding his recently compiled census of two thousand known urban "park areas" in the United States. Despite the register's thoroughness, Parker admitted that he lacked the "full information . . . of the Zoological parks of the world." He remembered meeting Baker during a visit to the National Zoo two years earlier and decided to ask his old acquaintance for a master list.[2] Baker responded by sending a compendium of "all the zoological collections of which I have definite knowledge," but he admitted that this knowledge was surely incomplete and asked Parker to "give me data regarding any [zoo] which I have not included."[3] Baker's register included seventy-seven zoos from places as distant as Constantinople, Sydney, Antwerp, and Giza. Thirteen were located in American cities—Baltimore, Buffalo, Chicago, Cincinnati, Detroit, New York City (Central Park and Brooklyn, breaking the one-per-city rule), Philadelphia, Pittsburgh, Providence, San Francisco, St. Louis, and Washington, DC.[4] Nonetheless, Baker seemed certain that elusive collections remained unaccounted for.

Three months later, Baker received a letter from Daniel Burns Dyer, former Cheyenne-Arapaho Indian agent and president of the Augusta Railway, wanting to know where he could "secure a book on the care and maintenance of birds, beasts, and reptiles." He assumed that the government—which funded a national zoo—published "such a book." Dyer also inquired about the cost of a black bear.[5] Apparently, Augusta, Georgia, possessed an unknown zoo, and Baker went into action immediately. He contacted the Census Office and requested a list of American cities with populations of at least twenty-five thousand.[6] Then, he sent a letter to each mayor: "I am endeavoring to secure . . . reliable data with regard to the different zoological collections in the United States, and I should be glad if you would inform me if there is one in or near your city." He also requested statistics on when they were established, whether they were funded by public or private parties, their acreage, the number of buildings they hosted, and the number and kind of species.[7]

The feedback he received surprised him. On November 16, 1901, Robert Wright, mayor of Denver, replied: "We have a small collection of Rocky Mountain fauna in our City Park. It was commenced about nine years ago . . . and, finding them very attractive to the public, we have added to them from time to time." This "small" zoo still maintained three bison, fourteen elk, four black-tailed deer, two white-tailed deer, eight pronghorn, one moose, one grizzly bear, two black bears, five cinnamon bears, two puma, three gray wolves, two swift foxes, three badgers, four raccoons, two minks, a prairie dog village, two golden eagles, two American swans, twenty-six wild ducks, three Mongolian pheasants, and six Canada geese.[8] Small indeed! In addition, Wright added that there was also a collection "in the Northern suburbs of our City known as 'Elitch's Garden.'" Impressed, Baker thanked Wright and stated that he was "pleased to know that your zoological collection has already reached such considerable proportions." And he asked if they had any pumas to spare.[9] J. M. Trimble, a park commissioner in Chattanooga, Tennessee, was the next to respond to Baker's letter, reporting that Chattanooga had an enormous collection (with species ranging from cockatoos, zebus, and macaws to baboons, Gila monsters, lions, and sea gulls) housed in two buildings in East Lake Park.[10] Baker mailed Trimble a long letter accompanied by the National Zoo's annual report.

Between the autumn of 1901 and the spring of 1902, letters from all over the nation poured in. Mayor Marshall Hicks of San Antonio informed Baker that the city's San Pedro Park hosted a private zoo of more than one hundred specimens, which was established in 1888. They even had jaguars and a prairie dog village.[11] S. N. Williams, mayor of Williamsport, Pennsylvania, spoke of a deer park in Clinton County that kept an astounding three hundred deer, nine Russian red deer, and fifteen elk.[12] William Clark, mayor of Manchester, New Hampshire, pointed to a collection at Corbin's Park, near Newport.[13] The mayor of Akron underscored an animal park owned by R. H. Lodge, just north of Cuyahoga Falls.[14] (Little did Baker know at the time that this Silver Lake Park would eventually become a world center for captive black bear breeding.)[15] An employee of Easton, Pennsylvania, directed attention to two wild animal preserves in northern New Jersey, one in Sussex County and one in Warren County.[16] A park commissioner of Omaha informed of a collection in Riverside Park.[17] The mayor of Grand Rapids told Baker about the John Ball Park zoo, established in the early 1890s.[18] The city auditor of Salt Lake City wrote about a deer park within the city and a buffalo herd on an island in the Great Salt Lake.[19] The mayor of Elmira, New York, informed Baker about a smallish zoo called Eldridge Park, established in the 1880s and equipped with a $4,000 bear pit.[20] And a park commissioner in Indianapolis informed of a small animal collection, including bears, deer, foxes, wolves, and birds.[21] As we saw in chapter 2, this zoo also possessed an elephant a year earlier, yet she was discarded after her downtown escapades.

The torrent of letters continued. Some offered information, and others simply stated that the city in question possessed no zoo at all. For example, the city clerk of Troy, New York, admitted that his city "has not and never had a collection of wild animals, excepting the time when the elephants belonging to Barnum broke loose and ran the streets."[22] Mayors, city clerks, and government officials all found time to respond to the superintendent of the National Zoo. Baker received letters from places as diverse as Wilmington, Oakland, Butte, and Spokane.[23] He opened correspondence from Pawtucket, Rhode Island; Schenectady, New York; Galveston, Texas; and Jackson, Michigan.[24] As Baker read and responded to these letters, he stitched together a zoo-networked nation. And he did so resiliently. When some city officials failed to respond to the first letter, Baker sent a gentle reminder in March 1902.

A secretary from a Scranton office told Baker about a small zoo with bears, monkeys, squirrels, raccoons, and an alligator.[25] The city clerk of Dayton wrote that his city possessed no wild animals but that the nearby National Soldiers Home had a living collection on site.[26] A Norfolk clerk told of an animal house and three-acre deer park in his Virginian port.[27] Also, a park chairman from Seattle updated Baker of three city parks that housed animal collections. One of these held deer, eagles, and an owl. Another displayed a wild cat, pheasant, and more owls. And the Leschi Park, a lovely steamboat stop owned by the Seattle Electric Railway Company, showcased seals, bears, wolves, monkeys, peacocks, coyotes, marten, and elk. This Seattle chairman added, as a good Progressive, "I am an 'ardent' on the matter of municipal parks and regard them . . . an essential necessity to the betterment of humanity."[28] These Seattle collections were not new. The first two were organized in 1897, and the Leschi collection was forged in the 1880s.[29] The United States not only hosted far more animal collections than Baker had ever imagined, but it seemed that many of these had emerged alongside the National Zoo as products of a nineteenth-century national zoo fever. Americans needed captive-wild animals in their growing cities.

Between 1902 and 1905, Baker continued to process data about zoos formed in the 1880s and 1890s. Mobile.[30] Binghamton.[31] Toledo.[32] Springfield, Massachusetts.[33] Portland, Oregon.[34] Portland, Maine.[35] Glen Island, New York.[36] In and through this communication, Baker frequently rehearsed a progressivist model for understanding zoos, one that still exists today, a model that envisioned a teleology stretching from rudimentary "menagerie" and small "urban animal collection" to "modern" zoological park.[37] Some of the collections that Baker discovered looked more like (what he believed constituted) "modern" zoos, grand projects of the City Beautiful movement,[38] than others did. However, the publics that visited these collections considered them all to be zoos. These local citizens, like the ones of Washington that frequented Hornaday's department, cherished their zoos and saw their animals as local symbols. Zoogoers around the country wished, however, that they could stroll through a place like the National Zoo, and they all hoped that their local collections might someday evolve into a premier, "modern" institution.

Baker's 1901 suspicions that he lacked knowledge about American zoos proved to be more than accurate. He admitted in an unofficial 1903 report titled "Zoological Collections of the United States" that "the lack of accurate . . . information with regard to zoological gardens of the United States has long been apparent and has made . . . it desirable to collect for publication all facts concerning these institutions which could be directly or indirectly obtained."[39] Originally, Baker told Parker that there were thirteen zoos in the United States. When his survey was complete, he counted no fewer than fifty-six living zoological collections.[40] He recorded the information he received into a makeshift chart that, while originally typed, was riddled with pencil markings, strike-throughs, corrections, and blank spaces. Baker never published the results of his survey. After three years of corresponding with local government officials, only a ragtag chart stood as a testament to Baker's work.

A makeshift database probably serves as the most accurate reflection of the American zoo movement. In the last three decades of the nineteenth century, living animal collections have become integral components of the American city, yet they were institutions in flux. Within a few short years, a collection of dead animals could transform into one of living animals. And these could quickly evolve into zoos. Through this lens, some collections looked like the menageries of a previous era, riddled with chains and cages. Others appeared to be something new, parks with larger enclosures and more developed routines of animal care. In truth, most collections were always both simultaneously. At the end of the day, despite differences within and between them, zoos always remained institutions of captivity and display where some animals received better care than others. Local circumstances, like funding, mattered to the destiny of individual collections. Some zoos faltered and collapsed. Others became landmark zoos. Most were somewhere in between. Yet all zoos, whether viewed as fledgling or grand, were unique, contingent on local circumstances, contexts, and cultures. They all deserve their own entangled histories. By corresponding with animal collections, Baker forged a network that both circulated knowledge about building, managing, and keeping a zoo and circulated resources that could support these endeavors. Between 1901 and 1910, Baker threw open the doors of the zoo-men gentleman's club, creating an infrastructure suitable for a nation that would host more than one hundred zoos by World War II.[41] Furthermore, through this process, Baker placed the National Zoo at the center of the American zoo movement and placed the federal government at the center of a growing federal system.

Baker did not create this network merely to improve his own zoo. On May 22, 1905, Dr. James Morgan from Wilmington, Delaware, scribbled Baker the following:

> As a lover of wild animals I write you relating our little Public Free Zoo asking you kind information regarding the possession of some small animals. . . . I know you can not sell your surplus, but will you advise me how we can obtain something for our Park. Our Senator will testify to our worthiness. . . . I hope you interest yourself in our behalf.

Baker responded dutifully, "We have at one time or another obtained mammals and birds from most of the people in the United States who supply such specimens and I shall be glad to send you the address of these dealers."[42] Three days later, Baker mailed Morgan a list of the premier dealers of the American animal trade.[43]

Assisting other zoos had always been important to Baker, but after extending his zoo network, this mission became easier. In 1906, George Clarke of East Liverpool, Ohio, corresponded with Baker about forming the "nucleus upon which to build . . . a small Zoo." He also asked for advice about designing a "Raccoon Tree," a common and inexpensive enclosure type.[44] That same year, MIT professor and bacteriologist William

Lyman Underwood, who was involved in the establishment of a zoo in Middlesex Falls, sent Baker an inquiry about bear enclosures.[45] Baker wrote back with photographs and a two-page detailed description of building materials and dimensions.[46] Otto Widmann of St. Louis wanted advice about the types of birds to place in the birdcages.[47] Baker replied by sending a dozen free night herons![48]

On April 11, 1906, William Stinchcomb, chief engineer of Parks in Cleveland, Ohio, sent a letter that truly demonstrated the centrality of the National Zoo to the American zoo movement. After informing Baker that Cleveland was "about to establish a new and modern zoo," Stinchcomb asked, with a sense of urgency, a long list of questions that inquired about acreage, building types, topography, enclosure design, deer parks, water supply, among other topics.[49] And, of course, Baker replied with thoughtful and thorough answers.[50] Thereafter, zoo men in Cleveland frequently contacted Baker as an expert. It is worth emphasizing here that Baker himself made this relationship possible, for five years earlier Baker had sent his original survey to the mayor of Cleveland.[51] This survey, then, ended up in the hands of the director of public service, Stinchcomb's superior. The information that Baker originally received, in 1901, was that Cleveland had a living collection organized in 1886.[52] Twenty years later, the menagerie was teeming with specimens, and the city was ready to formally establish a large zoo in the Old Brooklyn neighborhood, where the Cleveland Zoo still stands to this day. For advice, Stinchcomb (at the urging of his superior) turned to Frank Baker—the National Zoo superintendent who first expressed interest in Cleveland's animal collection.

Baker made himself the director of the nation's zoos, and he made his own zoo the template for zoo formation and management. The National Zoo inspired animal collections to make themselves in its image—to increase the number of animals, to establish walkways and buildings, to attract masses of zoogoers, to campaign for funding, to improve animal care, and, of course, to pursue popular zoology.

Toward Ecological Futures

Zoo animals were knotted, layered, and divided, always more than what we—an American zoogoing society—said they were. Zoos are teeming with stories. Runaway grizzlies. Angry congresspeople. Flying contraptions. Talking monkeys. Consumptive monkeys. Chained elephants. Absconding bighorns. Dead housecats. This book has shown a few ways in which the inhabitants of the National Zoo found unlikely homes in politics, print culture, science, ethics, conservation, and medicine. As the zoo movement spread, animals elsewhere similarly ran away from expectations, forging unlikely inroads into supposedly human endeavors. Popular zoology shaped nearly every facet of American science, culture, and conservation during the Gilded Age and Progressive Era. As millions of zoogoers began to visit zoos, some even began to think about animals in new ways. Americans, echoing environmental historian Jason Colby, needed first to experience animals for themselves before some could truly begin to appreciate them.[53]

In many ways, zoos themselves have changed immensely since the days of Frank Baker. In other ways, however, zoos have changed very little. By looking closely at the knots, layers, and animals (real and symbolic) of the National Zoo from 1887 to 1920, I hope we can better see the textures, ironies, complexities, and contradictions of the twenty-first century zoo. I hope we can better see its shortcomings, its stale histories, its violent structures, its ideological deceptions. And I hope we can better see its strengths, its compassions, its dedicated staffs, its interdisciplinary positionality, its unique place between professional science and popular science and between popular science and popular culture, and its stories of science, culture, and conservation. Most important, by looking closely at the historical zoo, I hope we can better see the animals that live with, around, because of, and in spite of us. Zoo historian Nigel Rothfels has written:

> Zoos have always argued that, while they might in the past have been bad, wasteful, poorly managed places, or destructive of nature, zoos today are better. . . . At a surface level the argument is true, but at a deeper level these institutions must forever struggle with the fundamental fact that while new missions—or what some would call justifications—have been added to zoos over the past couple of hundred years, the essential nature of the places as sites of human amusement that depend fundamentally on constraining the natural behaviours of other living creatures has not changed.[54]

Zoos, like their animals, are layered with histories that have settled over centuries. Zoos may always be stained by some of their deepest meanings, and they may always be indicted by their oldest colonial, hegemonic, anthropocentric structures that exploit animals for human amusement. Zoos have always *contained* animals—and still do. They have always presented captive animals as wild—and still do. Nonetheless, the assumption that zoos must be imprisoned by their histories may be just as teleological as the assumption that zoos have been moving toward Progress. Maybe colonial structures can become anti-colonial openings in the same way that overwrought humanistic displays and discourses can spin off in posthuman directions. Anthropologist and posthumanist Matthew Chrulew has written powerfully about the reintroduction programs of golden lion tamarins of the 1980s and 1990s, in which the National Zoo was involved alongside forty other networked institutions. Through careful and convincing analysis, Chrulew advances the argument that when the entire "ensemble of zoo-biological and wildlife-management techniques" (from breeding to feeding to management to surveillance) was directed "toward self-erasure and the production of wildness," zoo reintroduction programs could finally undo what the zoo has done; they could finally "dismantle" the "psychological cage" of the animal and lead "capable and independent, truly *self*-sustaining" tamarins into wild ecosystems. In other words, for golden lion tamarin reintroductions, the zoo "staged its own withdrawal," and only then was the zoo successful.[55]

In the twenty-first century, we need to unravel so many of the institutions around us, to push them to undo what they have done, to stage their withdrawal. Yet undoing is not dissolving, deconstruction is not nihilism. Rather, as queer theorist Jack Halberstam shows us, "unbuilding the world" can lead us to wildness, toward the less bounded worlds that we lost but now crave.[56] There can be "creative destruction" in taking apart, in directing overdetermined and exhausted monocultural logics toward more open terrains of ecological thinking. Indeed, as we unbuild and reimagine, we must—as field philosopher Thòm van Dooren writes in *The Wake of Crows: Living and Dying in Shared Worlds*—"remain *as open as possible* to others' lives, projects, and possibilities."[57] According to feminist science studies scholar Juno Salazar Parreñas, we must "reorient" ourselves toward others, "accept[ing] the risk of living together," if we are to decolonize conservation and truly imagine new ways of living and caring.[58] Why not start unbuilding our world *with* the zoological park? Rothfels states, "Show me a zoo that builds an enclosure for a 'boring' snake . . . that actually meets the natural conditions of that snake in the wild rather than being a simple, small glass box, and I will be surprised."[59]

To continue in this line of thought, show me a zoo that prioritizes amphibians before all else, and I will be surprised.[60] Show me a zoo that devotes all of its space and resources to keeping and caring for three or four species at a time, and I will be surprised. Show me a zoo that does not engage in breeding, cloning, and the forced reproduction of animals just for the continuation of a captive population, and I will be surprised.[61] Show me a zoo that treats its exotic captive animals as "companion animals" and "co-citizens" worthy of rights and personhood status rather than framing them as "wild animals," and I will be surprised.[62] Show me a zoo that encourages visitors to bring hiking boots or binoculars, and I will be surprised. Show me a zoo where plants are equal to animals, and I will be surprised. Show me a zoo where visitors observe patiently, rather than stroll/scroll through animals in rapid pornographic succession, and I will be surprised. Show me a zoo that critiques mega-corporations instead of seeking their sponsorships, and I will be surprised. Show me a zoo that teaches about the long histories of racism with which zoos are entangled, and I will be surprised. Show me a zoo that educates its visitors about the human exhibits of its past, and I will be even more surprised.[63] What recuperative zoos these could be! And what surprising encounters they could offer us as we try to create more ecological futures. Yet, zoos will not be able to remake themselves until we can remake ourselves. Everything and everyone is always in relation.

Notes

Abbreviations

CPLMRC	Chicago Public Library, Municipal Reference Collection
CPLSC	Chicago Public Library Special Collections
CPNC	Chicago Parks Newspaper Clippings
CR	*Congressional Record*
CZA	Cincinnati Zoo Archives
HUA	Harvard University Archives
LPDA	Lincoln Park District Archives
NAA	National Anthropological Archives, Smithsonian
NASMASA	National Air and Space Museum, Air and Space Archives
NUA	Northwestern University Archives
NYZS	New York Zoological Society
NZP	National Zoological Park
PZA	Philadelphia Zoo Archives
RLGP	Richard Lynch Garner Papers
RU	Record Unit—Anywhere RU is listed SIA is implied, since the Smithsonian Institution Archives, in part, organizes collections through the "Record Unit"
SIA	Smithsonian Institution Archives
TUSCRC	Temple University Special Collections Research Center
UCL	University of Chicago Library
WCSA	Wildlife Conservation Society Archives
WCSLA	Wildlife Conservation Society Library and Archives

Introduction: At the Entrance Gate

1. R. W. Shufeldt, "Some Washington Bird Notes," *Wilson Bulletin* 21 (June 1909): 95–96.

2. R. W. Shufeldt, "The Extermination of the Wild Turkey in the State of Virginia," *Auk* 28 (January 1911): 144–146; R. W. Shufeldt, "Contributions to Avian Paleontology," *Auk* 30 (January 1913): 29–39; R. W. Shufeldt, "An Introduction to the Study of the Eggs of North American Limicolae," *Condor* 15 (July 1913): 138–151.

3. Mark V. Barrow Jr., *A Passion for Birds: American Ornithology after Audubon* (Princeton: Princeton University Press, 1998), 63–64; David W. Johnston, *The*

History of Ornithology in Virginia (Charlottesville: University of Virginia Press, 2003), 73.

4. R. W. Shufeldt, "An Unusual Malady and Probable Cause of Death in a Toucan," *Auk* 30 (July 1913): 433–434; R. W. Shufeldt, "On the Comparative Osteology of Cereopsis novae-hollandiae," *Emu* 12 (April 1913): 209–237.

5. R. W. Shufeldt, "The National Zoological Garden," *Popular Science Monthly* 83, no. 5 (November 1913): 434–440.

6. Shufeldt, "National Zoological Garden," 434–440.

7. R. W. Shufeldt, *The Negro: A Menace to American Civilization* (Boston: Gorham Press, 1907).

8. Barrow, *A Passion for Birds*, 63–67; R. W. Shufeldt, "On the Medico-Legal Aspect of Impotency in Women," *Medico-Legal Journal* 14 (December 1896): 289–296.

9. Yuka Suzuki, *The Nature of Whiteness: Race, Animals, and Nation in Zimbabwe* (Seattle: University of Washington Press, 2017); Dorceta E. Taylor, *The Rise of the American Conservation Movement: Power, Privilege, and Environmental Protection* (Durham, NC: Duke University Press, 2016); Miles Powell, *Vanishing America: Species Extinction, Racial Peril, and the Origins of Conservation* (Cambridge, MA: Harvard University Press, 2016); Alexandra Minna Stern, *Eugenic Nation: Faults and Frontiers of Better Breeding in Modern America* (Oakland: University of California Press, 2015). I borrow "preserving the best" from Garland E. Allen, "'Culling the Herd': Eugenics and the Conservation Movement in the United States, 1900–1940," *Journal of the History of Biology* 46, no. 1 (2013): 36. For similar issues regarding Nazi Germany, see Frank Uekoetter, *The Green and the Brown: A History of Conservation in Nazi Germany* (Cambridge: Cambridge University Press, 2006).

10. Bénédicte Boisseron, *Afro-Dog: Blackness and the Animal Question* (New York: Columbia University Press, 2018), 10.

11. Sandra Swart, "African Studies," in *Handbook of Historical Animal Studies*, ed. Mieke Roscher, André Krebber, and Brett Mizelle (Berlin: De Gruyter, 2021), 89.

12. Though "posthuman" has a variety of entangled meanings, here and throughout this book I am referring to the constellation of cultural theory and philosophy that seeks to move beyond the (white, male, upper-class, European) anthropocentric assumptions of Renaissance humanisms that have forced the world into the conceptual prison we call "modernity." See, for example, Roberto Esposito, "Politics and Human Nature," *Angelaki* 16, no. 3 (September 2011): 77–84; Francesca Ferrando, *Philosophical Posthumanism* (London: Bloomsbury, 2019); and Timothy Morton, *Being Ecological* (Cambridge, MA: MIT Press, 2018).

13. Hilda Kean, "Challenges for Historians Writing Animal-Human History: What Is Really Enough?" *Anthrozoös* 25, no. 3 (August 2012): 65.

14. Ewa Domańska, "Posthumanist History," in *Debating New Approaches to History*, ed. Marek Tamm and Peter Burke (London: Bloomsbury, 2019), 337. See also Marek Tamm and Zoltán Boldizsár Simon, "More-Than-Human

History: Philosophy of History at the Time of the Anthropocene," in *Philosophy of History Twenty-First-Century Perspectives*, ed. Jouni-Matti Kuukkanen (London: Bloomsbury, 2021), 198–215; and Emily O'Gorman and Andrea Gaynor, "More-Than-Human Histories," *Environmental History* 25, no. 4 (August 2020): 711–735.

15. Zakiyyah Iman Jackson, *Becoming Human: Matter and Meaning in an Antiblack World* (New York: New York University Press, 2020), 16, 27–33.

16. Achille Mbembe, *On the Postcolony*, trans. Steven Rendell (Berkeley: University of California Press, 2001), 192.

17. Despite measurable and knowable surfaces, each of these "notions" are very much "entangled affairs." See, for example, Karen Barad, *Meeting the Universe Halfway: Quantum Physics and the Entanglement of Matter and Meaning* (Durham, NC: Duke University Press, 2007), 394.

18. Jacques Derrida, *Of Grammatology*, trans. Gayatri Chakravorty Spivak (Baltimore, MD: Johns Hopkins University Press, 1997), 296. All too often, (re)presenting another requires the subject (the one representing, the historian) to construct distance from the one being (re)presented (the object, the historicized subject). Derrida problematizes distances between insides and outsides, signs and signifiers, and thus criticizes representation as always the "loss of presence." By holding "the archive" near the text, I strive to minimize these inevitable losses.

19. Erica Fudge, "A Left-Handed Blow: Writing the History of Animals," in *Representing Animals*, ed. Nigel Rothfels (Bloomington: Indiana University Press, 2002), 5; Zeb Tortorici, "Animal Archive Stories: Species Anxieties in the Mexican National Archive," in *The Historical Animal*, ed. Susan Nance (Syracuse: Syracuse University Press, 2015), 96.

20. Donna J. Haraway, *Primate Visions: Gender, Race, and Nature in the World of Modern Science* (New York: Routledge, 1989), 35.

21. Donna J. Haraway, *Staying with the Trouble: Making Kin in the Chthulucene* (Durham, NC: Duke University Press, 2016), 14.

22. From Nietzsche, Beauvoir, and White to Said, Spivak, and bell hooks, a multitude of theorists and philosophers of history across linguistic, feminist, postcolonial, and critical race turns have thoroughly established the violences and ignorances enacted by timelines, forward-moving directionalities, and master narratives of progress or declension. While historiography has been influenced by these voices since the historical profession's cultural turn, the timeline (its cause-effect topographies, its contextualizations, its single temporality) still largely governs historical analysis. Nonetheless, in the twenty-first century, a diverse and growing body of leading interdisciplinary theorists (Donna J. Haraway, Eduardo Kohn, Bruno Latour, Anna Tsing, Isabelle Stengers, Eduardo Viveiros de Castro, Jane Bennett, to name a few) are pushing us all to think time and relations differently. This is a task we must strive for as we turn to face the global problems that simple narratives of "modernity," pressed upon a Cold War world, have created for all of us.

23. Steven Shaviro, *The Universe of Things: On Speculative Realism* (Minneapolis: University of Minnesota Press, 2014).

24. Sven Steinmo, "Historical Institutionalism," in *Approaches and Methodologies in the Social Sciences: A Pluralist Perspective*, ed. Donatella Della Porta and Michael Keating (Cambridge: Cambridge University Press, 2008), 166.

25. Haraway, *Staying with the Trouble*, 30–57.

26. Deborah Bird Rose, Thom van Dooren, and Matthew Chrulew, eds., *Extinction Studies: Stories of Time, Death, and Generations* (New York: Columbia University Press, 2017), 3.

27. Caroline Randall Williams, "1619 and Beyond: Explorations in Atlantic Slavery and Its American Legacy," Lecture at Center for Historical Research, The Ohio State University, November 5, 2021.

28. Hannah Arendt, *Men in Dark Times* (San Diego: Harcourt Brace, 1968), 105. For more on the importance of storytelling, see Hanna Meretoja, *The Ethics of Storytelling: Narrative Hermeneutics, History, and the Possible* (New York: Oxford University Press, 2018). In animal studies, see Susan McHugh, *Animal Stories: Narrating across Species Lines* (Minneapolis: University of Minnesota Press, 2011).

29. Randy Malamud, *Reading Zoos: Representations of Animals and Captivity* (New York: New York University Press, 1998), 5.

30. Harriet Ritvo, *The Animal Estate: The English and Other Creatures in the Victorian Age* (Cambridge: Harvard University Press, 1987); Bob Mullan and Garry Marvin, *Zoo Culture: The Book about Watching People Watch Animals*, 2nd edition (Urbana: University of Illinois Press, 1998). For other zoo history classics, see Eric Baratay and Elisabeth Hardouin-Fugier, *Zoo: A History of Zoological Gardens in the West* (London: Reaktion Books, 2002); Nigel Rothfels, *Savages and Beasts: The Birth of the Modern Zoo* (Baltimore, MD: Johns Hopkins University Press, 2002); and Elizabeth Hanson, *Animal Attractions: Nature on Display in American Zoos* (Princeton, NJ: Princeton University Press, 2002). For a recent review of zoo historiography and interdisciplinary zoo studies, see Tracy McDonald and Daniel Vandersommers, *Zoo Studies: A New Humanities* (Montreal: McGill-Queens University Press, 2019). See also Lisa Uddin, *Zoo Renewal: White Flight and the Animal Ghetto* (Minneapolis: University of Minnesota Press, 2015), 1–22.

31. Gilles Deleuze and Felix Guattari, *A Thousand Plateaus: Capitalism and Schizophrenia*, trans. Brian Massumi (Minneapolis: University of Minnesota Press, 1987, 2003), 202–206.

32. Andrew Flack, *The Wild Within: Histories of a Landmark British Zoo* (Charlottesville: University of Virginia Press, 2018), 14.

33. Gregg Mitman, *Reel Nature: America's Romance with Wildlife on Film* (Cambridge, MA: Harvard University Press, 1999), 13.

34. Mel Y. Chen, *Animacies: Biopolitics, Racial Mattering, and Queer Affect* (Durham, NC: Duke University Press, 2012), 98. As animal historian Boria Sax suggests, "Every animal is a tradition, and together they are a vast part of our heritage as human beings." I believe the zoo may help us measure the weight of

heritage. Boria Sax, *The Mythical Zoo: An Encyclopedia of Animals in World Myth, Legend, and Literature* (Santa Barbara, CA: ABC-CLIO, 2001), xx.

35. Yellowstone wildlife biologist Doug Smith asks this simple question, "Wolves at the pleasure of humans—is that how we should live?" Smith, "Should Humans Rule the World? Using Wolves to Connect with Nature," TED Talk at Big Sky, MO, January 26, 2019, https://www.youtube.com/watch?v=kCu34aAyBFc.

36. This book will help us see/feel how this universal, phenomenologic dynamic of power shapes lived, historical experiences within the zoo. Envisioning power as cleaved, incomplete, and open is theorized by Slavoj Žižek, who reworks Kantian philosophy through Hegel, Marx, and Lacan (each of whom also theorizes the unseen gaps produced by power's remainder). In *The Sublime Object of Ideology* (London, Verso Books, 1989), for example, Žižek writes, "But what is hidden behind the phenomenal appearance? Precisely the fact that there is nothing to hide. What is concealed is that the very act of concealing conceals nothing" (219). Nothing feels more Real in the zoo.

37. Bruno Latour, *Reassembling the Social: An Introduction to Actor-Network Theory* (Oxford: Oxford University Press, 2005), 79.

38. On authenticity, see Jennifer Price, *Flight Maps: Adventures with Nature in Modern America* (New York: Basic Books, 1999), 175–178, 181–182. As Bruno Latour acknowledges, "a deep source of *inauthenticity* is revealed every time we engage more thoroughly in the Anthropocene." Bruno Latour and Dipesh Chakrabarty, "Conflicts of Planetary Proportions—A Conversation," *Journal of the Philosophy of History* 14, no. 3 (November 2020): 3–4, 8.

39. I assume the following: that Subjects are always split (layered by different degrees and types of consciousness), that ideology/belief is always a "false consciousness," that the structuring principle of a system is always necessarily empty (and thus Real), that examining inwardness is of immense historical importance to cultural and intellectual history, and that "signifiers without a signified" (like "the animal") function as beacons toward the Real (structure) of an idea, culture, or system. I am very much indebted here to the work of Žižek and Timothy Morton, especially *Ecology Without Nature: Rethinking Environmental Aesthetics* (Cambridge, MA: Harvard University Press, 2007).

40. This notion of "the animal" is common in academic conversations in animal studies, human-animal studies, critical animal studies, and posthumanist philosophy. With its strange singularity, "the animal" refers to the larger category "Animal," its prerequisites, and its importance to discourse and culture. The concept invites you to pause and ask: What makes any individual animal—a bumblebee, tiger shark, condor, black rat, brontosaurus, border collie—a member of the category of "animal"? "The animal" refers to this constructed and vast category, and it operates similarly to notions of "the human." The term also encourages us to see double, so to speak. Anytime you encounter an animal, you encounter both "the animal" and that individual animal, as well as (dialectically) "the human" and yourself.

41. There is a robust theoretical literature on the "question of the animal," engaged especially with Continental philosophy. See, for example, Cary Wolfe, *Zoontologies: The Question of the Animal* (Minneapolis: University of Minnesota Press, 2003); and Matthew Calarco, *Zoographies: The Question of the Animal from Heidegger to Derrida* (New York: Columbia University Press, 2008).

42. Ralph Acampora, "Zoos and Eyes: Contesting Captivity and Seeking Successor Practices," *Society and Animals* 13, no. 1 (2005): 77, 69.

43. Cultural animal historian Kathleen Kete recognizes that the "image of sensibility," which so often absorbs or irons out the experience of dissonance, "deflects us from awareness of the violence between ourselves and others." Kathleen Kete, "Introduction: Animals and Human Empire," in *Cultural History of Animals in the Age of Empire*, ed. Kathleen Kete (New York: Berg, 2007), 15.

44. Antoine Traisnel, *Capture: American Pursuits and the Making of a New Animal Condition* (Minneapolis: University of Minnesota Press, 2020), 4.

45. Lynn Nyhart, *Modern Nature: The Rise of the Biological Perspective in Germany* (Chicago: University of Chicago Press, 2009), 20–21; William Coleman, *Biology in the Nineteenth Century: Problems of Form, Function, and Transformation* (New York: John Wiley & Sons, 1971), 1–3.

46. See Philip J. Pauly, *Biologists and the Promise of American Life* (Princeton, NJ: Princeton University Press, 2000).

47. Joel Dorman Steele and J. W. P. Jenks, *Popular Zoology* (New York: Chautauqua Press, 1887).

48. For an overview of studies concerning "popular science," see Andreas Daum, "Varieties of Popular Science and the Transformations of Public Knowledge: Some Historical Reflections," *Isis* 100, no. 2 (June 2009): 319–332; Jon Topham, "Rethinking the History of Science Popularization/Popular Science," in *Popularizing Science and Technology in the European Periphery, 1800–2000*, ed. Faidra Papanelopoulou, Augustí Nieto-Galan, and Enrique Perdiguero (Burlington, VT: Ashgate, 2009), 1–20; and Bernadette Bensaude-Vincent, "A Genealogy of the Increasing Gap between Science and Public," *Public Understanding of Science* 10, no. 1 (January 2001): 99–113. For a seminal work, see Bernard Lightman, *Victorian Popularizers of Science: Designing Nature for New Audiences* (Chicago: University of Chicago Press, 2007); and Aileen Fyfe and Bernard Lightman, eds., *Science in the Marketplace: Nineteenth-Century Sites and Experiences* (Chicago: University of Chicago Press, 2007).

49. Peter J. Bowler and Iwan Rhys Morus, *Making Modern Science: A Historical Survey* (Chicago: University of Chicago Press, 2005), 226. See also, for example, Roger Cooter, *The Cultural Meaning of Popular Science: Phrenology and the Organization of Consent in Nineteenth-Century Britain* (Cambridge: Cambridge University Press, 1984); Alison Winter, *Mesmerized: Powers of Mind in Victorian Britain* (Chicago: University of Chicago Press, 1998); and Aaron Sachs, *The Humboldt Current: Nineteenth-Century Exploration and the Roots of American Environmentalism* (New York: Viking, 2006).

50. See Thomas F. Gieryn, *Truth Spots: How Places Make People Believe*

(Chicago: University of Chicago Press, 2018). For classic works dispelling diffusionist models of science, see Roger Cooter and Stephen Pumfrey, "Separate Spheres and Public Places: Reflections on the History of Science Popularization and Science in Popular Culture," *History of Science* 32 (September 1994): 237–267; Anne Secord, "Science in the Pub: Artisan Botanists in Early Nineteenth-Century Lancashire," *History of Science* 32 (September 1994): 269–315; Stephen Hilgartner, "The Dominant View of Science Popularization: Conceptual Problems, Political Uses," *Social Studies of Science* 20, no. 3 (August 1990): 519–539; and Greg Myers, "Discourse Studies of Scientific Popularization: Questioning the Boundaries," *Discourse Studies* 5, no. 2 (May 2003): 265–279.

51. See James Secord, "Knowledge in Transit," *Isis* 95, no. 4 (December 2004): 664. See also his *Victorian Sensation: The Extraordinary Publication, Reception, and Secret Authorship of Vestiges of the Natural History of Creation* (Chicago: University of Chicago Press, 2000).

52. Daum, "Varieties of Popular Science and the Transformations of Public Knowledge," 320.

53. Clifford Geertz, "Thick Description: Towards an Interpretive Theory of Culture," in *The Interpretations of Cultures* (New York: Basic Books, 1973), 3–30. Regarding animals, see "Deep Play: Notes of the Balinese Cockfight," in *The Interpretations of Cultures*, 412–453.

54. I concur entirely with animal historian Violette Pouillard's observation that studies of historical animals share "close ties with history from below, which draws upon life experiences to highlight the history of subalterns—the poor, the working classes, the slaves—while avoiding both their victimization (the perception of them as passive beings) and, following the well-known Thompsonian formula, 'the enormous condescension of posterity.'" See her "Gust (ca 1952–1988), or A History from Below of the Changing Zoo," in *Zoo Studies*, 170.

55. Anna Tsing, *The Mushroom at the End of the World* (Princeton, NJ: Princeton University Press, 2015), 37.

56. Robert E. Kohler, *Inside Science: Stories from the Field in Human and Animal Science* (Chicago: University of Chicago Press, 2019), 13.

57. Donna Varga, "Babes in the Woods: Wilderness Aesthetics in Children's Stories and Toys, 1830–1915," *Society and Animals* 17 (January 2009): 188, 194, 197–198. See also "Winnie: Troubling the Idealization of the Bear as Childhood Innocent," in *Positioning Pooh: Edward Bear after 100 Years*, ed. J. Harrison (Jackson: University Press of Mississippi, 2021).

58. Susan Nance, *Animal Modernity: Jumbo the Elephant and the Human Dilemma* (London: Palgrave Macmillan, 2015).

59. For an introduction into animal history see the following edited volumes: Hilda Kean and Philip Howell, *The Routledge Companion to Animal-Human History* (New York: Routledge, 2018); Susan Nance, *The Historical Animal* (Syracuse: Syracuse University Press, 2015); Martha Few and Zeb Tortorici, *Centering Animals in Latin American History* (Durham, NC: Duke University Press, 2013); Dorothee Brantz, *Beastly Natures: Animals, Humans, and the Study of History*

(Charlottesville: University of Virginia Press, 2010); André Krebber and Mieke Roscher, *Animal Biography: Re-framing Animal Lives* (Cham, Switzerland: Palgrave Macmillan, 2018); Chiara Mengozzi, *Outside the Anthropological Machine: Crossing the Human-Animal Divide and Other Exit Strategies* (New York: Routledge, 2020); and Brett Mizelle, Mieke Roscher, and André Krebber, *Handbook of Historical Animal Studies* (Berlin: De Gruyter, 2021).

60. Daniel Gamito-Marques, "A Space of One's Own: Barbosa du Bocage, the Foundation of the National Museum of Lisbon, and the Construction of a Career in Zoology (1851–1907)," *Journal of the History of Biology* 51, no. 2 (June 2018): 224. This description of science describes "circulation," an important theme in science studies. See, for example, Kapil Raj, "Beyond Postcolonialism . . . and Postpositivism: Circulation and the Global History of Science," *Isis* 104, no. 2 (June 2013): 337–347.

61. The entirety of *Entangled Encounters* explores the importance of nonhuman agencies; however, for a brief overview, see Jamie Lorimer's succinct discussion in *Wildlife in the Anthropocene: Conservation after Nature* (Minneapolis: University of Minnesota Press, 2015), 25–28.

62. Susan Nance, "Animal History: The Final Frontier?" *American Historian* (November 2015): 29–34.

63. Drew Swanson, "Mountain Meeting Ground: History at an Intersection of Species," in *The Historical Animal*, ed. Susan Nance (Syracuse: Syracuse University Press, 2015), 240.

64. Michael McCormick, "Rats, Communications, and Plague: Toward an Ecological History," *Journal of Interdisciplinary History* 34, no. 1 (Summer 2003): 25.

65. This note, and the next six, will offer a few example titles (classics and new works) for each expanding genre of animal histories. Others will be cited throughout this book. For animal-environmental history, see, for example, Virginia DeJohn Anderson, *Creatures of Empire: How Domestic Animals Transformed Early America* (New York: Oxford University Press, 2004); Brett Walker, *The Lost Wolves of Japan* (Seattle: University of Washington Press, 2005); and Jason M. Colby, *Orca: How We Came to Know and Love the Ocean's Greatest Predator* (New York: Oxford University Press, 2018).

66. For animal-cultural histories, see, for example, Jon Coleman, *Vicious: Wolves and Men in America* (New Haven, CT: Yale University Press, 2004); Abel Alves, *The Animals of Spain: An Introduction to Imperial Perceptions and Human Interaction with Other Animals, 1492–1826* (Leiden: Brill, 2011); Karen Raber, *Animal Bodies, Renaissance Culture* (Philadelphia: University of Pennsylvania Press, 2013); Erica Fudge, *Perceiving Animals: Humans and Beasts in Early Modern English Culture* (Chicago: University of Illinois Press, 2002); and Sarah Cockram and Andrew Wells, eds., *Interspecies Interactions: Animals and Humans from the Middle Ages to Modernity* (London: Routledge, 2018). See, especially, the wonderfully short books of the Reaktion Books "Animal" series.

67. For animal-economic/commodity histories, see, for example, Alan

Mikhail, *The Animal in Ottoman Egypt* (New York: Oxford University Press, 2014); and John F. Richards, *The World Hunt: An Environmental History of the Commodification of Animals* (Berkeley: University of California Press, 2014).

68. For animal-urban histories, see, for example, Clemens Wischermann, Aline Steinbrecher, and Philip Howell, eds., *Animal History in the Modern City: Exploring Liminality* (London: Bloomsbury, 2018); Frederick L. Brown, *The City Is More Than Human: An Animal History of Seattle* (Seattle: University of Washington Press, 2016); Andrew Robichaud, *Animal City: The Domestication of America* (Cambridge, MA: Harvard University Press, 2019); Clay McShane and Joel Tarr, *The Horse in the City: Living Machines in the Nineteenth Century* (Baltimore: Johns Hopkins University Press, 2007); Catherine McNeur, *Taming Manhattan: Environmental Battles in the Antebellum City* (Cambridge, MA: Harvard University Press, 2014); and Dawn Day Biehler, *Pests in the City: Flies, Bedbugs, Cockroaches, and Rats* (Seattle: University of Washington Press, 2013).

69. For intersectional animal-historical works, see, for example, Carol Adams, *The Sexual Politics of Meat: A Feminist-Vegetarian Critical Theory* (New York: Continuum, 2010); Claire Jean Kim, *Dangerous Crossings: Race, Species, and Nature in a Multicultural Age* (New York: Cambridge University Press, 2015); Neel Ahuja, *Bioinsecurities: Disease Interventions, Empire, and the Government of Species* (Durham, NC: Duke University Press, 2016); Kalpana Rahita Seshadri, *HumAnimal: Race, Law, Language* (Minneapolis: University of Minnesota Press, 2012); Michael Lundblad, *The Birth of a Jungle: Animality in Progressive-Era U.S. Literature and Culture* (New York: Oxford University Press, 2013); and Radhika Govindrajan, *Animal Intimacies: Interspecies Relatedness in India's Central Himalayas* (Chicago: University of Chicago Press, 2018).

70. For animal-STM histories, see, for example, Etienne Benson, *Wired Wilderness: Technologies of Tracking and the Making of Modern Wildlife* (Baltimore, MD: Johns Hopkins University Press, 2010); Harriet Ritvo, *The Platypus and the Mermaid, and Other Figments of the Classifying Imagination* (Cambridge, MA: Harvard University Press, 1997); Karen Rader, *Making Mice: Standardizing Animals for American Biomedical Research, 1900–1955* (Princeton, NJ: Princeton University Press); and Robert Kohler, *Lords of the Fly: Drosophila Genetics and the Experimental Life* (Chicago: University of Chicago Press, 1994).

71. For animal histories on these macroscales, see, for example, Richard Bulliet, *Hunters, Herders, and Hamburgers: The Past and Future of Human-Animal Relationships* (New York: Columbia University Press, 2005); Brian Fagan, *The Intimate Bond: How Animals Shaped Human History* (New York: Bloomsbury, 2015); and Abraham Gibson, *Feral Animals in the American South: An Evolutionary History* (New York: Cambridge University Press, 2016).

72. On the concept "Anthropocene," see Anna L. Tsing, Jennifer Deger, Alder Keleman Saxena, and Feifei Zhou, "What Is the Anthropocene?" in *Feral Atlas: The More-Than-Human Anthropocene*, ed. Tsing, Deger, Saxena, Zhou (Published online: Stanford University Press, 2021), https://feralatlas.supdigital.org/?cd=true&bdtext=feral-atlas-and-the-more-than-human-anthropocene.

73. Haraway, *Staying with the Trouble*, 58–98. Haraway's use of "sympoiesis" is inspired by the work of feminist biologist Lynn Margulis on symbiosis and holobionts, which are biological assemblages, knots, systems, composed of many different organisms. Nearly all the "individuals" and self-contained systems that we traditionally use as common denominators in biology and political thought are, in fact, holobionts "all the way down." See, for example, Lynn Margulis, *Symbiotic Planet: A New Look at Evolution* (New York: Basic Books, 1998).

74. Etienne Benson, "Animal Writes: Historiography, Disciplinarity, and the Animal Trace," in *Making Animal Meaning*, ed. Linda Kalof and Georgina M. Montgomery (East Lansing: Michigan State University Press, 2011), 3.

75. Libby Robin and Will Steffen, "History for the Anthropocene," *History Compass* 5, no. 5 (July 2007): 1711–1712.

76. Baker to Hornaday, 1 August 1891, SIA, RU 74, Box 7, Folder 4. From here, onward, anywhere "RU" appears, SIA is implied.

77. "Entanglement" is a notion important to quantum and particle physics (first developed by Albert Einstein, Boris Podolsky, Nathan Rosen, and Erwin Schrödinger) to describe a state where two photons, electrons, or larger molecules can become conjoined in a larger system, despite physical distance, that collapses, or dissipates, upon measurement. Across the intersectional humanities, "entanglement" has become a productive catchphrase to describe the seemingly unlikely, paradoxical, repressed, and unforeseen interconnected relations among human-cultural histories, technological systems, and environments. I believe fully that quantum metaphors may be far more productive for us to use than the "laws" and "truths" that we, at least in the West, have grafted onto our everyday lives. As an introduction to the philosophical and physical properties of QE to contemporary physics, see F. J. Duarte, *Fundamentals of Quantum Entanglement* (Bristol: Institute of Physics Publishing, 2019).

78. Anita Guerrini, *The Courtiers' Anatomists: Animals and Humans in Louis XIV's Paris* (Chicago: University of Chicago Press, 2015).

Chapter 1. The Origins of a National Zoo

1. See Stefan Bechtel, *Mr. Hornaday's War: How A Peculiar Victorian Zookeeper Waged a Lonely Crusade for Wildlife That Changed the World* (New York: Beacon Press, 2012); and Gregory Dehler, *The Most Defiant Devil: William Temple Hornaday and His Controversial Crusade to Save American Wildlife* (Charlottesville: University of Virginia Press, 2013).

2. See Pamela Newkirk, *Spectacle: The Astonishing Life of Ota Benga* (New York: Harper Collins, 2015).

3. For more on Hornaday and the Smithsonian, zoos, and bison, see Mark V. Barrow Jr., *Nature's Ghosts: Confronting Extinction from the Age of Jefferson to the Age of Ecology* (Chicago: University of Chicago Press, 2009), 108–123; and Diane Smith, *Yellowstone and the Smithsonian: Centers of Wildlife Conservation* (Lawrence: University Press of Kansas, 2017), 59–93. For more on Ward and his Natural Science Establishment, see Mark V. Barrow Jr., "The Specimen Dealer:

Entrepreneurial Natural History in America's Gilded Age," *Journal of the History of Biology* 33, no. 3 (Winter 2000): 493–534.

4. Smith, *Yellowstone and the Smithsonian*, 14–25; Helen L. Horowitz, "The National Zoological Park: 'City of Refuge' or Zoo?" *Records of the Columbia Historical Society, Washington, D.C.* 49 (1973/1974): 408–409; Sally Gregory Kohlstedt, "History in a Natural History Museum: George Brown Goode and the Smithsonian Institution," *Public Historian* 10, no. 2 (Spring 1988): 11. For more on nineteenth-century Smithsonian history, see Nina Burleigh, *The Stranger and the Statesman: James Smithson, John Quincy Adams, and the Making of America's Greatest Museum: The Smithsonian* (New York: Harper Collins, 2003); Heather Ewing, *The Lost World of James Smithson: Science, Revolution, and the Birth of the Smithsonian* (New York: Bloomsbury, 2007); and Daniel Goldstein, "'Yours for Science': The Smithsonian Institution's Correspondents and the Shape of Scientific Community in Nineteenth-Century America," *Isis* 85, no. 4 (December 1994): 573–599.

5. Historian Elizabeth Hanson has described this trip as the reason for Hornaday's "conversion" from hunter to conservationist. See Hanson, *Animal Attractions: Nature on Display in American Zoos* (Princeton, NJ: Princeton University Press, 2002), 26; Smith, *Yellowstone and the Smithsonian*, 69–70. Historian James Dolph believed, however, this conversion occurred while writing *Extermination*. See Dolph, "Bringing Wildlife to the Millions: William Temple Hornaday, The Early Years, 1854–1896," PhD diss., University of Massachusetts, 1975, 503.

6. Hornaday to Baird, 21 December 1886, RU 305, Accession 18617, acquired at "Smithsonian Scrapbook: Letters, Diaries & Photographs from the Smithsonian Archives," accessed February 14, 2014: http://siarchives.si.edu/oldsite/history/exhibits/documents/21Dec1886Letter.htm.

7. Hanson, *Animal Attractions*, 134.

8. Nigel Rothfels, "The Antelope Collectors," in *Zoo Studies: A New Humanities*, ed. Tracy McDonald and Daniel Vandersommers (Montreal: McGill-Queens University Press, 2019), 61; Nigel Rothfels, "Catching Animals," in *Animals in Human Histories: The Mirror of Nature and Culture*, ed. Mary J. Henninger-Voss (Rochester, NY: University of Rochester Press, 2002), 182–228.

9. Susan Nance, *Animal Modernity: Jumbo the Elephant and the Human Dilemma* (London: Palgrave Macmillan, 2015), 63; Barrow, *Nature's Ghosts*, 108–110.

10. Horowitz, "National Zoological Park," 409–412; William T. Hornaday, "The Extermination of the American Bison with a Sketch of Its Discovery and Life History," in *Report of the United States National Museum under the Direction of the Smithsonian Institution*, 1887 (Washington, DC: Government Printing Office, 1889): 367–548. For more, see Andrew C. Isenberg, *The Destruction of the Bison: An Environmental History, 1750–1920* (New York: Cambridge University Press, 2000). For a deep history of human-bison relations, see Geoff Cunfer and Bill Waiser, eds., *Bison and People on the North American Great Plains: A Deep Environmental History* (College Station: Texas A&M University Press, 2016).

11. Hornaday to Goode, 2 December 1887, RU 201, Box 17, Folder 10.

12. Hornaday, "Extermination of the American Bison with a Sketch of Its Discovery and Life History," 528. See James A. Dolph and C. Ivar Dolph, "The American Bison: His Annihilation and Preservation," *Montana: The Magazine of Western History* 25, no. 3 (Summer 1975): 14–25.

13. Jay Kirk, *Kingdom Under Glass: A Tale of Obsession, Adventure, and One Man's Quest to Preserve the World's Greatest Animals* (New York: Picador, 2010), 150, 11–78.

14. For an excellent discussion of taxidermy, see Melissa Milgrom, *Still Life: Adventures in Taxidermy* (New York: Houghton Mifflin Harcourt, 2010). Also, while taxidermy was a field dominated by men, for a compelling argument that Hornaday (as well as American taxidermy in general) was influenced by the masterful work of Colorado taxidermist Martha Maxwell, see Smith, *Yellowstone and the Smithsonian*, 42–43, 49–55. On female taxidermists in ornithology, see Mark V. Barrow Jr., *A Passion for Birds: American Ornithology After Audubon* (Princeton, NJ: Princeton University Press, 1998), 43–45. For an animal studies perspective, see Giovanni Aloi, *Speculative Taxidermy: Natural History, Animal Surfaces, and Art in the Anthropocene* (New York: Columbia University Press, 2018). Finally, see Mary Anne Andrei, *Nature's Mirror: How Taxidermists Shaped America's Natural History Museums and Saved Endangered Species* (Chicago: University of Chicago Press, 2020).

15. W. T. Hornaday, "Report of the Department of Living Animals," 1–2, RU 158, Box 5, folder labeled "Department of Living Animals, 1887–1888."

16. Hornaday, "Report of the Department of Living Animals," 2–3; Smith, *Yellowstone and the Smithsonian*, 73.

17. Hornaday, "Report of the Department of Living Animals," 4–5.

18. Hornaday, "Report of the Department of Living Animals," 5–7.

19. For more on Hagenbeck, see Gary Bruce, *Through the Lion Gate: A History of the Berlin Zoo* (New York: Oxford University Press, 2017); Eric Ames, *Carl Hagenbeck's Empire of Entertainments* (Seattle: University of Washington Press, 2009); and Nigel Rothfels, *Savages and Beasts: The Birth of the Modern Zoo* (Baltimore, MD: Johns Hopkins University Press, 2002).

20. Hanson, *Animal Attractions*, 42.

21. Hornaday, "Report of the Department of Living Animals," 8–9.

22. David Harvey, *Paris, Capital of Modernity* (London: Taylor & Francis, 2003).

23. Louise E. Robbins, *Elephant Slaves and Pampered Parrots: Exotic Animals in Eighteenth-Century Paris: Animals, History, Culture* (Baltimore, MD: Johns Hopkins University Press, 2002), 37–67. See also Caroline Grigson, *Menagerie: The History of Exotic Animals in England, 1100–1837* (Oxford: Oxford University Press, 2016); and Helen Cowie, *Exhibiting Animals in Nineteenth-Century Britain: Empathy, Education, Entertainment* (London: Palgrave Macmillan, 2014).

24. Hornaday, "Report of the Department of Living Animals," 10.

25. Hornaday, "Report of the Department of Living Animals," 11.

26. Smith, *Yellowstone and the Smithsonian*, 76.

27. Hornaday, "Report of the Department of Living Animals," 11–12.

28. Hornaday, "Report of the Department of Living Animals," 12–14.

29. Hornaday, "Report of the Department of Living Animals," 14–15.

30. "Mr. Hornaday's New Pet" (*Washington Critic*, November 21, 1888), "Scrap book, vol. 2, 1888–1891," RU 74, Box 286, Folder 1.

31. Hornaday, "Report of the Department of Living Animals," 15–19; Branka Arsić, *Bird Relics: Grief and Vitalism in Thoreau* (Cambridge, MA: Harvard University Press, 2016), 164.

32. See John F. Reiger, *American Sportsmen and the Origins of Conservation*, 3rd ed. (Corvallis: Oregon State University Press, 2000).

33. Hornaday, "Report of the Department of Living Animals," 20–21.

34. Thomas Veltre, "Menageries, Metaphors and Meanings," in *New Worlds, New Animals: From Menagerie to Zoological Park in the Nineteenth Century*, ed. R. J. Hoage and William A. Deiss (Baltimore, MD: Johns Hopkins University Press, 1996), 19–29.

35. Hornaday, "Report of the Department of Living Animals," 16.

36. Hornaday, "Report of the Department of Living Animals," 20.

37. S. P. Langley, "Report, Secretary of the Smithsonian Institution, For Year Ending June 30, 1889," "The Department of Living Animals," RU 74, Box 285, "Scrapbook, 1887–1902."

38. "An Embryo Zoo" (*Evening Star*, June 2, 1888), "Scrap book, vol. 2, 1888–1891." Over the course of the nineteenth century, especially after the Civil War, naturalists, scientists, animal rights activists, hunters, anglers, and Romantics drew attention to the plight of American wildlife. See Mark Barrow, *Nature's Ghosts* and *A Passion for Birds*; Kimberly K. Smith, *The Conservation Constitution: The Conservation Movement and Constitutional Change, 1870–1930* (Lawrence: University Press of Kansas, 2019); Lisa Mighetto, *Wild Animals and American Environmental Ethics* (Tucson: University of Arizona Press, 1991); and Richard W. Judd, *Common Lands, Common People: The Origins of Conservation in New England* (Cambridge, MA: Harvard University Press, 1997).

39. Martin Qvarnström, Ahlberg, Per E., Niedźwiedzki, Grzegorz, "Tyrannosaurid-like Osteophagy by a Triassic Archosaur," *Scientific Reports* 9, no. 925 (January 2019).

40. "Embryo Zoo."

41. "Novelties at the 'Zoo'" (*Washington Star*, June 6, 1888), "Scrap book, vol. 2, 1888–1891." On "zoo stories" in literature, see Randy Malamud, *Reading Zoos: Representations of Animals and Captivity* (New York: New York University Press, 1998).

42. "The National Zoo" (*Evening Lamp*, July 14, 1888), "Scrap book, vol. 2, 1888–1891." On Jefferson and animals, see Lee Alan Dugatkin, *Mr. Jefferson and the Giant Moose: Natural History in Early America* (Chicago: University of Chicago Press, 2009). On animal diplomacy, see, for example, Erik Ringmar, "Audience for a Giraffe: European Expansionism and the Quest for the Exotic," *Journal of World History* 17, no. 4 (December 2006): 375–397.

43. John H. Eicher and David J. Eicher, *Civil War High Commands* (Stanford, CA: Stanford University Press, 2001), 239–240.

44. Forney, September 12, 1888, 50th Cong., 1st sess., *Congressional Record* 19, pt. 9: 8545; Horowitz, "National Zoological Park," 412–415.

45. Horowitz, "National Zoological Park," 413.

46. S. 2752, April 23, 1888, 50th Cong., 1st sess., *ProQuest Congressional*; Langley, "Zoological Park in the District of Columbia," Report to accompany H.R. 11810, 2 February 1889, RU 74, Box 285, "Scrapbook, 1887–1902."

47. Bland, September 12, 1888, 50th Cong., 1st sess., *CR* 19, pt.9: 8547.

48. McMillin, February 27, 1889, 50th Cong., 2nd sess., *CR* 20, pt.3: 2422; Schlup, Leonard, "Benton McMillin," Tennessee Encyclopedia, https://tennesseeencyclopedia.net/entries/benton-mcmillin/.

49. Stockdale, August 13, 1888, 50th Cong., 1st sess., *CR* 19, pt.8: 7498; February 5, 1891, 51st Cong., 2nd sess., *CR* 22, pt.3: 2211–2212.

50. Blount, April 9, 1890, 51st Cong., 1st sess., *CR* 21, pt.4: 3208.

51. Morse, 3210, Clements, 3213.

52. Dockery, 3214–3215.

53. For more on the intertwined discourses of race and animal, see Claire Jean Kim, *Dangerous Crossings: Race, Species, and Nature in a Multicultural Age* (New York: Cambridge University Press, 2015); Erica Tom, "Humanizing Animals: Talking about Second Chances, Horses, and Prisoners," in *Racial Ecologies*, ed. Leilani Nishime and Kim D. Hester Williams (Seattle: University of Washington Press, 2018), 123–138; Sujit Sivasundaram, "Imperial Transgressions: The Animal and Human in the Idea of Race," *Comparative Studies of South Asia, Africa and the Middle East* 35, no. 1 (May 2015): 156–172; and Marianna Szczygielska, "Pandas and the Reproduction of Race and Heterosexuality in the Zoo," in *Zoo Studies*, 211–236.

54. McCreary, April 18, 1890, 51st Cong., 1st sess., *CR* 21, pt.4: 3489.

55. February 28, 1889, 50th Cong., 2nd sess., *CR* 20, pt.3: 2456.

56. "Fooled the Watch-Dog," June 1, 1890; "The Park Zoo," October 30, 1890, *Dallas Morning News*.

57. Gregory A. Bryant and Athena C. Aktipis, "The Animal Nature of Spontaneous Human Laughter," *Evolution and Human Behavior* 35, no. 4 (July 2014): 327–335.

58. Tom Tyler, *CIFERAE: A Bestiary in Five Fingers* (Minneapolis: University of Minnesota Press, 2012), 29.

59. Matthew Senior, "The Souls of Men and Beasts, 1637–1764," in *A Cultural History of Animals in the Age of Enlightenment*, ed. Matthew Senior (Oxford: Berg Publishers, 2007), 23–46.

60. Drew Gilpin Faust, *This Republic of Suffering: Death and the American Civil War* (New York: Vintage Books, 2008), xiv.

61. David W. Blight, *Race and Reunion: The Civil War in American Memory* (Cambridge, MA: Harvard University Press, 2001), 1, 5.

62. Caroline E. Janney, *Remembering the Civil War: Reunion and the Limits of Reconciliation* (Chapel Hill: University of North Carolina Press, 2013), 5.

63. Goode to Purcell, 13 September 1890; Baker to Adair, 15 September 1890; Baker to Langley, 12 January 1891; Baker to Rerdell, 2 October 1891; Baker to Palmer, 18 May 1891; Baker to Tracy, 21 July 1891, RU 74, Box 7, Folders 1, 2, 4.

64. "Zoological Collections of U.S.," Compiled by Baker in 1903, RU 74, Box 100, Folder 5; "Statistics of Zoological Collections," RU 74, Box 100, Folder 6.

65. Baker to Parker, 15 July 1901, "List of Zoological Gardens," RU 74, Box 100, Folder 1.

66. Flack, *The Wild Within*, 14.

67. "The Zoological Park Proposition, and the Voice of the Press in Regard to It," RU 74, Box 285, "Scrapbook, 1887–1902."

68. On nostalgia, see Dolly Jørgensen, *Recovering Lost Species in the Modern Age: Histories of Longing and Belonging* (Cambridge, MA: MIT Press, 2019), 6–7, 51–52.

69. "Zoological Park Proposition, and the Voice of the Press in Regard to It."

70. Judith Hamera, *Parlor Ponds: The Cultural Work of the American Home Aquarium, 1850–1970* (Ann Arbor: University of Michigan Press, 2012), 3.

71. For more on the public park movement and zoos, see Hanson, *Animal Attractions*, 16–28. For more on "middle landscape," see Leo Marx, *The Machine in the Garden: Technology and the Pastoral Ideal in America* (New York: Oxford University Press, 1964). See also David Schuyler, *The New Urban Landscape: The Redefinition of City Form in Nineteenth-Century America* (Baltimore, MD: Johns Hopkins University Press, 1986); and Galen Cranz, *The Politics of Park Design: A History of Urban Parks in America* (Cambridge, MA: MIT Press, 1982).

72. "Zoological Park Proposition, and the Voice of the Press in Regard to It."

73. "Zoological Park Proposition, and the Voice of the Press in Regard to It." On science in the early history of the London Zoo, see Takashi Ito, *London Zoo and the Victorians, 1828–1859* (Cambridge: Cambridge University Press, 2014), 107–137.

74. Chris Otter, *The Victorian Eye: A Political History of Light and Vision in Britain, 1800–1910* (Chicago: University of Chicago Press, 2008), 131.

75. "The Value of Zoological Gardens to Education," Editorial Department, RU 74, Box 6, Folder 3.

76. "Zoological Park Proposition, and the Voice of the Press in Regard to It."

77. "Zoological Park Proposition, and the Voice of the Press in Regard to It."

78. Langley, "National Zoological Park," Report to accompany S. 2284, 19 February 1890, RU 74, Box 285, "Scrapbook, 1887–1902."

79. Untitled article (*Public Opinion*, August 4, 1888), "Scrap book, vol. 2, 1888–1891."

80. Horowitz, "National Zoological Park," 412.

81. "The Zoological Park" (*Evening Star*, July 25, 1889), "Scrap book, vol. 2, 1888–1891."

82. "The Park Appropriation" (*Washington Post*, July 27, 1889), "Scrap book, vol. 2, 1888–1891."

83. "Money for the Park" (*Washington Post*, July 26, 1889), "Scrap book, vol. 2, 1888–1891."

84. "The Zoological Park" (*Evening Star*, August 3, 1889), "The District Pays One-Half" (*Washington Post*, August 4, 1889), "Scrap book, vol. 2, 1888–1891."

85. "The Zoological Park" (*Evening Star*, February 11, 1890), "Scrap book, vol. 2, 1888–1891."

86. Baker to Langley, 2 January 1891, RU 74, Box 7, Folder 2.

87. Smith, *Yellowstone and the Smithsonian*.

88. Baker to W.R.C. Johnstone, to Sleah and Sloap, to Henry Sloan, to Max Ferche, to Secretary of Treasury; 31 August 1892, 6 September 1892, 8 September 1892, RU 74, Box 8, Folder 2.

89. William T. Hornaday, "The Right Way to Teach Zoölogy," *Outlook* 95 (June 1910): 258.

90. This relationship between museums and zoos was not unique. An 1890 circular concerning the proposed zoological gardens and aquarium for Boston, for example, stated, "If dead creatures and their bare skeletons can attract multitudes of visitors, of how much deeper interest will living creatures prove." "Circular About the Proposed Zoological Gardens + Aquarium," 7, BSNH III-1.7, Museum of Science, Boston, Archives.

91. Steve Conn, *Museums and American Intellectual Life, 1876–1926* (Chicago: University of Chicago Press, 1998), 4–5. See also Catherine A. Nichols, *Exchanging Objects: Nineteenth-Century Museum Anthropology at the Smithsonian Institution* (New York: Berghahn, 2021).

92. For more on museum studies approaches to display, see, for example, Sharon Macdonald, ed., *The Politics of Display: Museums, Science, Culture* (New York: Routledge, 1998); Karen A. Rader and Victoria E. M. Cain, eds., *Life on Display: Revolutionizing U.S. Museums of Science and Natural History in the Twentieth Century* (Chicago: University of Chicago Press, 2014); and Liv Emma Thorsen, "Animal Matter in Museums: Exemplifying Materiality," in *The Routledge Companion to Animal-Human History*, ed. Hilda Kean and Philip Howell (London: Routledge, 2019), 171–193.

93. On the cultural transformations of the Progressive Era, see, for example, Michael McGerr, *A Fierce Discontent: The Rise and Fall of the Progressive Movement in America, 1870–1920* (New York: Oxford University Press, 2003); Marshall Berman, *All That Is Solid Melts into Air: The Experience of Modernity* (New York: Penguin, 1988); John F. Kasson, *Amusing the Million: Coney Island at the Turn of the Century* (New York: Hill and Wang, 1978); Christopher Lasch, *The New Radicalism in America, 1889–1963: The Intellectual as a Social Type* (New York: Norton, 1997); Daniel T. Rodgers, *Atlantic Crossings: Social Politics in a Progressive Age* (Cambridge, MA: Harvard University Press, 1998); and Nell Irvin Painter, *Standing at Armageddon: A Grassroots History of the Progressive Era* (New York: Norton, 2008).

94. On the philosophy of encapsulation, see Peter Sloterdijk, *Foams*, trans. Wieland Hoban (Cambridge, MA: MIT Press, 2016).

95. "The Land for the Zoo" (*Washington Post*, March 13, 1889) and "The Zoological Commission" (*Evening Star*, March 13, 1889), "Scrap book, vol. 2, 1888–1891." For more on Powell's legacy, see Donald Worster, *A River Running West: The Life of John Wesley Powell* (New York: Oxford University Press, 2002).

96. "The Zoological Park" (*Evening Star*, March 26, 1889), "Scrap book, vol. 2, 1888–1891." A column with the same title ran on multiple days in these early months.

97. "The Zoological Park" (March 16, May 18, May 29, 1889).

98. "The Zoological Park" (August 19, September 20, 1889).

99. Starkweather to Commissioners, 22 April 1889, RU 74, Box 6, Folder 3.

100. "Land for the New Zoo" (*Washington Post*, November 21, 1889); "The Zoological Park" (*Evening Star*, November 22, 1889); "The Zoological Park" (*Evening Star*, November 25, 1889); "The Land for the Zoo" (*Washington Post*, December 3, 1889); "The Zoological Park" (*Evening Star*, December 3, 1889); and "The Zoological Park" (*Evening Star*, January 27, 1891), "Scrap book, vol. 2, 1888–1891."

101. "The Zoological Park" (*Evening Star*, March 14, 1889), "Scrap book, vol. 2, 1888–1891."

102. "The 'Zoos' of the World" (*Evening Star*, March 30, 1889), "Scrap book, vol. 2, 1888–1891."

103. "Rock Creek National Park: Information for Public," (Washington, DC: Judd and Detweiler), 3–5, in "NZP Commission Papers," RU 74, Box 6, Folder 8.

104. Though Olmsted toured and surveyed the site and drew up original plans, the Smithsonian could not afford him for long. Olmsted's sons, however, were intermittently involved with the National Zoological Park across the 1890s, until 1905.

105. "Rock Creek," 5–26. In 1933, Rock Creek Park and the National Capital Parks were transferred to the National Park Service (NPS). The zoo, though, to this day, is not a part of the NPS. It remains under the auspices of the Smithsonian. See William Bushong, *Rock Creek Park, District of Columbia: Historic Resource Study* (Washington, DC: Department of the Interior, National Park Service, 1990).

106. "At Work on the Zoo" (*Washington Post*, August 27, 1890), "Scrap book, vol. 2, 1888–1891."

107. "The Zoological Park" (*Evening Star*, February 11, 1890), "Scrap book, vol. 2, 1888–1891."

108. "The Zoological Park" (February 11, 1890).

109. David Edgerton, *Shock of the Old: Technology and Global History since 1900* (London: Profile Books, 2011).

110. "The Zoological Park" (*Evening Star*, February 11, 1890), "Scrap book, vol. 2, 1888–1891."

111. "The Zoological Park" (February 11, 1890).

112. "Memoranda Relative to Work," "Diaries, Ledgers, and Memoranda . . . 1890–1931," RU 74, Box 4.

113. On occasion, common laborers could escape their confinement in the underclasses and work their way through the ranks. Harry Dean, for example, was hired by the National Zoo in 1899 as a part-time laborer. He was paid $1.50 per day. By 1915, he was promoted to zookeeper, making $840 per year (approximately $21,300 in 2022). By 1928, he was earning a salary of $1,680. Laborers Charles Lewis (hired in 1890), Albert Barnes (1898), Rodney Rose (1900), and water boy John Turner (1890) had similar work histories, all eventually becoming national zookeepers. These individuals show us that zookeeping knowledge was always assembled from below as the zoo was simultaneously established from above, which is why gamekeepers, foresters, farmers, ranchers, and pet-owners of all backgrounds felt qualified to apply for keeper positions. Retirement cards, RU 74, Box 150, Folder 9; Job applications, RU 74, Box 41.

114. "A New Suburban Attraction" (*Evening Star*, August 16, 1890), "Scrap book, vol. 2, 1888–1891."

115. Baker to Langdale, 10 November 1890, RU 74, Box 7, Folder 2; Sybil Hamlet, "History of the National Zoological Park, 1805–1987," unpublished manuscript, undated, RU 365, Box 37, Folder 1, "1894," 6–7.

116. Hamlet, Folder 1, "1892," 5. Also, as a frame of reference, for the first eight months of its existence, the Philadelphia Zoo attracted 227,557 visitors. In the next twelve months, 419,776 people attended, followed by 657,295 over the following twelve months. These attendance figures are for July 1, 1874–February 25, 1875; March 1, 1875–February 28, 1876; and March 1, 1876–February 28, 1877, respectively. "Attendance Grand Totals," Box #ZHA1, Item #15, PZA.

Chapter 2. Runaway Animals

1. "Escaped from the Zoo" (*Evening Star*, May 24, 1892), "Scrapbook, 1887–1902," RU 74, Box 285.

2. "A Zoo Bear At Large" (*Washington Post*, May 25, 1892), "Scrapbooks, Vol. 3, 1890–1896" RU 74, Box 286, Folder 2.

3. "Escaped from the Zoo."

4. "At Work on The Zoo" (*Washington Post*, August 21, 1890), "Scrapbook, 1887–1902."

5. Damir Ugarković, Nera Fabijanić, Kristijan Tomljanović, Davor Krmpotić, and Nikolina Kelava Ugarković, "Microhabitat Characteristics of Brown Bear Den Areas," *Baltic Forestry* 26, no. 2 (2020): 176–184. Other types include nest, stump, trunk, root excavation, and anthill dens.

6. "Escaped from the Zoo."

7. Baker to Blackburne, 17 July 1890 & 14 January 1891, RU 74, Box 7, Folders 1 & 2. William H. Blackburne was a highly respected figure of the National Zoo, working as head keeper for fifty-three years. With presidential approval, Blackburne worked long past the mandatory retirement age, until he was eighty-seven.

Supposedly, as his story went, the love of his life would not marry him unless he stopped his traveling circus life. Blackburne was known to have "quit the circus, joined the zoo, gotten married, and given up drinking all on the same day." Sybil Hamlet, "History of the National Zoological Park, 1805–1987," unpublished manuscript, undated, RU 365, Box 37, Folder 1, "1891," 5–6.

8. "Escaped from the Zoo."

9. "Escaped from the Zoo." This search party also conjured the ritual and symbolisms of "the wolf hunt." See Jon Coleman, *Vicious: Wolves and Men in America* (New Haven, CT: Yale University Press, 2006), 123–146.

10. "Zoo Bear At Large."

11. "Escaped from the Zoo."

12. "Escaped from the Zoo."

13. "The Dead Grizzly" (*Evening Star*, May 25, 1892), "Scrapbook, 1887–1902."

14. Katherine Tweed, "Rock-Climbing Bears," *Science World* 67, no. 3 (October 18, 2010); Lech Stempniewicz, "Polar Bears Observed Climbing Steep Slopes to Graze on Scurvy Grass in Svalbard," *Polar Research* 36, no. 1 (June 15, 2017).

15. "Dead Grizzly."

16. "'B'ar' Meat for Clubmen" (unknown newspaper, unknown day, 1892), "Scrapbook, 1887–1902."

17. "Dead Grizzly."

18. "The Big Bear Hunt" (*Washington Post*, May 26, 1892), "Scrapbook, 1887–1902."

19. "The Reporter and the Bear" (unknown newspaper, unknown date), "Scrapbook, 1887–1902."

20. "Big Bear Hunt."

21. "Zoo Bear At Large."

22. "Zoo Bear At Large."

23. Beasts, monsters, and predators hunting Civilization is a story line that has deeply shaped cultural artifacts about human-animal relations. A large swathe of environmental and animal history chronicles the historical formation and impact of these plots. There is also a significant literary scholarship about how these narratives shape contemporary horror—from Spielberg's *Jaws* to Stephen King's *Pet Sematary* to Michael Crichton's *Jurassic Park*. See, for example, Nichole Neff, "The Belly of the Beast: The Uncanny Shark," *Gothic Studies* 18, no. 2 (November 2016): 52–61; Ruth Heholt and Melissa Edmundson, eds., *Gothic Animals: Uncanny Otherness and the Animal With-Out* (Cham, Switzerland: Palgrave, 2020).

24. Coleman, *Vicious*, 117; "Concerning Bears" (*Evening Star*, June 4, 1892), "Scrapbook, 1887–1902."

25. "Concerning Bears."

26. See Giorgio Agamben, *The Open: Man and Animal*, trans. Kevin Attell (Stanford, CA: Stanford University Press, 2003).

27. Karen Raber, *Animal Bodies, Renaissance Culture* (Philadelphia: University of Pennsylvania Press, 2013), 134.

28. "Fierce Grip of the Panther" (*New-York Tribune*, June 27, 1890), "Scrapbook, 1887–1902."

29. "Broke Out the Den," *Grand Rapids Herald*, March 4, 1898, 5.

30. "Bear Hunt in Bronx Park" (*Evening Star*, July 17, 1900), "Scrapbook, 1887–1902."

31. "Black-Tailed Prairie Dog," Smithsonian's National Zoo and Conservation Biology Institute, https://nationalzoo.si.edu/animals/black-tailed-prairie-dog; "Black-tailed Prairie Dogs," Scotts Bluff National Monument, National Park Service, https://www.nps.gov/scbl/planyourvisit/upload/Prairie-Dogs.pdf. Amazingly, the largest prairie dog colony ever recorded covered 25,000 square miles and housed 400 million individuals, a true metropolis!

32. "An Animal Garden in Fairmount Park," Zoological Society of Philadelphia, 1988, 5, PZA.

33. Marie Howland, "The Philadelphia Zoo," *Harper's New Monthly Magazine* 58 (April 1879): 708–709; M. Howland, "Prairie-Dogs in the Philadelphia Zoo," *New England Farmer, and Horticultural Register* 58 (April 19, 1879): 4.

34. "In the Zoo Garden by Day and Night" (*Press*, November 29, 1891), "Scrapbook, 1887–1902."

35. "Ostrich Bent on Suicide," *New Haven Evening Register*, July 24, 1899, 5.

36. Mary Louise Kete, *Sentimental Collaborations: Mourning and Middle-Class Identity in Nineteenth-Century America* (Durham, NC: Duke University Press, 2000).

37. For example, see J. R. Shorter and O. Rueppell, "A Review on Self-destructive Defense Behaviors in Social Insects," *Insectes Sociaux* 59, no. 1 (December 2011): 1–10; or Samir Okasha, "Biological Altruism," in *The Stanford Encyclopedia of Philosophy*, ed. Edward N. Zalta (Summer 2020 edition), https://plato.stanford.edu/archives/sum2020/entries/altruism-biological/.

38. I mean for this term to echo the "instinctive" valences of both Schopenhauer's "will to life" and Nietzsche's "will to power." Both concepts would be enriched if freed from humanist straightjackets.

39. Lori Gruen, *Entangled Empathy: An Alternative Ethic for Our Relationships with Animals* (Brooklyn: Lantern Books, 2015). See also Mary Midgley's classic "What Is Anthropomorphism," in *Animals and Why They Matter* (Athens: University Press of Georgia, 1983), 125–133; Chengcheng You, "Representing Zoo Animals: The Other-than-Anthropocentric in Anthony Brown's Picture Books," *Lion and the Unicorn* 43, no. 1 (January 2019): 22–41; Chengcheng You, "The Necessity of an Anthropomorphic Approach to Children's Literature," *Children's Literature in Education* 52, no. 2 (June 2021): 183–199.

40. Brian C. R. Bertram, *The Ostrich Communal Nesting System* (Princeton, NJ: Princeton University Press, 1992).

41. "Foxy, the Mischievous Lemur, Leads Keepers a Merry Chase," *Philadelphia Inquirer*, July 15, 1900, 2.

42. "Buffaloes Riot Through Chicago," *Philadelphia Inquirer*, September 16, 1900, 4.

43. "Wildcat Escaped," *Sun* (Baltimore), May 4, 1900, 10.

44. "Wildcat in a Hotel," *Evening Press* (Grand Rapids), January 22, 1903, 4.

45. "An Elephant on a Tear," *Daily Picayune* (New Orleans), March 19, 1901, 10.

46. "Panther Captured," *San Antonio Daily Express*, July 31, 1902, 8.

47. "Eighty Snakes Escape from a Box in Zoo," *Evening News* (San Jose), April 11, 1904, 6.

48. "Wolverine Saw New York," *Plain Dealer* (Cleveland), February 13, 1906, 2.

49. "Catch Sea Lion in River," *Plain Dealer*, July 24, 1910, 1.

50. "Leopards Cause Big Panic in Audience," *Fort Worth Telegram*, May 1, 1908, 5.

51. "Monkeys Elope from Lincoln Zoo," *Grand Forks Daily Herald* (North Dakota), April 27, 1909, 4.

52. "Bull Buffalo Wins in Battle with Automobile," *Bellingham Herald* (Washington), January 16, 1912, 1.

53. "The Zoo: Incidents of Animal Life" (*Evening Telegraph*, March 10, 1877), *Scrap Book* (1866–1903), Item #PUB R1, PZA.

54. "Open to Public," *Press*, July 1, 1874, 8, Folio "Articles on Zoo Opening, July 1, 1874," Item #ZHA1#4, PZA.

55. "The 'Zoo': How the Animals Live" (*Evening Telegraph*, March 9, 1877), *Scrap Book*.

56. "Novelties of the Zoo" (*Evening Telegraph*, March 25, 1879), *Scrap Book*.

57. "'Don' Put in Chains" (unknown newspaper, November 2, 1882), *Scrap Book*.

58. "Tale of a Lion" (*Practical Farmer*, September 15, 1877), *Scrap Book*.

59. "In the Lion's Mouth," *Columbus Enquirer-Sun*, June 3, 1891, 1.

60. "In the Zoo Garden By Day and Night" (*Press*, November 29, 1891), "Scrapbook, 1887–1902."

61. "Bear's Claws on Him," *Evening Press* (Grand Rapids), August 1, 1904, 5.

62. "Attacked by Deer," *Evening Press* (Grand Rapids), December 20, 1907, 13.

63. "Attacked by a Park Deer," *Trenton Evening Times*, December 27, 1896, 1.

64. Susan Nance, *Animal Modernity: Jumbo the Elephant and the Human Dilemma* (London: Palgrave Macmillan, 2015), 20.

65. See, for example, Lisa Guenther, *Solitary Confinement: Social Death and Its Afterlives* (Minneapolis: University of Minnesota Press, 2013).

66. "Sunday at the 'Zoo'" (*Washington Post*, June 23, 1889), "Scrapbook, Vol. 2, 1888–1891," RU 74, Box 286, Folder 1.

67. Baker to S. W. Curtin, 15 November 1890; Baker to Dr. Lloyd Smith, 28 July 1891, RU 74, Box 7, Folders 2 & 4.

68. "At the Zoo" (*Evening Star*, May 20, 1892), "Scrapbook, Vol. 3, 1890–1896."

69. "Battled with a Deer" (*Washington Post*, February 15, 1892), "Scrapbook, Vol. 3, 1890–1896."

70. I borrow this phrase from Wilfrid Blunt, *The Ark in the Park: The Zoo in the Nineteenth Century* (London: Hamish Hamilton, 1976).

71. Susan Nance, *Entertaining Elephants: Animal Agency and the Business of the American Circus* (Baltimore, MD: Johns Hopkins University Press, 2013), 120.

72. Cleveland Moffett, "Wild Beasts and Their Keepers: How the Animals in a Menagerie Are Tamed, Trained, and Cared For," 545, "Scrapbook 1887–1902."

73. "Battled with a Deer."

74. Table of Contents for "Scrapbook, 1900–1919," RU 74, Box 287.

75. "Coyote Attacks Keeper Edwards" (*Times*, May 16, 1904) "Scrapbook, 1900–1919."

76. See Dan Flores, *Coyote America: A Natural and Supernatural History* (New York: Basic Books, 2016). On the urban coyote, see Gavin Van Horn, *The Way of Coyote: Shared Journeys in Urban Wilds* (Chicago: University of Chicago Press, 2018).

77. "The Lost Beaver" (*Evening Star*, April 3, 1897), "Scrapbook, 1887–1902." The founding beaver colony in the National Zoo was formed in 1894, consisting of nine beavers acquired from Yellowstone National Park. Elwood (Billy) Hofer, a famous hunter and trapper, was given instructions by Captain George Anderson to capture these beavers for the zoo. When these animals arrived in Washington, they were placed in a part of Rock Creek that was labeled the "Missouri Valley." Their enclosure was 100 x 200 feet. Immediately, they began felling small trees and built three dams. One of the dams reached four feet in height. Hamlet, Folder 1, "1894," 1–2. There is a vast historiography on beavers and the fur trade, but for a wonderful overview, see Rachel Poliquin, *Beaver* (London: Reaktion Books, 2015).

78. "Fox Makes Its Escape" (*Evening Star*, November 23, 1899), "Scrapbook, 1887–1902."

79. "Anxious for Return of Fox" (*Evening Star*, November 24, 1899), "Scrapbook, 1887–1902."

80. I am borrowing this term from Jeffrey Nugent Hyson, "Urban Jungles: Zoos and American Society," PhD diss., Cornell University, 1999.

81. "Report of the Superintendent," For Year Ending June 30, 1900, N.Z.P. Correspondence to the S.I., Volume 1, April 2, 1900–January 28, 1901, RU 74, Box 27.

82. "Report," Year Ending June 30, 1904, Vol. 5, November 17, 1903–February 21, 1905, RU 74, Box 28.

83. "Report," Year Ending June 30, 1905, Vol. 6, February 25, 1905–June 2, 1906, RU 74, Box 28.

84. "Report," Year Ending June 30, 1906, Vol. 7, June 2, 1906–January 13, 1908, RU 74, Box 28.

85. "Report," Year Ending June 30, 1908, Vol. 8, January 15, 1908–September 2, 1910, RU 74, Box 29, Folder 1.

86. "Report," Year Ending in June 30, 1909, Vol. 8.

87. "Report," Year Ending June 30, 1911, Vol. 9, September 6, 1910–April 5, 1913, RU 74, Box 29, Folder 2.

88. "Report," Year Ending June 30, 1912, Vol. 9.

89. "Report," Year Ending June 30, 1913, Vol. 10, April 7, 1913–January 27, 1916, RU 74, Box 29, Folder 3.

90. "Wild Animals in Captivity" (*Evening Star*, November 13, 1897), "Scrapbook, 1896–1907," RU 74, Box 286, Folder 3. For a powerful, more recent illustration of this point concerning Tilikum, an orca held in SeaWorld, see *Blackfish*, directed by Gabriela Cowperthwaite (Manny O. Productions, 2013). Also see Randy Malamud, "Al Gore, *Blackfish*, and Me: Eco-activist Progress and Prospects for the Future," in *Zoo Studies: A New Humanities*, ed. Tracy McDonald and Daniel Vandersommers (Montreal: McGill-Queens University Press, 2019), 262–275.

91. "Wild Animals in Captivity."

92. The only literature on the following episode is Abe Gibson, "Dog Days at the Zoo, Part II," Smithsonian Institution Archives (September 2012), Accessed May 25, 2020, http://www.siarchives.si.edu/blog/dog-days-zoo-part-ii.

93. "Mad Wolf from Zoo" (*Washington Post*, October 19, 1902), "Scrapbook, 1896–1907."

94. "Mad Wolf from Zoo."

95. "Mad Wolf from Zoo."

96. "Zoo's Truant Wolf" (*Washington Post*, October 20, 1902), "Cleveland Park in Arms" (unknown newspaper, October 21, 1902), "Scrapbook, 1896–1907."

97. "Zoo's Truant Wolf."

98. For more on the entangled notions of stray, domestic, wild, and feral, see Abe Gibson, *Feral Animals in the American South: An Evolutionary History* (Cambridge: Cambridge University Press, 2016).

99. "Zoo's Truant Wolf."

100. "Mad Dog Appears" (*Evening Star*, October 20, 1902), "Police Believe Dog Bit Jackson Children" (*Times*, October 21, 1902), "Scrapbook, 1896–1907."

101. "It Was Not a Wolf" (*Evening Star*, October 20, 1902), "Scrapbook, 1896–1907."

102. "Zoo's Truant Wolf."

103. Article title unknown (*Times*, October 21, 1902), "Scrapbook, 1896–1907."

104. *Times*, October 21, 1902.

105. "Admiral Dewey's Dog Now Under Suspicion" (unknown newspaper, unknown date, 1902), "Scrapbook, 1896–1907."

106. "Dewey Says Kill the Dog" (*Washington Post*, October 25, 1902), "Scrapbook, 1896–1907."

107. "Admiral Dewey's Dog Now Under Suspicion."

108. "Dewey Says Kill the Dog."

109. "Dewey Says Kill the Dog."

110. For more on hydrophobia, see Harriet Ritvo, *The Animal Estate: The English and Other Creatures in the Victorian Age* (Cambridge, MA: Harvard University Press, 1987), 167–204; Bill Wasik and Monica Murphy, *Rabid: A Cultural History of the World's Most Diabolical Virus* (New York: Viking, 2012); Neil Pemberton and Michael Worboys, *Mad Dogs and Englishmen: Rabies in Britain, 1830–2000* (London: Palgrave Macmillan, 2013); and Jessica Wang, *Mad Dogs and Other New Yorkers: Rabies, Medicine, and Society in an American Metropolis, 1840–1920* (Baltimore, MD: Johns Hopkins University Press, 2019).

111. "Mad From Wolf's Bite" (*Washington Post*, October 26, 1902), "Scrapbook, 1896–1907."

112. "Mixed as to the 'Bites,'" *Trenton Times*, October 24, 1902, 7.

113. "Was It Admiral Dewey's Dog?" (*Washington Post*, October 27, 1902), "Scrapbook, 1896–1907."

114. "Was Zoo Wolf Mad?" (*Washington Post*, January 3, 1903), "Scrapbook, 1896–1907."

115. "Mr. Einstein, Many Thanks!" (*Washington Post*, November 9, 1902), "Scrapbook, 1896–1907."

116. "Was Zoo Wolf Mad?"

117. "Truant Wolf Is Alive" (*Washington Post*, January 16, 1902), "Scrapbook, 1896–1907."

118. "Was Zoo Wolf Mad?"

119. "Wolf Case in Court" (*Washington Post*, June 24, 1903), "Scrapbook, 1896–1907."

120. "Another Wolf Damage Suit" (*Washington Post*, July 15, 1903), "Scrapbook, 1896–1907."

121. Colin Dayan, *The Law Is a White Dog: How Legal Rituals Make and Unmake Persons* (Princeton, NJ: Princeton University Press, 2011). See also her *With Dogs at the Edge of Life* (New York: Columbia University Press, 2016).

122. "Wolf Case in Court."

123. Juno Salazar Parreñas, *Decolonizing Extinction: The Work of Care in Orangutan Rehabilitation* (Durham, NC: Duke University Press, 2018), 132–134.

Chapter 3. The Crossroads of Science and Popular Culture

1. On the importance of understanding science as a form of communication, see James Secord, "Knowledge in Transit," *Isis* 95, no. 4 (December 2004): 654–672.

2. "The Zoological Park Proposition, and the Voice of the Press in Regard to It," RU 74, Box 285, "Scrapbook, 1887–1902."

3. "Prof. Hornaday Resigns" (*Washington Post*, May 20, 1890); "Mr. Hornaday Resigns" (*Evening Star*, May 20, 1890), "Scrapbook, 1887–1902."

4. Helen L. Horowitz, "The National Zoological Park: 'City of Refuge' or Zoo?" *Records of the Columbia Historical Society, Washington, D.C.* 49 (1973/1974): 420–424; Sybil Hamlet, "History of the National Zoological Park, 1805–1987," unpublished manuscript, undated, RU 365, Box 37, Folder 1, 1890, 16–26; J. Gordon Vaeth, *Langley: Man of Science and Flight* (New York: The Ronald Press Company, 1966), 54–62.

5. Baker to Iles, 29 July 1890, RU 74, Box 7, Folder 3; Mark Collins Jenkins, "Dr. Frank Baker: If Only He Had Been Allowed to Treat President Garfield," *National Geographic*, blog, Accessed March 1, 2013, http://newswatch.nationalgeographic.com/2012/07/12/dr-frank-baker-if-only-he-had-been-allowed-to-treat-president-garfield/.

6. Baker to Alpheus Hyatt, 5 December 1890, RU 74, Box 7, Folder 3.

7. Frank Baker, "The Ascent of Man," *American Anthropologist* 3, no. 4 (October 1890): 297–320; Frank Baker, "Primitive Man," *American Anthropologist* 11, no. 12 (December 1898): 357–66.

8. As an active Lamarckian, Baker can be associated with Julian Huxley's "eclipse of Darwinism," a term that describes the trend in which many scientists at the beginning of the twentieth century refused to subscribe to Darwin's evolutionary theory. For more, see Peter Bowler, *The Eclipse of Darwinism: Anti-Darwinian Evolutionary Theories in the Decades Around 1900* (Baltimore, MD: Johns Hopkins University Press, 1983) and "Revisiting the Eclipse of Darwinism," *Journal of the History of Biology* 38, no. 1 (Spring 2005): 19–32; and Mark A. Largent, "The So-Called Eclipse of Darwinism," *Transactions of the American Philosophical Society* 99, no. 1 (2009): 3–21. In "Primitive Man," Baker strayed from pure Lamarckianism to a more nuanced recapitulation theory based on Ernst Haeckel's contention that "ontogeny recapitulates phylogeny." See Stephen Jay Gould, *Ontogeny and Phylogeny* (Cambridge, MA: Harvard University Press, 1985).

9. Baker, "Ascent of Man," 298–299. Histories of racism and zoos are intimately entangled. For more, see Lisa Uddin, *Zoo Renewal: White Flight and the Animal Ghetto* (Minneapolis: University of Minnesota Press, 2015); and Marianna Szczygielska, "Pandas and the Reproduction of Race and Heterosexuality in the Zoo," in *Zoo Studies: A New Humanities*, ed. Tracy McDonald and Daniel Vandersommers (Montreal: McGill-Queens University Press, 2019), 211–236.

10. Baker to Keen, 1 October 1891, RU 74, Box 7, Folder 3. His salaries as professor, curator, and zoo director combined for a whopping $3,500 per year (almost $100,000 today).

11. Baker to Manigault, 2 December 1890, RU 74, Box 7, Folder 2.

12. Baker to William Fewsmith, 3 December 1890, RU 74, Box 7, Folder 2.

13. Baker to Funk Wagnalls, 4 June 1892, RU 74, Box 7, Folder 3.

14. Baker to H. C. Jones, 1 October 1891, RU 74, Box 7, Folder 3.

15. Baker to Dorsey, 4 October 1892, RU 74, Box 7, Folder 3.

16. Baker to J. B. Bullane, 2 July 1891, RU 74, Box 7, Folder 4. Fascinatingly, experiments concerning venom have a rich history. See Jutta Schickore, *About Method: Experimenters, Snake Venom, and the History of Writing Scientifically* (Chicago: University of Chicago Press, 2017).

17. Baker to A. C. Downs, 22 July 1891; Baker to G. O. Ghenault, 26 August 1891, RU 74, Box 7, Folder 4. For more on nineteenth-century beetle-collecting and classifying crazes, see J. F. M. Clark, *Bugs and the Victorians* (New Haven, CT: Yale University Press, 2009).

18. Baker to B. Westerman & Co., 28 July 1891, RU 74, Box 7, Folder 4.

19. Baker to G. Brown Goode, 12 August 1891, RU 74, Box 7, Folder 4. On the diffusion of Joseph Lister's antisepsis, see M. Anne Crowther, "Lister at Home and Abroad: A Continuing Legacy," *Notes and Records of the Royal Society of London* 67, no. 3 (September 2013): 281–294; Michael Worboys, "Joseph Lister and the Performance of Antiseptic Surgery," *Notes Rec. R. Soc.*, same issue, 199–209.

20. Baker to anonymous, undated; Baker to G. Brown Goode, 2 & 12 September 1891; Baker to Langley, 13 September 1891, RU 74, Box 7, Folder 4.

21. Baker to Stiles, 29 September 1891; Baker to King, 6 October 1890, RU 74, Box 7, Folder 4; Baker to Macmillan & Company, 10 July 1890, RU 74, Box 7, Folder 3.

22. Eric Baratay and Elisabeth Hardouin-Fugier, *Zoo: A History of Zoological Gardens in the West* (London: Reaktion Books, 2002), 99.

23. Untitled Article (Unnamed Newspaper, 1874?), *Scrap Book*, Item #PUB R1, PZA.

24. "Zoological Garden" (Unnamed Newspaper, 1867), *Scrap Book*.

25. "Cash Wanted" (Unnamed Newspaper, 1874?), *Scrap Book*.

26. Bruno Latour, *Science in Action: How to Follow Scientists and Engineers Through Society* (Cambridge, MA: Harvard University Press, 1988).

27. Paul Israel, *Edison: A Life of Invention* (New York: John Wiley, 1998), 142–156, 277–291. See also David Morton, *Off the Record: The Technology and Culture of Sound Recording in America* (New Brunswick: Rutgers University Press, 2000).

28. "Do Monkeys Talk?" (*Washington Star*, 1891), "Scrapbook, 1887–1902."

29. "Do Monkeys Talk?"

30. Gregory Radick, "Primate Language and the Playback Experiment, in 1890 and 1980," *Journal of the History of Biology* 38, no. 3 (Autumn 2005): 466–467; R. L. Garner, "The Speech of Animals," unpublished manuscript, RLGP, Box 6, "Writings: 'Sanctuary to Women—'Spider Webs,'" NAA; R. L. Garner, "The Simian Tongue," *New Review* 4 (1891): 555–562.

31. R. L. Garner, "Do Monkeys Express Themselves in Words?" (unknown newspaper, June 8, 1891), "Scrapbook, 1887–1902"; John P. Harrington, "He Spoke: Biography and Scientific Work of Richard Lynch Garner" (Washington, DC: Government Printing Office, 1941): 10, RLGP, Box 5, "Biographer's Papers," NAA. These papers contain vast material concerning Garner in Africa and his writings on primates. Garner was interested in a wide range of topics, and he wrote voluminously—scientific analyses, poetry, short fictional stories, diary entries, accounts of African societies, political commentaries, and bizarre musings on myriad topics.

32. Baker to Langley, 3 May 1892, RU 365, Box 12, Folder 25.

33. Baker to Brooks, 22 April 1892, RU 365, Box 12, Folder 25.

34. Noah Cincinnati, "Too Sullen for Survival: Historicizing Gorilla Extinction, 1900–1930," in *The Historical Animal*, ed. Susan Nance (Syracuse, NY: Syracuse University Press, 2015), 175.

35. John Sorenson, *Ape* (London: Reaktion, 2009), 48–62.

36. Linda Kerber, "Separate Spheres, Female Worlds, Woman's Place: The Rhetoric of Women's History," *Journal of American History* 75, no. 1 (June 1988): 9–39.

37. Radick, "Primate Language and the Playback Experiment," 464. On the history of playbacks, see also J. B. Falls, "Playback: A Historical Perspective," in *Playback and Studies of Animal Communication*, ed. P. K. McGregor (New York:

Plenum Press, 1992), 11–33. For another take on Garner that contextualizes him within the field of comparative psychology and alongside zoologist Conwy Lloyd Morgan, see Gregory Radick, "Morgan's Canon, Garner's Phonograph, and the Evolutionary Origins of Language and Reason," *British Journal for the History of Science* 33, no. 1 (March 2000): 3–23.

38. Garner, "Do Monkeys Express Themselves in Words?"

39. Radick, "Primate Language and the Playback Experiment," 462–463; William James, *The Principles of Psychology*, in *The Works of William James*, ed. Frederick H. Burkhardt, Fredson Bowers, and Ignas K. Skrupskelis, vol. 8–10 (Cambridge, MA: Harvard University Press, 1981), 1470.

40. The classic Cartesian text that established this difference of Reason between human and animal is his "Discourse on Method," published in 1637 and most famous for the phrase "I think, therefore I am." A voluminous literature, of course, exists on the philosophical importance of this work. Animal studies scholars, though, have been systematically dismantling Cartesian conceptions of "the animal" since the early works of Peter Singer. For two important animal studies critiques on Descartes' "animal," see Erica Fudge, *Brutal Reasoning: Animals, Rationality and Humanity in Early Modern England* (Ithaca, NY: Cornell University Press, 2006); and Karen Raber, *Animal Bodies, Renaissance Culture* (Philadelphia: University of Pennsylvania Press, 2013).

41. Garner, "Do Monkeys Express Themselves in Words?"

42. Jeremy Rich, *Missing Links: The African and American Worlds of R. L. Garner, Primate Collector* (Athens: University of Georgia Press, 2012), especially 88–124.

43. R. L. Garner, *The Speech of Monkeys* (London: William Heinemann, 1892).

44. Scientific expeditions like Garner's became quite common in the early twentieth century. For discussion of William M. Mann (the fifth NZP superintendent) and his expeditions to Tanganyika (1926), the Dutch East Indies (1937), and Liberia (1940), see Elizabeth Hanson's chapter "Zoo Expeditions," in *Animal Attractions: Nature on Display in American Zoos* (Princeton, NJ: Princeton University Press, 2002).

45. Baker to R. L. Garner, 8 December 1890, RU 74, Series 6, Box 7, Folder 2.

46. Donna J. Haraway, *Primate Visions: Gender, Race, and Nature in the World of Modern Science* (New York: Routledge, 1989), 42, 45. In the context of taxidermist and nature photographer Carl Akeley (1864–1926), Haraway frames early photography as "hunting with a camera." See also Finis Dunway, "Hunting with the Camera: Nature Photography, Manliness, and Modern Memory, 1890–1930," *Journal of American Studies* 34, no. 2 (August 2000): 207–230. On empire-making and hunting "with a gun," see Angela Thompsell, *Hunting Africa: British Sport, African Knowledge and the Nature of Empire* (New York: Palgrave Macmillan, 2015). For a brilliant analysis of the moving image and Eadweard Muybridge, see Antoine Traisnel, *Capture: American Pursuits and the Making of a New Animal Condition* (Minneapolis: University of Minnesota Press, 2020), 153–190. And for a classic on vision and power in postwar zoos and wildlife parks, see Gregg

Mitman, "When Nature *Is* the Zoo: Vision and Power in the Art and Science of Natural History," *Osiris* 11 (1996): 117–143.

47. "To Talk with Gorillas" (unknown newspaper, 1891), "Scrapbook, 1887–1902."

48. Daniel E. Bender, *The Animal Game: Searching for Wildness at the American Zoo* (Cambridge, MA: Harvard University Press, 2016), 52.

49. "Talking with Gorillas" (*Buffalo Courier*, December 13, 1891), "Scrapbook, 1887–1902."

50. "Phonographing the Language of the Apes," *Current Literature* 10, no. 2 (June 1892): 207.

51. "Monkey Talk," *Youth's Companion* 75, no. 23 (June 6, 1901): 296.

52. "Simian Speech and Simian Thought," *Cosmopolitan; A Monthly Illustrated Magazine* 13, no. 1 (May 1892): 72.

53. "Monkeys Can Count," *Daily Picayune*, May 29, 1892, 16.

54. "Richard L. Garner," *Salt Lake Weekly Tribune*, August 11, 1892, 7.

55. "A Love-Sick Monkey," *Wheeling Sunday Register* 9, issue 314 (May 22, 1892): 9.

56. Radick, "Primate Language and the Playback Experiment," 463.

57. Baker to Brooks, 22 April 1892, RU 365, Box 12, Folder 25.

58. Baker to Guthridge, 8 June 1891, RU 74, Box 7, Folder 4.

59. Baker to Skehan, 12 February 1891, RU 74, Box 7, Folder 2.

60. Baker to Sterling Ruffin, 12 April 1892, RU 365, Box 12, Folder 25.

61. Jon Cohen, *Almost Chimpanzee: Searching for What Makes Us Human, in Rainforests, Labs, Sanctuaries, and Zoos* (New York: Henry Holt, 2010), 101–109.

62. "Monkey Talk," *Bismarck Daily Tribune*, March 27, 1894, 1.

63. "Monkey Talk."

64. Gregory Radick, *The Simian Tongue: The Long Debate about Animal Language* (Chicago: University of Chicago Press, 2008), 147, 156–158.

65. Henry Morton Stanley, for example, invented an entirely fictionalized biography for himself. See Adam Hochschild, *King Leopold's Ghost: A Story of Greed, Terror, and Heroism in Colonial Africa* (New York: Houghton Mifflin, 1998), 23–25, 31–32.

66. Radick, *Simian Tongue*, 155; R. L. Garner, *Gorillas and Chimpanzees* (London: Osgood, McIlvaine, 1896).

67. Similarly, see Donna J. Haraway, "Primatology Is Politics by Other Means," *PSA: Proceedings of the Biennial Meeting of the Philosophy of Science Association* 1984, no. 2 (Chicago: University of Chicago Press, 1984), 489–524. For more on the history of primatology, see Georgina M. Montgomery, "Place, Practice and Primatology: Clarence Ray Carpenter, Primate Communication and the Development of Field Methodology, 1931–1945," *Journal of the History of Biology* 38, no. 3 (Autumn 2005): 495–533; and Amanda Rees, "Reflections on the Field: Primatology, Popular Science and the Politics of Personhood," *Social Studies of Science* 37, no. 6 (December 2007): 881–907.

68. Haraway, *Primate Visions*, 13–14.

69. "He Knows Six Gorilla Words," *Daily Charlotte Observer* 7, issue 733 (April 1, 1894): 4.

70. "Talks to Gorillas," *Knoxville Journal* 11, issue 139 (July 15, 1895): 8.

71. "The New Philology," *Kansas Weekly Capital and Farm Journal* 16, issue 23 (June 7, 1894): 10; "The New Philology," *Trenton Times*, April 6, 1894, 6; "The New Philology," *Bismarck Daily Tribune*, April 18, 1894, 4.

72. Untitled blurb, *Arizona Weekly Journal-Miner* 31, issue 40 (January 22, 1896): 4.

73. Tracy McDonald, "Sculpting Dinah with the Blunt Tools of the Historian," in *Zoo Studies*, 93–118. On Garner's animal trading, see Rich, *Missing Links*, 45–66. See Harrington, "He Spoke," 12–28.

74. Richard Lynch Garner, *Apes and Monkeys* (Boston: Ginn and Company's The Athenaeum Press, 1900). *Apes and Monkeys* is a rich three-hundred-page account of Garner's experiments. Composed of twenty-four chapters, the book balances Garner's observations of specific primates with general analysis of the implications of these observations. While scientists like Baker would attack the integrity of Garner's work because of its colloquial style, anthropomorphizing tendencies, and lack of rigor and discipline, these shortcomings make this book an interesting window into popular conceptions of primates at the turn of the twentieth century.

75. Radick, "Primate Language and the Playback Experiment," 463; Richard L. Garner, "A Monkey's Academy in Africa," *New Review* 7 (September 1892): 282–292.

76. Rich, *Missing Links*, 4–6; Register to RLGP, "Introduction" and "Chronology," NAA.

77. For an incredible account, see Radick, *Simian Tongue*.

78. "Project CETI," Cetacean Translation Initiative, 2021, https://www.projectceti.org/.

79. Haraway, *Primate Visions*, 5.

80. Radick, "Primate Language and the Playback Experiment," 468–469; Julian Huxley and Ludwig Koch, *Animal Language* (London: Country Life, 1938).

81. For more on Yerkes see Haraway, *Primate Visions*, 59–83.

82. Yerkish required primates to use a keyboard that possessed keys containing lexigrams, symbols that represent larger concepts.

83. Radick, "Primate Language and the Playback Experiment," 463, 465.

84. The space surrounding the Castle functioned as an ad-hoc "science center," peppered with quickly and cheaply constructed buildings that supported the heterogeneous scientific endeavors of the Smithsonian.

85. "The Zoological Park" (*Washington Post*, January 30, 1891), "Scrapbook 1887–1902."

86. For more on nineteenth-century geology and astrophysics, see John Lankford, *American Astronomy: Community, Careers, and Power, 1859–1940* (Chicago: University of Chicago Press, 1997); and Deborah R. Coen, *The Earthquake*

Observers: Disaster Science from Lisbon to Richter (Chicago: Chicago University Press, 2012).

87. "Sunday at the 'Zoo'" (*Washington Post*, June 23, 1889), "Scrap book 1888–1891, vol. 2," RU 74, Box 286, Folder 1.

88. See Armand Marie Leroi, *The Lagoon: How Aristotle Invented Science* (New York: Bloomsbury, 2014).

89. "This Is Our Own Park" (*Evening Star*, October 18, 1890), "Scrapbooks 1890–1896, vol. 3," RU 74, Box 286, Folder 2. On the obsession of classification, see Harriet Ritvo, *The Platypus and the Mermaid: And Other Figments of the Classifying Imagination* (Cambridge, MA: Harvard University Press, 1997); and Nicholas Jardine, James A. Secord, and E. C. Spary, eds., *Cultures of Natural History* (Cambridge: Cambridge University Press, 1996).

90. "An Immense Zoo" (*Times-Democrat*, New Orleans, October 29, 1899); "Nation's New Zoo" (*Press*, Grand Rapids, November 9, 1899), "Scrapbook 1887–1902."

91. Baker to F. E. Kitamiller, unknown day January 1891, RU 74, Series 6, Box 7, Folder 2.

92. For more on capturing gorillas, see McDonald, "Sculpting Dinah with the Blunt Tools of the Historian"; Violette Pouillard, "Gust (1952–1988), or a History from Below of the Changing Zoo," in *Zoo Studies*, 167–190; and Noah Cincinnati, "Too Sullen for Survival," 166–183. For an overview on gorillas in human culture, especially concerning fears of race, sexuality, and criminality, see Kathryn Weir and Ted Gott, *Gorilla* (London: Reaktion Books, 2012).

93. "Uncle Sam, His Show" (unknown newspaper, Section title "Illustrated Magazine Section," November (?), 1899), "Scrapbook 1887–1902."

94. Lynn Nyhart, "Civic and Economic Zoology in Nineteenth-Century Germany: The 'Living Communities' of Karl Mobius," *Isis* 89, no. 4 (December 1998): 605–630.

95. Walt Disney, "Educational Values in Factual Nature Pictures," *Educational Horizons* 33, no. 2 (Winter 1954): 82–84. See also Susan G. Davis, *Spectacular Nature: Corporate Culture and the Sea World Experience* (Berkeley: University of California Press, 1997).

96. Zoo babies play a particularly important role in zoos, both past and present. See Guro Flinterud, "The Rise and Fall of Polar Bear Knut," in *Zoo Studies*, 191–210.

97. On cuteness, see Joshua Paul Dale, Joyce Goggin, Julia Leyda, Anthony P. McIntyre, and Diane Negra, eds., *The Aesthetics and Affects of Cuteness* (New York: Routledge, 2017).

98. "An Odd Infantile Pet" (unknown newspaper, Winter 1891), "Scrapbooks, 1887–1902."

99. Susan Hunter, "A Nation's Zoo: Government Zoological Park at Washington" (*Home Magazine*, July 1900), "Scrapbook 1900–1919" RU 74, Box 287.

100. See Ernest Seton-Thompson's two-part article: "The National Zoo at Washington: A Study of Its Animals in Relation to Their Natural Environment,"

Century Illustrated Magazine 59/60, no. 5 (March & May 1900). Seton became famous for his animal fiction (especially his 1898 *Wild Animals I Have Known*), but he also authored works in natural history concerning wild animals, environment, and conservation. Many artists during the decades around 1900 wanted to study zoo animals. Charles R. Knight, Louis Agassiz Fuertes, and J. M. Gleeson similarly drew animals in the National Zoo. Gleeson even composed the only painting from life of a living thylacine and her cubs. Hamlet, 1900, 12. See also Thomas R. Dunlap, "The Realistic Animal Story: Ernest Thompson Seton, Charles Roberts, and Darwinism," *Forest & Conservation History* 36, no. 2 (April 1992): 56–62.

101. "Some Queer Freaks Among the Cubans" (*New York Times*, November 10, 1900), "Scrapbook 1900–1919."

102. "Spring at the Zoo" (*Evening Star*, April 1, 1901), "Scrapbook 1900–1919."

103. "A Night Visit to the Zoo" (*Washington Times*, May 26, 1901), "Scrapbook 1900–1919." On nocturnality, see Andy Flack, "Dark Trails: Animal Histories Beyond the Light of Day," *Environmental History* 27, no. 2 (April 2022), 215–241.

104. "Largest of All Carnivora" (*Washington Post*, March 2, 1902), "Scrapbook 1900–1919."

105. "Zoo May Have a Musk Ox" (*Washington Post*, March 2, 1902), "Scrapbook 1900–1919."

106. "Battled with a Deer" (*Washington Post*, February 15, 1892), "Scrapbooks, 1890–1896, vol. 3," RU 74, Box 286, Folder 2.

107. "Christening at the Zoo" (*Washington Post*, July 5, 1892), "Scrapbooks, 1890–1896, vol. 3."

108. "Fraternalism at the Zoo" (*Washington Post*, August 22, 1892), "Scrapbooks, 1890–1896, vol. 3."

109. "Spring Opening at the Zoo" (*Washington Post*, April 3, 1893), "Scrapbooks, 1890–1896, vol. 3."

110. "Tiger That Eats Rats" (*Washington Post*, September 14, 1896), "Scrapbooks, 1896–1907," RU 74, Box 286, Folder 3.

111. "Sea Water for the Zoo" (unknown newspaper, 1902), "Scrapbooks, 1896–1907."

112. "Arctic Foxes at Zoo" (*Washington Post*, December 12, 1902), "Scrapbooks, 1896–1907."

113. "Patients at the Zoo" (unknown newspaper, March 23, 1903), "Scrapbooks, 1896–1907."

114. "Painting Animals' Eyes" (*Washington Post*, May 30, 1903), "Scrapbook 1900–1919."

115. "Meat Contract for Zoo Animals" (*Washington Post*, July 14, 1903), "Scrapbook 1900–1919."

116. "Big Lion Chloroformed" (*Washington Times*, August 19, 1907), "Scrapbooks, 1896–1907."

117. "At the National Zoo" (*Washington Post*, September 6, 1903), "Scrapbooks 1896–1907."

118. The pronunciation of animal names did not first surface in the National Zoo. A decade earlier, for example, a byline of a Philadelphia newspaper stated the following: "The Parrots Are Growing Cross-Eyed and Very Weak from Watching Visitors Trying to Pronounce Their Latin Names." This byline suggested that zoogoers needed to improve their pronunciation, whereas the Washington article of 1903 bragged about the already established pronunciating abilities of zoogoers. Maybe the pronunciation of zoogoers nationwide improved between 1892 and 1903. Or, maybe these articles were merely emphasizing two different sides of zoo booster rhetoric—zoo as an elitist display of refinement and zoo as an educating medium for the people. For either case, however, the issue of the democratization and popularization of scientific language was noteworthy in zoo journalism. On similar issues regarding the "botanical vernacular" associated with gardening, see Amy M. King, *Bloom: The Botanical Vernacular in the English Novel* (Oxford: Oxford University Press, 2003).

119. "Information Concerning Zoölogical Department," 4–7, HUC 8598.2, HUA. That year, the new department offered Zoölogy, Morphology of Animals, Comparative Anatomy of Vertebrates, and Microscopical Anatomy to undergraduates. Graduate students could take Comparative Osteology and Anatomy and Development in Animals. On Harvard zoology, see Mary P. Windsor, *Reading the Shape of Nature: Comparative Zoology at the Agassiz Museum* (Chicago: University of Chicago Press, 1990).

120. Nancy Fraser, "Rethinking the Public Sphere: A Contribution to the Critique of Actually Existing Democracy," in *Habermas and the Public Sphere*, ed. Craig Calhoun (Cambridge, MA: MIT Press, 1992), 110.

121. Charles D. Walcott's *Biographical Memoir of Samuel Pierpont Langley, 1834–1906* (Washington, DC: National Academy of Science, April 1912) is the first biographical account of Langley. At Harvard, Langley was employed as an assistant in the College Observatory. At the Naval Academy, technically employed as an assistant professor of mathematics, Langley also worked on establishing an observatory. At Western, Langley worked as a professor of astronomy and physics. While there, he created the Allegheny system—the first standard time employed by railroads for the purpose of shipping goods efficiently. Simultaneously, he studied "the physical nature and functions of . . . celestial bodies" (248). He was particularly known for his examination of an 1873 sunspot and the total eclipses of 1869, 1870, 1878, and 1900. Most influentially, before beginning his work on flight, Langley worked on researching and revising the principles of solar radiation.

122. "Historical Note," Samuel P. Langley Collection, NASMASA, Accession No. XXXX-0494; Walcott, *Biographical Memoir of Samuel Pierpont Langley*, 252–254. For more on Langley and other experimenters preceding the Wright Brothers, see John D. Anderson, *Inventing Flight: The Wright Brothers and Their Predecessors* (Baltimore, MD: Johns Hopkins University Press, 2004) and *A History of Aerodynamics and Its Impacts on Flying Machines* (Cambridge: Cambridge University Press, 1998); Tom D. Crouch, *A Dream of Wings: Americans and the*

Airplane (New York: W. W. Norton, 2002); Richard P. Hallions, *Taking Flight: Inventing the Aerial Age, from Antiquity through the First World War* (New York: Oxford University Press, 2003); and James Tobin, *To Conquer the Air: The Wright Brothers and the Great Race for Flight* (New York: Free Press, 2004).

123. Langley to Baker, 14 March 1901, RU 74, Box 99, Folder 7.

124. Baker to Langley, 15 March 1901, RU 74, Box 99, Folder 7.

125. Baker to Langley, 30 April 1901; Baker to Hagenbeck, 25 March 1901; Baker to Hatfield, 30 April 1901; Baker to Ayson (fisheries commissioner for New Zealand), 30 April 1901; Baker to General Whitside (Commanding Department of Santiago), 30 April 1901, RU 74, Box 99, Folder 7.

126. Madeleine Moore has shown that between 1899 and 1908, the National Zoo acquired more than five hundred specimens (of more than one hundred species) from at least twenty-six different countries and territories using diplomatic and military connections. Madeleine Moore, "Red Blood and Red Tape: Building the Collections of the National Zoological Park, 1887–1908," Thesis, Georgetown University, 2017, 36, 35–71.

127. Baker to Langley, 30 April 1901.

128. Langley to Baker, 14 May 1901, RU 74, Box 99, Folder 7, 1–2.

129. Langley to Baker, 14 May 1901, 2.

130. Soaring bird sketch, RU 74, Box 99, Folder 7.

131. Langley to Baker, 14 May 1901, RU 74, Box 99, Folder 7, 3.

132. Langley to Baker, 14 May 1901, 4.

133. Langley sketches to Ridgeway, 29 March 1900, RU 74, Box 99, Folder 7. Birds #1 and #2 ("John crows") were done by an individual with the last name Holmes, with the assistance of Langley. Bird #3 ("John crow") and Bird #4 (gull) was sketched by Langley. Langley had communicated with Ridgeway about "soaring" and "hovering" birds prior to his interest in studying flight in the National Zoo. A few of these letters are contained in "Notes on Bird Flight—Langley," Samuel P. Langley Collection, NASMASA, Accession No. XXXX-0494 Box 44. In this collection, there is data (compiled by either Langley or Ridgeway) concerning American bald eagles, American white pelicans, common turkey buzzards, South American condors, humming birds, English kestrels, and frigate birds.

134. Otto Lilienthal, *Birdflight as the Basis of Aviation: A Contribution Towards a System of Aviation* (London: Longmans, Green, and Co., 1889, 1911).

135. Peter L. Jakab, *Visions of a Flying Machine* (Washington, DC: Smithsonian Institution Press, 1990); Tom Crouch, *The Bishop's Boys: A Life of Wilbur and Orville Wright* (New York: Norton, 2003).

136. Anderson, *History of Aerodynamics and Its Impacts on Flying Machines*, 21–25. Leonardo's classic text on flight is *Codex on the Flight of Birds*, written at the same time he painted the *Mona Lisa*. Prior to *Codex*, Leonardo sketched a series of ornithopters, contraptions that would hypothetically use flapping wings to lift and carry a person through the air.

137. Langley to Ridgeway, 29 March 1900, RU 74, Box 99, Folder 7, 3–6.

138. On Descartes's anatomical-philosophical theories on the differences between human and animal bodies, especially in regard to the materialization of the soul, see Matthew Senior, "The Souls of Men and Beasts, 1630–1764," in *A Cultural History of Animals in the Age of Enlightenment*, ed. Matthew Senior (Oxford: Berg Publishers, 2007), 23–45.

139. Walcott, *Biographical Memoir of Samuel Pierpont Langley*, 251–252.

140. S. P. Langley, *Experiments in Aerodynamics* (Washington, DC: Smithsonian Institution Press, 1891), 107.

141. Langley, *Experiments in Aerodynamics*, 12. Navier's calculations were published in Gay-Lussac, Flourens, Navier, "Rapport sur un Mémoire de M. Chabrier concernant les moyens de voyager dans l'air et de s'y diriger, contenent une nouvelle théorie des mouvements progressifs" (Paris: Mém. Acad. Sci. XI, 1832): 61–118.

142. Langley, *Experiments in Aerodynamics*, 4. For more on Navier, physicist George Gabriel Stokes, and their joint work on the mathematics of friction, see Anderson, *History of Aerodynamics and Its Impacts on Flying Machines*, 88–92.

143. Langley, *Experiments in Aerodynamics*, 12.

144. Langley, *Experiments in Aerodynamics*, 34.

145. S. P. Langley, "The Internal Work of the Wind" (Washington, DC: Smithsonian Institution Press, 1893): 1.

146. Walcott, *Biographical Memoir of Samuel Pierpont Langley*, 253.

147. Langley, "Internal Work of the Wind," 22. Langley borrows the nautical term "see the wind" from innovator Louis Pierre Mouillard, who experimented with gliders prior to Langley's research career in aeronautics. See Louis Pierre Mouillard, *L'empire De L'air: Essai D'ornithologie Appliquée A L'aviation* (Paris: Ulan Press, 2012).

148. Walcott, *Biographical Memoir of Samuel Pierpont Langley*, 253–254.

149. Rathbun to Baker, 8 October 1901; Baker to Blackburne, 9 October 1901, RU 74, Box 99, Folder 8.

150. Baker to Langley, 9 October 1901, RU 74, Box 99, Folder 8.

151. Note on scaffolding costs, 6 November 1901, RU 74, Box 99, Folder 8.

152. Langley to Baker, 8 February 1902; Baker to Johnson, 11 February 1902, RU 74, Box 99, Folder 8. The entire world of the Smithsonian was at Langley's fingertips. For example, Langley requested that A. B. Johnson, chief clerk of the Treasury Department's Light-House Board, make the compass cards. This demonstrates the power that Langley commanded within the greater Smithsonian community and underscores the rootedness of the National Zoo within the bureaucracies of federal government.

153. A. B. Baker to Frank Baker, 26 November 1901, RU 74, Box 99, Folder 8; Langley to Hodge, 13 December 1901, RU 74, Box 99, Folder 9.

154. Langley to Smillie, 8 February 1902; Langley to Baker, 12 February 1902, RU 74, Box 99, Folder 8; Langley to Bausch and Lomb Optical Company, 29 April 1902 & 7 May 1902; Langley to Rochester Optical and Camera Company, 5 June 1902; Scrawled note, 29 April 1902, RU 74, Box 99, Folder 9.

155. I use "technics," twice in this chapter, instead of "technology" to suggest subtly—with posthumanist philosopher Bernard Stiegler in mind—that the cameras and recording devices here are indeed part of a much larger philosophical anthropology in which the technics *of* humans (not humans themselves) forged "the human" (its Being, its Time) and thus "the animal" as incidental byproducts of their primary function as tools of memory and repression. As I suggest in the introduction, the love affair between these two opposites is integral to the structure of the zoo. See Bernard Stiegler, *Technics and Time, 1: The Fault of Epimetheus*, trans. Richard Beardsworth and George Collins (Stanford, CA: Stanford University Press, 1998).

156. Langley to Manly, 8 October 1902, RU 74, Box 99, Folder 9.

157. Rene Bache, "Photographing Wild Birds on the Wing to Find Out How to Make a Flying Machine for Man" (June 19, 1902), "Scrapbook, 1896–1907," RU 74, Box 286, Folder 3, 1.

158. Bache, "Photographing Wild Birds on the Wing to Find Out How to Make a Flying Machine for Man," 2–3.

159. Bache, "Photographing Wild Birds on the Wing to Find Out How to Make a Flying Machine for Man," 3–5.

160. Beebe to Baker, 27 May 1908, RU 74, Box 99, Folder 9.

161. Baker to Beebe, 22 May 1908, RU 74, Box 99, Folder 9.

162. Baker to Pearson, 28 May 1908, RU 74, Box 99, Folder 9.

163. Baker to Beebe, 11 June 1908, RU 74, Box 99, Folder 9. On Pearson, see Oliver H. Orr Jr., *Saving American Birds: T. Gilbert Pearson and the Founding of the Audubon Movement* (Gainesville: University Press of Florida, 1992).

164. Biomimetics is a growing and diverse field of thought that places ecological and sustainable thinking at the center of human endeavors that would have once been examples of pure anthropocentrism—construction, locomotion, medicine, optics, agriculture, robotics, and so on. For an introduction to this expansive field, see Amina Khan, *Adapt: How Humans Are Tapping into Nature's Secrets to Design and Build a Better Future* (New York: St. Martin's Press, 2017). In this genre of biomimetics, though in a much different spirit from Langley, see also Robin Wall Kimmerer, *Braiding Sweetgrass: Indigenous Wisdom, Scientific Knowledge and the Teachings of Plants* (Minneapolis, MN: Milkweed Editions, 2013).

165. Raber, *Animal Bodies, Renaissance Culture*, 38.

166. Thomas Nagel, "What Is It Like to Be a Bat?" *Philosophical Review* 83, no. 4 (October 1974): 435–450; Jacques Derrida, *The Animal That Therefore I Am*, ed. Marie-Louise Mallet, trans. David Wills (New York: Fordham University Press, 2008).

167. It should not come as a surprise that these experiments failed. As historian of science Oliver Hochadel acknowledges, many scientific observers across zoo history (from Alfred Edmund Brehm to Heini Hediger) complained that "the zoo did not live up to expectations." Oliver Hochadel, "Science at the Zoo: An Introduction," *Centaurus: Journal of the European Society for the History of*

Science 64, no. 3 (2022): 564–565. For more on science in zoo history—regarding zoos in Barcelona, Berlin, London, Munich, New York, Paris, and Poznań—see the special issue introduced by this article.

168. Ian Jared Miller, *The Nature of the Beasts: Empire and Exhibition at the Tokyo Imperial Zoo* (Berkeley: University of California Press, 2013), 2–3.

169. Bruno Latour, *We Have Never Been Modern*, trans. Catherine Porter (Cambridge, MA: Harvard University Press, 1993).

170. Timothy Morton, *Dark Ecology: For a Logic of Coexistence* (New York: Columbia University Press, 2016), 5.

171. Achille Mbembe, *On the Postcolony*, trans. Steven Rendell (Berkeley: University of California Press, 2001), 6; Michel Foucault, *The Order of Things: An Archeology of the Human Sciences* (New York: Vintage Books, 1994), 16.

172. On Levinas and the "faces" of animals, see Peter Atterton and Tamra Wright, eds., *Face to Face with Animals: Levinas and the Animal Question* (Albany: SUNY Press, 2019).

Chapter 4. Animal Activism and the Zoo-Networked Nation

1. On animal-centered activism prior to World War II, in the British context, see Harriet Ritvo, *The Animal Estate: The English and Other Creatures in the Victorian Age* (Cambridge, MA: Harvard University Press, 1987), 125–166; Moira Ferguson, *Animal Advocacy and Englishwomen, 1780–1900: Patriots, Nation, and Empire* (Ann Arbor: University of Michigan Press, 2001); and Chien-hui Li, *Mobilizing Traditions in the First Wave of the British Animal Defense Movement* (London: Palgrave Macmillan, 2019). For the French context, see Kathleen Kete, *The Beast in the Boudoir: Petkeeping in Nineteenth-Century Paris* (Berkeley: University of California Press, 1994). For the US context, see Diane Beers, *For the Prevention of Cruelty: The History and Legacy of Animal Rights Activism in the United States* (Athens: Ohio University Press, 2006); and Gary Jarvis, "The Road Not Taken: Humanitarian Reform and the Origins of Animal Rights in Britain and the United States, 1883–1919," PhD diss., University of Iowa, 2009. For expansive histories, see James Turner, *Reckoning with the Beast: Animals, Pain, and Humanity in the Victorian Mind (Baltimore, MD: Johns Hopkins University Press, 2000);* and Gerald Carson, *Men, Beasts, and Gods: A History of Cruelty and Kindness to Animals* (New York: Charles Scribner's Sons, 1972). On animals and the ethics of the turn-of-the-century laboratory, especially on vivisection, see Theodore G. Obenchain, *The Victorian Vivisection Debate: Frances Power Cobbe, Experimental Science and the "Claim of Brutes"* (New York: McFarland, 2012).

2. For discussions about zoos and animal rights, later in the twentieth century, see Elizabeth Hanson, *Animal Attractions: Nature on Display in American Zoos* (Princeton, NJ: Princeton University Press, 2002), 179–184; Stephen St. C. Bostock, *Zoos and Animal Rights: The Ethics of Keeping Animals* (New York: Routledge, 1993); and Bryan G. Norton, Michael Hutchins, Elizabeth F. Stevens, and Terry L. Maple, eds., *Ethics on the Ark: Zoos, Animal Welfare, and Wildlife Conservation* (Washington, DC: Smithsonian Institution Press, 1995). Most important,

see Dale Jamieson's "Against Zoos," in *In Defense of Animals*, ed. Peter Singer (New York: Blackwell, 1984), 132–143. For conversations within zoos, see Jesse Donahue, ed., *Increasing Legal Rights for Zoo Animals: Justice on the Ark* (Lanham, MD: Lexington Books, 2017).

3. Janet Davis, *The Gospel of Kindness: Animal Welfare and the Making of Modern America* (Oxford: Oxford University Press, 2016). See also her "Cockfight Nationalism: Blood Sport and the Moral Politics of American Empire and Nation Building," *American Quarterly* 65, no. 3 (September 2013): 549–574.

4. "Uncle Sam's Elephant" (*Washington Post*, April 21, 1891), "Scrapbooks, Vol. 3, 1890–1896," RU 74, Box 286, Folder 2.

5. "Elephant for the Zoo" (*Evening Star*, April 21, 1891), "Scrapbooks, Vol. 3."

6. "Presented to the Zoo" (*Evening Star*, April 30, 1891), "Scrapbooks, Vol. 3"; Sybil Hamlet, "History of the National Zoological Park, 1805–1987," unpublished manuscript, undated, RU 365, Box 37, Folder 1, 1891, 11; Guro Flinterud, "The Rise and Fall of Polar Bear Knut," in *Zoo Studies: A New Humanities*, ed. Tracy McDonald and Daniel Vandersommers (Montreal: McGill-Queens University Press, 2019), 201.

7. Untitled article (*Washington Post*, May 1, 1891), "Scrapbooks, Vol. 3."

8. "An Elephant at the Zoo" (*Evening Star*, May 1, 1891), "Scrapbooks, Vol. 3."

9. Hamlet, Folder 1, 1891, 11.

10. Hanson, *Animal Attractions*, 44, 59–67; Hanson also presents an interesting discussion about how the elephants of Boston's Franklin Park Zoo promoted "civic unity." See also Marianna Szczygielska, "Elephant Empire: Zoos and Colonial Encounters in Eastern Europe," *Cultural Studies* 34, no. 5 (September 2020): 789–810.

11. "An Elephant at the Zoo" (*Evening Star*, May 1, 1891), "A March to the Zoo" (*Evening Star*, May 2, 1891), "Scrapbooks, Vol. 3"; Untitled blurb (unknown newspaper, undated), "Scrapbook, 1887–1902," RU 74, Box 285; Hamlet, Folder 1, 1891, 9–11.

12. "The Zoo Elephants" (*Evening Star*, May 19, 1891), "Scrapbooks, Vol. 3."

13. Nigel Rothfels, *Elephant Trails: A History of Animals and Cultures* (Baltimore, MD: Johns Hopkins University Press, 2021).

14. Jim Mason, "Animals: From Souls and the Sacred in Prehistoric Times to Symbols and Slaves in Antiquity," in *A Cultural History of Animals in Antiquity*, ed. Linda Kalof (New York: Berg, 2007), 17–45.

15. For more on elephants in America, see Susan Nance, *Entertaining Elephants: Animal Agency and the Business of the American Circus* (Baltimore, MD: Johns Hopkins University Press, 2013). According to Nance, viewing elephants as "troublesome brutes" was especially common during the Gilded Age. Earlier in the century, however, elephants were seen as "sagacious." See also Sujit Sivasundaram, "Trading Knowledge: The East India Company's Elephants in India and Britain," *Historical Journal* 48, no. 1 (March 2005): 27–63; Alix Heintzman, "E Is for Elephant: Jungle Animals in Late Nineteenth-Century Picture Books," *Environmental History* 19, no. 3 (July 2014): 553–563; Jonathan Saha, "Colonizing

Elephants: Animal Agency, Undead Capital and Imperial Science in British Burma," *BJHS Themes* 2 (April 2017): 169–189; and Kim Stallwood, "Topsy: The Elephant We Must Never Forget," in *Animal Biography: Re-framing Animal Lives*, ed. André Krebber and Mieke Roscher (Cham, Switzerland: Palgrave Macmillan), 227–242.

16. Daniel E. Bender, *The Animal Game: Searching for Wildness at the American Zoo* (Cambridge, MA: Harvard University Press, 2016), 44–50; "Tip at His Old Tricks" (unknown newspaper), "Scrapbook, 1887–1902." The report that Tip had killed eight keepers is most surely an exaggeration. As director William Conklin commented (in "Tip's Character Is Bad," below), Tip was responsible for the deaths of only two keepers, both of whom were "careless in the way they were handling him."

17. "Tip Smashes Furniture" (*New York Herald Tribune*, June 20, 1891); "Tip's Character Is Bad" (*New York Herald Tribune*, June 29, 1891); "Naughty Miss Fanchon" (*Boston Herald*, February 21,1891); "Damaged by an Elephant" (*New York Herald Tribune*, June 9, 1891); "Vicious Elephant Shot" (unknown newspaper, December 1891?), "Scrapbook, 1887–1902."

18. "Keeper and His Pets" (*Washington Post*, January 12, 1892), "Scrapbook, 1887–1902."

19. "Gold Dust on the Rampage" (*Evening Star*, September 1, 1891), "Scrapbooks, Vol. 3."

20. "Spring Opening at the Zoo" (*Washington Post*, April 3, 1893), "Scrapbooks, Vol. 3."

21. "Dunk's Little Outing" (*Washington Post*, July 12, 1898), Scrapbooks, Vol. 3."

22. "More Room for the Elephants" (*Evening Star*, June 3, 1891), "Scrapbooks, Vol. 3."

23. A. M. Riyas Ahamed, "Activity Time Budget of the Asian Elephant (Elephas Maximus Linn.) in the Wild," *Trends in Biosciences* 8, no. 12 (January 2015): 3024–3028.

24. "Gold Dust Passes Away" (*Evening Star*, November 7, 1898); and "Elephant Gold Dust Dead" (*Washington Post*, November 7, 1898), Scrapbooks, Vol. 3."

25. "Pleads for Elephant" (*Washington Post*, April 16, 1902), "Scrapbooks, 1900–1919," RU 74, Box 287.

26. Beers, *For the Prevention of Cruelty*, 116.

27. "Pleads for Elephant" and "In Behalf of Dunk" (*Evening Star*, April 16, 1902), Scrapbooks, Vol. 3"; Hamlet, Folder 2, 1902, 5.

28. Beers, *For the Prevention of Cruelty*, 75.

29. Louise E. Robbins, *Elephant Slaves and Pampered Parrots: Exotic Animals in Eighteenth-Century Paris: Animals, History, Culture* (Baltimore, MD: Johns Hopkins University Press, 2002), 193.

30. "Space, How Much Is Enough," Global Sanctuary for Elephants, 2017, Accessed June 3, 2019, https://globalelephants.org/space-much-enough/.

31. "A Plea for Dunk" (*Evening Star*, November 8, 1902), Scrapbooks, Vol. 3."

32. "Plea for Dunk."

33. It is astonishing how sometimes "thinking ahead" involves harkening back to old truths erased, suppressed, or ignored. Pliny the Elder in his *Natural History* (ca. 77 CE) wrote this of the elephant: "Its intelligence approaches the nearest to man. It understands the language of its country, it obeys commands, and it remembers all the duties which it has been taught. It is sensible alike of the pleasures of love and glory, and, to a degree that is rare among men even, possesses notions of honesty, prudence, and equity; it has a religious respect also for the stars, and a veneration for the sun and the moon."

34. Ron Broglio, *Surface Encounters: Thinking with Animals and Art* (Minneapolis: University of Minnesota Press, 2011). For a wonderful volume on elephants, see Christen Wemmer and Catherine Christen, eds., *Elephants and Ethics: Toward a Morality of Coexistence* (Baltimore, MD: Johns Hopkins University Press, 2008). See also primatologist Frans de Waal's landmark *Are We Smart Enough to Know How Smart Animals Are?* (New York: Norton, 2016).

35. "But Two Have Escaped" (*Evening Star*, November 22, 1902), Scrapbooks, Vol. 3."

36. Hamlet, Folder 2, 1903, 1.

37. "The Pet of the Zoo" (*Evening Star*, Summer, 1903), Scrapbooks, Vol. 3."

38. On the "human" (and Western humanism) "in disguise" as such, see Friedrich Nietzsche, *The Gay Science*, trans. Walter Kaufmann (New York: Penguin, 1974), 259.

39. These "gaps" are in part historically and discursively constructed, as an enormous swathe of environmental and intellectual history has demonstrated and as I demonstrate in this book. Yet, in my view, they are also derivative of what philosopher Slavoj Žižek theorizes as parallax gaps, separations created by an ontological "cut that separates two incompatible dimensions." This cut *within* incommensurability causes a "doubling back" of "opposites," which forces opposed or paradoxical objects to become conjoined in quirky ways. Slavoj Žižek, *Sex and the Failed Absolute* (London: Bloomsbury, 2019), 430. For more on the psychoanalytic-ontological inspirations that lie deep within this book, see also Slavoj Žižek, *The Parallax View* (Cambridge, MA: MIT Press, 2006); and Alenka Zupančič, *What Is Sex?* (Cambridge, MA: MIT Press, 2017).

40. Jason M. Colby, *Orca: How We Came to Know and Love the Ocean's Greatest Predator* (New York: Oxford University Press, 2018), 5

41. "Dunk, Zoo Elephant, Hurt, Is Shot, And Children Mourn Death of Pet," *Washington Post*, March 31, 1917.

42. Historian Gary Bruce makes a similar observation regarding German zoos of the same period, specifically calling attention to zoologist Alfred Brehm's ten-volume *Life of Animals*. He also emphasizes the importance of elephants (specifically, Chunee and Rostom) to the rise of animal activism in Europe. Gary Bruce, *Through the Lion Gate: A History of the Berlin Zoo* (New York: Oxford University Press, 2017), 105–113.

43. For more on these questions as they shaped the history of science, see

Robert J. Richards, *Darwin and the Emergence of Evolutionary Theories of Mind and Behavior* (Chicago: University of Chicago Press, 1987).

44. W. T. Hornaday, "Report of the Department of Living Animals," 20–21, RU 158, Box 5, folder labeled "Department of Living Animals, 1887–1888"; "Cruelty to Animals at the 'Zoo'" (*Washington Star*, January 10, 1889), "Scrapbook vol. 2, 1888–1891," RU 74, Box 286, Folder 1.

45. "Sunday at the 'Zoo'" (*Washington Post*, June 23, 1889), "Scrapbook vol. 2."

46. "Tossed by a Buffalo Cow" (*Washington Post*, June 9, 1890), "Scrapbooks, 1887–1900." The National Zoo before 1920 was a less openly discriminatory place than other locations in the city. This zoo, unlike others, welcomed African American zoogoers. The zoo of course, though, factored into the broader racist cartography of the city. For example, in 1906, the National Zoo allowed African American children to take part in an Easter egg–rolling event since they were excluded from the traditional, whites-only egg-rolling that took place at the White House. In this way, the zoo, a place for animals, was viewed as an appropriate place for those pushed out of the national public sphere. Hamlet, Folder 2, 1906, 6–7.

47. "Don't Bother the Animal" (*Washington Post*, June 11, 1890), "Scrapbooks, 1887–1900."

48. Randy Malamud, *Reading Zoos: Representations of Animals and Captivity* (New York: New York University Press, 1998), 179.

49. "He Spat in the Bear's Eye" (*Evening Star*, August 6, 1890), "Scrapbooks, 1887–1900."

50. "Zoo's Receiving Day" (*Washington Post*, November 13, 1893), "Scrapbooks, vol. 3."

51. "Diana Is Dead" (*Evening Star*, June 1, 1895) and "Diana and the Laurel" (*Outlook*, June 22, 1895), "Scrapbooks, 1887–1900."

52. "Dynamite Made Them Roar" (*Washington Post*, October 24, 1895) and "Killed by Explosion" (*Evening Star*), "Scrapbooks, vol. 3."

53. Scribbled Note, 29 November 1913, RU 74, Box 102, Folder 3.

54. "Keeping the Animals Cool" (*Washington Post*, July 28, 1892), "Scrapbooks, vol. 3."

55. "Animals at Zoo" (*Evening Star*, August 13, 1892), "Scrapbooks, vol. 3."

56. "Baby Elk Born" (*Washington Post*, June 26, 1893), "Scrapbooks, vol. 3."

57. "Fraternalism at Zoo" (*Washington Post*, August 22, 1892), "Scrapbooks, vol. 3."

58. "Zoo in Winter Quarters" (*Washington Post*, October 17, 1892), "Scrapbooks, vol. 3."

59. Baker to Diehl, 29 October 1900, RU 74, Box 96, Folder 3.

60. Blackburne to Baker, 2 January 1906, RU 74, Box 96, Folder 3.

61. "Big Crowds Throng Zoo" (*Washington Post*, October 15, 1906), "Scrap books, volume 4, 1896–1907," RU 74, Box 286, Folder 3.

62. "The National Zoological Park" (*Our Animal Friends*, 1890), "Scrapbooks, 1887–1902." Even though *Our Animal Friends* was the magazine of the

ASPCA, by no means was this article an attack on the zoo. In fact, Adney at times spoke almost like a zoo booster, explicitly distancing zoos from their circus and menagerie predecessors.

63. Mulhall to Baker, 1 February 1905, RU 74, Box 102, Folder 1.

64. Baker to Mulhall, 3 February 1905, RU 74, Box 102, Folder 1; Gregory A. Franckowiak, Marlo Perdicas, and Gregory A. Smith, "Spatial Ecology of Coyotes in the Urbanizing Landscape of the Cuyahoga Valley, Ohio," *PloS One* 14, no. 12 (December 30, 2019): e0227028.

65. Frederick L. Brown, *The City Is More Than Human: An Animal History of Seattle* (Seattle: University of Washington Press, 2016), 41–42, 118–120, 182–183. See especially Carol Adams, *The Sexual Politics of Meat: A Feminist-Vegetarian Critical Theory* (New York: Continuum, 1990). See also Davis, *Gospel of Kindness*; Ferguson, *Animal Advocacy and Englishwomen*; Susan Pearson, *The Rights of the Defenseless: Protecting Animals and Children in Gilded Age America* (Chicago: University of Chicago Press, 2011); and Emily Gaarder, *Women and the Animal Rights Movement* (New Brunswick: Rutgers University Press, 2011).

66. Miller to Baker, (?) February 1905, RU 74, Box 102, Folder 1.

67. Barber to Gentlemen, 8 October 1907, RU 74, Box 102, Folder 1.

68. Miller to Baker, (?) December 1914, RU 74, Box 102, Folder 3. After receiving this letter, Baker asked William Blackburne to wash the Eskimo dogs' eyes out. Blackburne was not sure if this would remove the greenish matter. Nonetheless, Miller's complaint did cause the zoo to take action. Blackburne to Baker, 13 May 1914, RU 74, Box 102, Folder 3.

69. Williams to Superintendent, 7 July 1914, RU 74, Box 102, Folder 3.

70. Raub to Walcott, 18 December 1909, RU 74, Box 102, Folder 2.

71. Baker to Secretary, 20 December 1909, RU 74, Box 102, Folder 2. Feral dogs also wandered through the zoo. Occasionally, they killed and injured zoo animals, and they often frightened zoogoers. References to the "wild dogs" of the zoo appear often in correspondence. For more on feral dogs, see Abe Gibson, "Dog Days at the Zoo, Part I," Smithsonian Institution Archives (September 2012), accessed May 15, 2020, http://siarchives.si.edu/blog/dog-days-zoo-part-i. For more on squirrels in urban parks, see Etienne Benson, "The Urbanization of the Eastern Gray Squirrel in the United States," *Journal of American History* 100, no. 3 (December 2013): 691–710. In 1902, the National Zoo received eight black squirrels and fifty gray squirrels from Crown Lands, Ontario, and the zoo released them in the park, allowing them to live on the grounds and propagate. This set the precedent. Throughout its early years, the National Zoo released thirty-six mountain chipmunks; 205 black, gray, and albino squirrels; thirty-two fox squirrels; and nine prairie hens. Hamlet, Folder 2, 1902, 3.

72. Raub to Walcott, 18 December 1909, RU 74, Box 102, Folder 2.

73. Baker to Secretary, 2 December 1913, RU 74, Box 102, Folder 3. In this letter, Baker explained that "the Park has been considerably troubled by vagrant cats and guns are kept at hand to dispose of them. They nest and breed in the gutter pipes and are as wild as any animals of the woods. . . . The neighbors are

aware of our difficulty as we have had several complaints because of the shooting of pet cats."

74. Bacon to Walcott, 6 December 1913, RU 74, Box 102, Folder 3.

75. Andrew Robichaud, interview by Catherine McNeur, "Animal City: The Domestication of America," Boston University Initiative on Cities, February 9, 2021, https://www.bu.edu/ioc/2021/02/05/animal-city-the-domestication-of-america/; Andrew Robichaud, *Animal City: The Domestication of America* (Cambridge, MA: Harvard University Press, 2019), 248. On "sacrifice zones" in an environmental justice context, see Steve Lerner, *Sacrifice Zones: The Front Lines of Toxic Chemical Exposure in the United States* (Cambridge, MA: MIT Press, 2010). On pests, urban spaces, and invisibility, see Dawn Day Biehler, *Pests in the City: Flies, Bedbugs, Cockroaches, and Rats* (Seattle: University of Washington Press, 2013).

76. Raub to Walcott, 6 November 1914; Baker to Walcott, 9 November 1914, RU 74, Box 102, Folder 3.

77. Malamud, *Reading Zoos*, 15–23. For an analysis of the animality of Marianne Moore, see Hatley Clifford, "A More-Than-Human Green: National Parks and Animality in Marianne Moore's 'An Octopus,'" in *Modernism in the Green: Public Greens in Modern Literature and Culture*, ed. Julia E. Daniel and Margaret Konkol (New York: Routledge, 2020).

78. Susan Nance, *Animal Modernity: Jumbo the Elephant and the Human Dilemma* (London: Palgrave Macmillan, 2015), 76–77, 3–5, 81.

79. Karen Iacobbo and Michael Iacobbo, *Vegetarian America: A History* (Westport, CT: Praiger, 2004), 119–121; Roderick Frazier Nash, *The Rights of Nature: A History of Environmental Ethics* (Madison: University of Wisconsin Press, 1989), 52–54.

80. Gunderson to Baker, 28 July 1910, RU 74, Box 102, Folder 2, 1–2. Read Gunderson's entire letter at Daniel Vandersommers, "Animal Activism and the Zoo-Networked Nation," *Humanimalia* 6, no. 2 (March 2015): 111–165.

81. Rousseau contended in *Discourse on Inequality* that "being destitute of intelligence and liberty, [animals] cannot recognize that [natural] law; as they partake, however, in some measure of our nature, in consequence of the sensibility wherewith they are endowed, *they ought to partake of natural right*; so that mankind is subjected to a *kind of obligation* even toward the brutes." Kant, on the other hand, in his *Groundwork for the Metaphysics of Morals*, argued that humans did not have direct responsibilities to animals but should nonetheless refrain from acting violently toward them, since violence would bring out the bad qualities of mankind.

82. Chien-hui Li, *Mobilizing Traditions in the First Wave of the British Animal Defense Movement* (London: Palgrave Macmillan, 2019), 121–127; Timothy Morton, *Shelley and the Revolution in Taste: The Body and the Natural World* (New York: Cambridge University Press, 1994), 13–56.

83. Li, *Mobilizing Traditions in the First Wave of the British Animal Defense Movement*, 120.

84. Gunderson to Baker, 28 July 1910, 2–3.

85. Gunderson to Baker, 28 July 1910, 2–3.

86. Gunderson to Baker, 28 July 1910, 4–5.

87. Gunderson to Baker, 28 July 1910, 6.

88. David Von Drehle, *Triangle: The Fire That Changed America* (New York: Grove Press, 2003), 32.

89. Peter Harrison, "Descartes on Animals," *Philosophical Quarterly* 42, no. 167 (April 1992): 219. The Cartesian system, especially animal mechanism, was not accepted completely and without question by scientists; it still competed with other systems. Nonetheless, Cartesian language certainly gained power as animals became industrialized in the nineteenth century. See Anita Guerrina, *The Courtiers' Anatomists: Animals and Humans in Louis XIV's Paris* (Chicago: University of Chicago Press, 2015), 5, 121–122.

90. See Pearson, *Rights of the Defenseless.*

91. See Peter Singer, *Animal Liberation: A New Ethics for Our Treatment of Animals* (New York: Harper Collins, 1975), Tom Regan, *The Case for Animal Rights* (Berkeley: University of California Press, 1983), and Giorgio Agamben, *The Open: Man and Animal*, trans. Kevin Attell (Stanford, CA: Stanford University Press, 2003).

92. French to Baker, 8 November 1908, RU 74, Box 102, Folder 1.

93. W. W. B. to Baker, 31 May 1911, RU 74, Box 102, Folder 3.

94. Baker to Bride, 9 June 1911, RU 74, Box 102, Folder 3.

95. See, for example, Heidi Ahonen, Robert G Harcourt, Adam J Stow, and Isabelle Charrier, "Geographic Vocal Variation and Perceptual Discrimination Abilities in Male Australian Sea Lions," *Animal Cognition* 21, no. 2 (March 2018): 235–243.

96. W. W. B. to Baker, 3 June 1911, RU 74, Box 102, Folder 3.

97. "Zoo Appropriation" (*Evening Star*, April 15, 1892), "Scrapbooks, 1887–1900."

98. Untitled Article (*Bulletin,* July 16, 1874), *Scrap Book* (1866–1903), Item #PUB R1, PZA.

99. "The Zoological" (*Bulletin,* July 20, 1874), *Scrap Book.*

100. "More Foolish Legislation Coming" (unknown newspaper, 1876?), *Scrap Book.*

101. See James McGrath Morris, *Pulitzer: A Life in Politics, Print, and Power* (New York: Harper Collins, 2010).

102. The zoo and the asylum share a common intellectual-architectural history. See Matthew Senior, "Psychotic Humans, Psychotic Animals: The Zoo and the Mental Hospital, 1656–1664," in *Zoo Studies*, 19–44.

103. On "Nellie Bly," see Brooke Kroeger, *Nellie Bly: Daredevil, Reporter, Feminist* (Pittsburgh, PA: Three Rivers Press, 1995); and Jean Marie Lutes, "Into the Madhouse with Nellie Bly: Girl Stunt Reporting in Late Nineteenth-Century America," *American Quarterly* 54, no. 2 (June 2002): 217–253.

104. e. e. cummings, "The Secret of the Zoo Exposed," in *E. E. Cummings: A Miscellany*, ed. George J. Firmage (New York: Argophile Press, 1958).

105. "Nellie Bly's Inferno: Sufferings of Central Park Animals: A Legalized Institution of Torture" (*World*, August 27, 1894), "Scrapbooks, 1887–1900."

106. "Nellie Bly's Inferno."

107. "Nellie Bly's Inferno."

108. "Why These Cruelties?" (*World*, September 13, 1894), "Scrapbooks, 1887–1900."

109. "Gov. Altgeld at the Zoo" (*World*, September 16, 1894), "Scrapbooks, 1887–1900."

110. "Poor Beasts, If You Knew!" (*World*, September 30, 1894), "Scrapbooks, 1887–1900." For more on the mutual dehumanization of animals and Black Americans, see Joshua Bennett, *Being Property Once Myself: Blackness and the End of Man* (Cambridge, MA: Harvard University Press, 2020).

111. On Bergh, see Ernest Freeberg, *A Traitor to His Species: Henry Bergh and the Birth of the Animal Rights Movement* (New York: Basic Books, 2020).

112. "Poor Beasts, If You Knew!"

113. I am riffing here on Friedrich Nietzsche, *Beyond Good and Evil: Prelude to a Philosophy of the Future*, trans. Walter Kaufmann (New York: Vintage, 1989).

114. John Berger, "Why Look at Animals?" in *About Looking* (New York: Pantheon Books, 1980), 4–5.

Chapter 5. Zoo Conservation and Its Discontents: Chasing Bighorn Sheep

1. See, for example, Timothy Sweet, *Extinction and the Human: Four American Encounters* (Philadelphia: University of Pennsylvania Press, 2021), 103–129.

2. Rob Roy Ramey II, "Evolutionary Genetics and Systematics of North American Mountain Sheep: Implications for Conservation," PhD diss., Cornell University, 1993. It is possible that mountain sheep migrated from Eurasia to North America during any of the glacial periods, along with other mammals like mammoths, bears, moose, and musk ox.

3. The understanding that North American mountain sheep fall into three distinct species was forged in 1940, and, for the most part, this understanding has remained uncontested despite continuing debate about subspecies. See Ian McTaggart Cowan, "Distribution and Variation in the Native Sheep of North America," *American Midland Naturalist* 24, no. 3 (November 1940): 506–507.

4. John D. Wehausen, Vernon C. Bleich, Rob R. Ramey II, "Correct Nomenclature for Sierra Nevada Bighorn Sheep," *California Fish and Game* 91, no. 3 (June 2005): 216–218; John D. Wehausen and Rob Roy Ramey II, "Cranial Morphometric and Evolutionary Relationships in the Northern Range of *Ovis Canadensis*," *Journal of Mammalogy* 81, no. 1 (February 2000): 145–161.

5. Ellen Meloy, *Eating Stone: Imagination and the Loss of the Wild* (New York: Vintage Books, 2005), 128.

6. Rick Graetz and Susie Graetz, *Crow Country: Montana's Crow Tribe of Indians* (Billings: Northern Rockies Publishing Company, 2000).

7. With "contact zones," I am referencing Donna J. Haraway, building upon

Mary Louise Pratt, in *When Species Meet* (Minneapolis: University of Minnesota Press, 2007), 205–247.

8. For magnificent bighorn art, see David S. Whitley, *The Art of the Shaman: Rock Art of California* (Salt Lake City: University of Utah Press, 2000); and Kenneth B. Castleton, *Petroglyphs and Pictographs of Utah, Vol. 1 and Vol. 2* (Salt Lake City: Utah Museum of Natural History, 2002).

9. Stephen Greenblatt, *Marvelous Possessions: The Wonder of the New World* (Chicago: University of Chicago Press, 1991).

10. Robert Jameson and Thomas Laurie, "Art. XXXVII.—On the Rocky Mountain Sheep of the Americans: Remarks for the Wernerian Society on the Skin of the Rocky Mountain Sheep," *The Boston Journal of Philosophy and the Arts, Intended to Exhibit a View of the Progress of Discovery in Natural Philosophy, Mechanics, Chemistry, Geology and Mineralogy, Natural History, Comparative Anatomy and Physiology, Geography, Statistics and the Fine and Useful Arts* 1 (May 1, 1824): 340. In this article, Jameson quoted Jesuit Miguel Venegas's *Noticia de la California*, published in 1757, which synthesized the reports of numerous missionaries across Sinaloa, Sonora, and Pimería Alta.

11. Duncan McGillivray, "Article I: Account of the Wild North-American Sheep. Memorandum Respecting the Mountain Ram of North America," *The Medical Repository of Original Essays and Intelligence, Relative to Physic, Surgery, Chemistry, and Natural History* 6, no. 3 (August 1, 1803): 239.

12. J. A. Allen, "Historical and Nomenclatorial Notes on North American Sheep," in *Bulletin of the American Museum of Natural History*, ed. J. A. Allen, vol. 31 (New York: Trustees of the AMNH, 1912): 2–9. This publication represents the most complete account of North American mountain sheep until the mid-twentieth century. In this article, J. A. Allen (zoologist, ornithologist, and former student of Louis Agassiz) wrote the first detailed intellectual genealogy of wild American sheep, mostly concerned with issues of classification and naming.

13. Allen, "Historical and Nomenclatorial Notes on North American Sheep," 15.

14. Jameson and Laurie, "Art. XXXVII.—On the Rocky Mountain Sheep of the Americans: Remarks for the Wernerian Society on the Skin of the Rocky Mountain Sheep," 341.

15. Helen Cowie, *Llama* (London: Reaktion, 2017), 110–111; Robert Marks, *The Origins of the Modern World: A Global and Ecological Narrative from the Fifteenth to the Twenty-First Century*, 2nd ed. (New York: Rowman & Littlefield, 2007), 108.

16. PR, "Mountain Sheep," *Spirit of the Times: A Chronicle of the Turf, Agriculture, Field Sports, Literature and the Stage* 28, no. 16 (May 29, 1858): 188.

17. Thinking with the topic of biomimetics raised in chapter 3, the porous bone architecture of mountain sheep horns is currently being studied to better understand energy absorption and to develop successful approaches to brain injury mitigation. Luca H. Fuller and Seth W. Donahue, "Material Properties of Bighorn Sheep (Ovis canadensis) Horncore Bone with Implications for Energy

Absorption Impacts," *Journal of the Mechanical Behavior of Biomedical Materials* 114 (February 2021).

18. As historian Edmund Russell argues, hunters' obsession over large antlers created an evolutionary selective advantage for sheep with (genetic predispositions for) smaller horns (and thus smaller bodies). Therefore, as European-Americans began hunting big game (including sheep), that "big" game became smaller, even within just a few generations. Edmund Russell, *Evolutionary History: Uniting History and Biology to Understand Life on Earth* (Cambridge: Cambridge University Press, 2011), 19–21.

19. "Big Horns, or Rocky Mountain Sheep, Surprised," *Frank Leslie's Popular Monthly* 3, no. 3 (March 1877): 280–281.

20. Dunraven, "Sheep-Hunting in the Mountains," *Littell's Living Age* 151, no. 1954 (December 3, 1881): 535–546.

21. "Stalking the Mountain Sheep," *Forest and Stream: A Journal of Outdoor Life, Travel, Nature Study, Shooting, Fishing, Yachting* 23, no. 1 (July 31, 1884): 4.

22. Howard Oliver, "Capturing Rocky Mountain Sheep," *Overland Monthly and Out West Magazine* 17, no. 102 (June 1891): 634–640.

23. Mary Austin, "The Rocky Mountain Sheep," *St. Nicholas; an Illustrated Magazine for Young Folks* 27, no. 11 (September 1900): 954.

24. Baker to Langley, 2 October 1890, RU 74, Box 7, Folder 1. This early bighorn did not live long. An autopsy revealed a brain clot and congestion of the cortex and ventricles, the signature marks of apoplexy. The Philadelphia Zoo and Lincoln Park Zoo also had previously, and briefly, held mountain sheep in captivity. Baker to Sleeper, 8 January 1901, RU 74, Box 81, Folder 2.

25. William T. Hornaday, "The Buffalo, Musk-Ox, Mountain Sheep, and Mountain Goat," *St. Nicholas; an Illustrated Magazine for Young Folks* 22, no. 8 (June 1895): 680–681.

26. Elizabeth Hanson, *Animal Attractions: Nature on Display in American Zoos* (Princeton, NJ: Princeton University Press, 2002), 136–137.

27. Mark V. Barrow Jr., "The Specimen Dealer: Entrepreneurial Natural History in America's Gilded Age," *Journal of the History of Biology* 33, no. 3 (December 2000): 493–532.

28. Ten years later, Elrod will play a central role in the bison conservation projects organized by Hornaday. Andrew C. Isenberg, "The Returns of the Bison: Nostalgia, Profit, and Preservation," *Environmental History* 2, no. 2 (April 1997): 187–188.

29. Baker to Elrod, 1 March 1899, RU 74, Box 80, Folder 10. Elrod's papers are held at the Archives and Special Collections of the Maureen and Mike Mansfield Library at The University of Montana-Missoula. His papers are quite voluminous and would be illuminating to anyone interested in the American West, especially topics concerning Montana wildlife or the history of Glacier National Park.

30. Baker to Rock, 4 March 1899, RU 74, Box 80, Folder 10.

31. Baker to Elrod, 13 March 1899, RU 74, Box 80, Folder 10.

32. Baker to Smith, 22 September 1899; Baker to Cotton, October 19, 1899, RU 74, Box 80, Folder 10.

33. Langley to Smith, 13 November 1899; "Articles of Agreement," RU 74, Box 80, Folder 10.

34. Wolcott to Langley, 15 February 1900; Langley to Wolcott, 11 April 1900, RU 74, Box 80, Folder 11.

35. Langley to Roosevelt, 18 April 1900, RU 74, Box 80, Folder 11.

36. Roosevelt to Langley, 18 April 1900, RU 74, Box 80, Folder 11. For more on Theodore Roosevelt and his passion for nature, wildlife, and the outdoors, see Douglas Brinkley, *The Wilderness Warrior: Theodore Roosevelt and the Crusade for America* (New York: Harper Collins, 2009); and Timothy Egan, *The Big Burn: Teddy Roosevelt and the Fire That Saved America* (New York: Mariner Books, 2010).

37. Swift to Baker, 19 April 1900; Baker to Ferguson, 18 April 1900; Baker to Root, 18 April 1900; Baker to Langley, 19 April 1900; Baker to Langley, 20 April 1900, RU 74, Box 80, Folder 11.

38. Langley to Baker, 16 April 1900, RU 74, Box 80, Folder 11.

39. Baker to Langley, 17 April 1900, RU 74, Box 80, Folder 11.

40. Theodore Roosevelt, "Mountain Sheep," in *Hunting Trips of a Ranchman: Sketches of Sport on the Northern Cattle Plains* (New York: Putnam, 1886), 242.

41. Like Elrod, Jones would also play an important (yet reckless) role in bison conservation and breeding. See Mackenzie Brown and Robert Easton, *Lord of Beasts: The Saga of Buffalo Jones* (Tucson: University of Arizona Press, 1961). See also Isenberg, "The Returns," 183, 186–189; and Andrew C. Isenberg, *The Destruction of the Bison: An Environmental History* (Cambridge: Cambridge University Press, 2000), 172–185. Later in life, Jones acted as the first game warden of Yellowstone National Park.

42. Baker to Langley, 20 April 1900, RU 74, Box 80, Folder 11.

43. "Articles of Agreement."

44. Jones to Baker, 20 May 1900, RU 74, Box 80, Folder 12.

45. Jones to Langley, 22 May 1900, RU 74, Box 80, Folder 12.

46. Baker to Jones, 1 June 1900, RU 74, Box 80, Folder 13.

47. Some evidence exists that Jones's party did not catch these two lambs themselves. Instead, they may have been caught by local resident Charles Smith. If this rumor is true, Jones, then, bought the lambs from Smith and pretended that he caught them himself. Unknown newspaper blurb, RU 74, Box 80, Folder 12.

48. "Wild Big Horns," from unnamed Topeka newspaper, RU 74, Box 80, Folder 13.

49. Jones to Baker, 27 May 1900, RU 74, Box 80, Folder 14.

50. Jones to Baker, 19 June 1900, RU 74, Box 80, Folder 13.

51. Jones to Baker, 19 June 1900.

52. Jones to Baker, 26 June 1900, RU 74, Box 80, Folder 13.

53. Jones to Baker, 26 June 1900.

54. Redieske to Baker, 24 September 1900, RU 74, Box 80, Folder 14.

55. Jones to Baker, 7 July 1900, RU 74, Box 80, Folder 13.

56. Jones to Doctor, 8 August 1900, RU 74, Box 80, Folder 13.

57. Jones to Baker, 16 August 1900, RU 74, Box 80, Folder 13.

58. Jones to Baker, 16 August 1900, telegram, RU 74, Box 80, Folder 14.

59. Jones to Baker, 7 September 1900, RU 74, Box 80, Folder 13.

60. Baker to Jones, 19 September 1900, RU 74, Box 80, Folder 13.

61. Jones to Baker, 19 September 1900, RU 74, Box 80, Folder 13.

62. Jones to Baker, 24 September 1900, RU 74, Box 80, Folder 13.

63. Goodnight note, undated, RU 74, Box 80, Folder 13.

64. Baker to Goodnight, 26 September 1900, RU 74, Box 80, Folder 13.

65. Jones to Baker, 30 September 1900, RU 74, Box 80, Folder 14.

66. Beaman to Baker, 18 December 1901, RU 74, Box 81, Folder 8.

67. John Webster, "Sentient Animals Do Not Just Live in the Present," *Animal Sentience* 25, no. 10 (2019); Keith M. Kendrick, "Social Cognition in Sheep," *Animal Sentience* 25, no. 40 (2019).

68. Haraway, *When Species Meet*, 60.

69. Peter Novick, *That Noble Dream: The "Objectivity Question" and the American Historical Profession* (Cambridge: Cambridge University Press, 1988), 1–18.

70. While this chapter follows Colorado's *wild* sheep, for more about the state's domestic sheep industry, see Andrew Gulliford, *The Woolly West: Colorado's Hidden History of Sheepscapes* (College Station: Texas A&M University Press, 2018).

71. Baker to Langley, 4 December 1900, RU 74, Box 81, Folder 1.

72. Baker to Langley, 18 March 1901, RU 74, Box 81, Folder 3.

73. Baker to Shaffer, 20 September 1900, RU 74, Box 81, Folder 1; Theodore Whaley Cart, "The Lacey Act: America's First Nationwide Wildlife Statute," *Forest History* 17, no. 3 (October 1973): 4–13; Mark V. Barrow Jr., *Nature's Ghosts: Confronting Extinction from the Age of Jefferson to the Age of Ecology* (Chicago: University of Chicago Press, 2009), 105.

74. Langley to Teller, 17 January 1901, RU 74, Box 81, Folder 2.

75. Langley to Teller, 22 January 1901, RU 74, Box 81, Folder 2; "*Ovis canadensis*," Animal Diversity Web, University of Michigan Museum of Zoology, https://animaldiversity.org/accounts/Ovis_canadensis/.

76. Langley to Moninger, 21 January 1901, RU 74, Box 81, Folder 2.

77. Harms to Langley, 28 January 1901, RU 74, Box 81, Folder 2.

78. Moninger to Baker, 5 February 1901, RU 74, Box 81, Folder 3.

79. Americans had long viewed mountain lions as nuisances. Like wolves, coyotes, and eagles, mountain lions were exterminated in great numbers, largely to protect the domesticated animals of farmers and ranchers. While Americans reduced the numbers of mountain sheep by hunting, collecting, and transforming their habitats, they also pursued vicious lion-extermination programs that eliminated mountain sheep's primary predator. Though mountain lions, themselves, prefer the slower mule deer as prey, certainly Americans' turn-of-the-century

hatred for predators contributed to the survival of mountain sheep when their hooves were on the brink. On similar hatreds for other predators, see Jon Coleman, *Vicious: Wolves and Men in America* (New Haven, CT: Yale University Press, 2004); and Isabel Hofmeyr, "J is for Jackal (and Dingo)," in *Animalia: An Anti-Imperial Bestiary for Our Times*, ed. Antoinette Burton and Renisa Mawani (Durham, NC: Duke University Press, 2020), 96–100.

80. John Mionczynski, *The Pack Goat* (Boulder, CO: Pruett Publishing, 1992).

81. Brian Fagan, *The Intimate Bond: How Animals Shaped Human History* (New York: Bloomsbury, 2015), 48.

82. Baker to Langley, 26 April 1901, RU 74, Box 81, Folder 4.

83. Sharp to Baker, 25 March 1901, RU 74, Box 81, Folder 3.

84. Sharp to Baker, 19 April 1901, RU 74, Box 81, Folder 4.

85. Sharp to Baker, 31 August 1901, RU 74, Box 81, Folder 7.

86. Jones to Baker, 19 April 1901, RU 74, Box 81, Folder 4.

87. Baker to Jones, 23 April 1901, RU 74, Box 81, Folder 4.

88. Cosley to Baker, 18 December 1900, RU 74, Box 81, Folder 1.

89. Baker to Langley, 26 April 1901.

90. Cosley to Baker, 5 March 1901, RU 74, Box 81, Folder 3.

91. Marco Festa-Bianchet, "Bighorn Sheep," in *The Smithsonian Book of North American Mammals*, eds. Don E. Wilson and Sue Ruff (Washington, DC: Smithsonian Institution Press, 1999), 348–350.

92. Baker to Jones, 24 May 1901, RU 74, Box 81, Folder 5.

93. Jones to Baker, 21 May 1901, RU 74, Box 81, Folder 5.

94. Baker to Shafer, 24 June 1901, RU 74, Box 81, Folder 6.

95. Baker to Fisher, 1 July 1901, RU 74, Box 81, Folder 7.

96. Baker to Monroe, 10 June 1901, RU 74, Box 81, Folder 6.

97. Langley to the Governor of Montana, 11 May 1901, RU 74, Box 81, Folder 5.

98. Scott to Baker, 30 May 1901, RU 74, Box 81, Folder 5.

99. Langley to McKay, 8 March 1901; Baker to Hitt, 9 February 1901, RU 74, Box 81, Folder 3; Baker to Johnstone, 30 August 1901; Baker to Bissell, 30 September 1901; Baker to Chandler, 22 July 1901, RU 74, Box 81, Folder 7.

100. Baker to Hornaday, 21 May 1901, RU 74, Box 81, Folder 5; Baker to Brown, 6 June 1902, RU 74, Box 81, Folder 12.

101. Baker to Ham, 4 June 1901, RU 74, Box 81, Folder 6.

102. Baker to Root, 16 July 1902, RU 74, Box 81, Folder 12.

103. Baker to Frakes, 25 June 1902, RU 74, Box 81, Folder 12.

104. Walker to Director US Zoo, 2 February 1903, RU 74, Box 81, Folder 14.

105. Beaman to Baker, 18 December 1901.

106. Hornaday to Baker, 9 May 1901, RU 74, Box 81, Folder 5.

107. Baker to Field, 12 June 1901, RU 74, Box 81, Folder 6; Baker to Graham, 5 February 1902, RU 74, Box 81, Folder 9.

108. Periolat to Baker, 22 June 1901, RU 74, Box 81, Folder 6.

109. Baker to Burns & Daily, 9 August 1902, RU 74, Box 81, Folder 12.

110. Burns and Daily to Baker, 16 August 1902, RU 74, Box 81, Folder 13.

111. Baker to Chandler, 22 July 1901.

112. Fullerton to Baker, 26 January 1902, RU 74, Box 81, Folder 9.

113. Root to Baker, 21 July 1902, RU 74, Box 81, Folder 12.

114. Phillips to Baker, 17 March 1903, RU 74, Box 81, Folder 14.

115. Frakes to Baker, 15 August 1902, RU 74, Box 81, Folder 12.

116. Moninger to Baker, 12 February 1902, RU 74, Box 81, Folder 9.

117. Baker to Sharp, 5 February 1902, RU 74, Box 81, Folder 9.

118. Baker to Harris, 12 April 1902, RU 74, Box 81, Folder 10.

119. Harris to Baker, 28 March 1902, RU 74, Box 81, Folder 10.

120. Nowlin to Baker, 22 April 1902, RU 74, Box 81, Folder 11.

121. Langley to Clark, 14 February 1903, RU 74, Box 81, Folder 14.

122. Baker to Benson, 12 April 1902, RU 74, Box 81, Folder 10.

123. Baker to Benson, 18 February 1902, RU 74, Box 81, Folder 9.

124. Baker to Underwood, 31 May 1902, RU 74, Box 81, Folder 11.

125. On the more-than-human geographies of westward expansion and their role in Manifest Destiny narratives, see Dominik Ohrem, "The Ends of Man: The Zooanthropological Imaginary and the Animal Geographies of Westward Expansion in Antebellum America," in *American Beasts: Perspectives on Animals, Animality and U.S. Culture, 1776–1920*, ed. Dominik Ohrem (Berlin: Neofelis Verlag, 2017), 259–270.

126. Barrow, "Specimen Dealer," 493–498.

127. Victor J. Evans, *Illustrated Guide Book to the National Zoological Park* (Washington, DC: Henry E. Wilkens Printing Company, 1902), 10, RU 74, Box 120, Folder 10. There is so much scholarly work—historical, literary, and intersectional—to be done on the topic of zoo guidebooks. See Andrew Flack, *The Wild Within: Histories of a Landmark British Zoo* (Charlottesville: University of Virginia Press, 2018), 66–67, 95–96, 111–112, 123; and Daniel E. Bender, *The Animal Game: Searching for Wildness at the American Zoo* (Cambridge, MA: Harvard University Press, 2016), 27–28.

128. Baker to Jones, 4 December 1903, RU 74, Box 81, Folder 14.

129. Rathbun to Sheppard, 5 October 1917; Sheppard to Hollister, 1 October 1917, RU 74, Box 82, Folder 4.

130. Baker to Duclos, 27 July 1917, RU 74, Box 82, Folder 4.

131. Jeremy Vetter, *Field Life: Science in the American West during the Railroad Era* (Pittsburgh: University of Pittsburgh Press, 2016), 76–137.

132. Cudahy to Hollister, 31 July 1922, RU 74, Box 82, Folder 4.

133. Hollister to Cudahy, 29 July 1922, RU 74, Box 82, Folder 4.

134. Anna Tsing, *The Mushroom at the End of the World* (Princeton, NJ: Princeton University Press, 2015), 63–66, 296.

135. Mieke Roscher, "Animals as Signifiers: Re-Reading Michel Foucault's *In the Order of Things* as a Genealogical Working Tool for Historical Human-Animal Studies," in *Beyond the Human-Animal Divide: Creaturely Lives in*

Literature and Culture, ed. Dominik Ohrem and Roman Bartosch (New York: Palgrave Macmillan, 2017), 200.

136. This literature is vast. In addition to the work of Bruno Latour previously cited, see, for example, works as diverse as Susan Scott Parrish, *American Curiosity: Cultures of Natural History in the Colonial British Atlantic World* (Chapel Hill: University of North Carolina Press, 2006); Thomas Gieryn, *Cultural Boundaries of Science: Credibility on the Line* (Chicago: University of Chicago Press, 1999); Sandra Harding, *Sciences from Below: Feminisms, Postcolonialities, and Modernities* (Durham, NC: Duke University Press, 2008); Andrew Pickering, *Constructing Quarks: A Sociological History of Particle Physics* (Chicago: University of Chicago Press, 1984); and Shigehisa Kuriyama's simply incredible *The Expressiveness of the Body and the Divergence of Greek and Chinese Medicine* (New York: Zone Books, 2002).

137. Richard W. Bulliet, *The Camel and the Wheel* (New York: Columbia University Press, 1990); Jason M. Colby, *Orca: How We Came to Know and Love the Ocean's Greatest Predator* (New York: Oxford University Press, 2018); Nancy J. Jacobs, *Birders of Africa: History of a Network* (New Haven, CT: Yale University Press, 2016); Brett Mizelle, *Pig* (London: Reaktion, 2011).

138. See Barrow, *Nature's Ghosts*; Karl Jacoby, *Crimes against Nature: Squatters, Poachers, Thieves, and the Hidden History of American Conservation* (Berkeley: University of California Press, 2001); Lisa Mighetto, *Wild Animals and American Environmental Ethics* (Tucson: University of Arizona Press, 1991); James A. Tober, *Who Owns the Wildlife?: The Political Economy of Conservation in Nineteenth-Century America* (Westport, CT: Greenwood Press, 1981); and Richard W. Judd, *Common Lands, Common People: The Origins of Conservation in New England* (Cambridge, MA: Harvard University Press, 1997).

139. Jedediah Purdy, *After Nature: A Politics for the Anthropocene* (Cambridge, MA: Harvard University Press, 2015), 153–187.

140. For an excellent discussion about this irony, see Louis S. Warren, *The Hunter's Game: Poachers and Conservationists in Twentieth-Century America* (New Haven, CT: Yale University Press, 1999).

141. Thom van Dooren, *Flight Ways: Life and Loss on the Edge of Extinction* (New York: Columbia University Press, 2014), 87–122.

142. Philosopher Slavoj Žižek has asked recently: "What if science will demonstrate that trees and other plants really somehow communicate and react in a panicky way when they are threatened? What if animals in industrial farms live a life of unbearable suffering ignored by us, humans? *But, even worse, what if our blindness for this immense suffering is the obverse, the necessary backside, of our progress? What if humanity with all its great cultural achievements grew out of this blindness? What if, in order to see all that we see, we had to ignore the immense domain of these silent murmurs?*" [Emphasis mine.] I cannot help but feel, in this chapter, how the cultural institution of the zoo "grew out of" silenced bleats. Slavoj Žižek, *Sex and the Failed Absolute* (London: Bloomsbury, 2019), 58.

143. Defenders of Wildlife, "Basic Facts About Bighorn Sheep," accessed May 18, 2015, http://www.defenders.org/bighorn-sheep/basic-facts.

144. On mountain lions and sheep, see Eric M. Rominger, Heather A. Whitlaw, Darrel L. Weybright, William C. Dunn, and Warren B. Ballard, "The Influence of Mountain Lion Predation on Bighorn Sheep Translocations," *Journal of Wildlife Management* 68, no. 4 (October 2004): 993–999.

145. The ecological, conservationist, biological, and wildlife management literatures concerning human encroachment on mountain sheep are large and varied. See, for example, Brian D. Jansen, Paul R. Krausman, Kirby D. Bristow, James R. Heffelfinger, and James C. deVos, "Surface Mining and Ecology of Desert Bighorn Sheep," *Southwestern Naturalist* 54, no. 4 (December 2009): 430–438; Kathleen Longshore, Chris Lowrey, and Daniel B. Thompson, "Detecting Short-Term Responses to Weekend Recreation Activity: Desert Bighorn Sheep Avoidance of Hiking Trails," *Wildlife Society Bulletin* 37, no. 4 (December 2013): 698–706; and Victoria M. Donovan, Samantha P. H. Dwinnell, Jeffrey L. Beck, Caleb P. Roberts, Justin G. Clapp, Greg S. Hiatt, Kevin L. Monteith, and Dirac Tidwell, "Fire-Driven Landscape Heterogeneity Shapes Habitat Selection of Bighorn Sheep," *Journal of Mammalogy* 102, no. 3 (June 2021): 757–771.

146. Karen A. Sausman, "Survival of Captive-Born *Ovis canadensis* in North American Zoos," *Zoo Biology* 3, no. 2 (1984): 111–121.

147. World Association of Zoos and Aquariums, accessed July 19, 2016: http://www.waza.org/en/zoo/visit-the-zoo/sheep-goats-chamois-and-musk-ox-1254385523/ovis-canadensis.

Chapter 6. The *Zoo*notic Nature of Tuberculosis

1. "Global Tuberculosis Report 2020" (Geneva: World Health Organization): xiii; Tom Paulson, "Epidemiology: A Mortal Foe," *Nature* 502 (October 2013).

2. Sharon Levy, "The Evolution of Tuberculosis: Genetic Analysis Offers New Insight on the Spread of an Ancient Disease," *BioScience* 62, no. 7 (July 2012): 625–629.

3. Pere-Joan Cardona, Martí Català, and Clara Prats, "Origin of Tuberculosis in the Paleolithic Predicts Unprecedented Population Growth and Female Resistance," *Scientific Reports* 10, no. 42 (January 2020).

4. Dolly Jørgensen, "Tracking Animals in a Pandemic," and John McNeill, "Bats, Battiness, and the COVID-19 Pandemic," in "Reflections: Environmental History in the Era of COVID-19," *Environmental History* 25, no. 4 (October 2020): 626–631, 635–639.

5. On airborne transmission and the 1910–1911 development of early masks regarding a different contagion, plague, in Manchuria, see Christos Lynteris, "Plague Masks: The Visual Emergence of Anti-Epidemic Personal Protection Equipment," *Medical Anthropology* 37, no. 6 (August 2018): 442–457.

6. "Tuberculosis at the Zoo" (*Evening Star*, January 6, 1905), Scrapbook "1896–1907," RU 74, Box 286, Folder 3.

7. "Tuberculosis at the Zoo."

8. "Tuberculosis at the Zoo."

9. For a classic history of veterinarians, see Susan D. Jones, *Valuing Animals: Veterinarians and Their Patients in Modern America* (Baltimore, MD: Johns Hopkins University Press, 2003). See also Irus Braverman, "Saving Species, One Individual at a Time: Zoo Veterinarians Between Welfare and Conservation," *Humanimalia* 9, no. 1 (September 2018): 1–27.

10. Abigail Woods, "Doctors in the Zoo: Connecting Human and Animal Health in British Zoological Gardens, 1828–1890," in *Animals and the Shaping of Modern Medicine: One Health and Its Histories*, ed. Michael Bresalier Woods, Angela Cassidy, and Rachel Mason Dentinger (London: Palgrave Macmillan, 2018), 31, 39–46.

11. On tuberculosis and rickets, see Woods, "Doctors in the Zoo," 54–60.

12. Knowledge about tuberculosis at the turn of the century was contested; experts around the world disagreed about some of the disease's fundamental qualities. Therefore, the ability to detect and diagnose the disease evolved piecemeal. Many zoo animals undoubtedly lived and died with undetected tuberculosis, especially before 1900. Yet the historian can only guess at numbers. Ultimately, undiagnosed diseases must remain undiagnosed and lost to history. Sometimes, the infected lived with the disease for decades. Other times, they died within weeks of contraction. While the detectability of the tubercle bacillus under microscope lenses eventually became indisputable evidence for the presence of the disease, many individuals with lifelong tuberculosis died from other maladies that masked their systemic tuberculosis. The "cause of death" in an autopsy report, for example, could be listed as "peritonitis" or "endocarditis" without acknowledging the presence of a long-lasting tuberculosis that exacerbated the symptoms of the "primary cause of death." Furthermore, before autopsies could be performed, funding for autopsies, medical buildings, and medical equipment had to be allocated and infrastructures that handled dead animals needed to be established.

13. On improved conditions of zoos in the 1930s, see Jesse Donahue and Erik Trump, *American Zoos During the Depression: A New Deal for Animals* (Jefferson, NC: McFarland Press, 2010).

14. Herbert L. Ratcliffe, "Contribution of a Zoo to an Ecology of Disease," *Proceedings of the American Philosophical Society* 112, no. 4 (August 1968): 235–244.

15. Sheila M. Rothman, *Living in the Shadow of Death: Tuberculosis and the Social Experience of Illness in American History* (New York: Basic Books, 1994), 3–6, 180–193. There is a robust historiography of tuberculosis in American and European society, exploring the shift from "consumption" to "tuberculosis," as well as examining issues of class, public policy, state power, popular reaction, germ theory, bacteriology, microscope technology, and food safety. This chapter centers the zoo and will discuss this literature only as illuminated by zoo sources. See Michael Worboys, *Spreading Germs: Disease Theories and Medical Practice in Britain, 1865–1900* (New York: Cambridge University Press, 2000); David S. Barnes, *The Making of a Social Disease: Tuberculosis in Nineteenth-Century France*

(Berkeley: University of California Press, 1995); Linda Bryder, *Below the Magic Mountain: A Social History of Tuberculosis in Twentieth-Century Britain* (New York: Oxford University Press, 1988); Peter Baldwin, *Contagion and the State in Europe, 1830–1930* (Cambridge: Cambridge University Press, 1999); Katherine McCuaig, *The Weariness, the Fever and the Fret: The Campaign Against Tuberculosis in Canada, 1900–1950* (Montreal: McGill-Queens University Press, 1999); Katherine Ott, *Fevered Lives: Tuberculosis in American Culture since 1870* (Cambridge, MA: Harvard University Press, 1996); Barbara Bates, *Bargaining for Life: A Social History of Tuberculosis, 1876–1938* (Philadelphia: University of Pennsylvania Press, 1992); and Thomas Dormandy, *The White Death: A History of Tuberculosis* (New York: New York University Press, 1999).

16. "Prevention of Tuberculosis Among Government Employees," Prepared for Executive Order of December 7, 1905, Folder "Animal Research: Correspondence—NZP and similar organizations—Tuberculosis among animals and employees 1905–1916," RU 74, Box 99, Folder 12, 3–4.

17. Thomas M. Daniel, *Captain of Death: The Story of Tuberculosis* (Rochester, NY: University of Rochester Press, 1997), 17. While the precise etymology of "pthisis" is unknown, it likely stems from the classical Greek verbs for either "to spit" or "to consume." The word "consumption," though, is still used in modern Bibles to translate *schanhepheth*.

18. "Prevention of Tuberculosis Among Government Employees," 5–6.

19. "Prevention of Tuberculosis Among Government Employees," 6–7.

20. Thomas Kuhn, *The Structure of Scientific Revolutions* (Chicago: University of Chicago Press, 1962).

21. Daniel, *Captain of Death*, 92–93.

22. "Prevention of Tuberculosis Among Government Employees," 7–8.

23. Notions of "fresh air" and "the natural" pervades the history of tuberculosis. See, for example, Pyrs Gruffudd, "'A Crusade Against Consumption': Environment, Health and Social Reform in Wales, 1900–1939," *Journal of Historical Geography* 21, no. 1 (January 1995): 39–54.

24. "Prevention of Tuberculosis Among Government Employees," 8–9.

25. "Prevention of Tuberculosis Among Government Employees," 9–11.

26. "Prevention of Tuberculosis Among Government Employees," 11–12.

27. Theodore Roosevelt, typed by "White House," "Executive Order," February 28, 1906, RU 74, Box 99, Folder 12.

28. Rathbun to Baker, 26 May 1906; Baker to Rathbun, 22 June 1906, RU 74, Box 99, Folder 12; Sybil Hamlet, "History of the National Zoological Park, 1805–1987," unpublished manuscript, undated, RU 365, Box 37, "1905," 2–3.

29. Baker to Born, 22 June 1906; Baker to Morgal, 22 June 1906; Baker to Blackburne, 22 June 1906, RU 74, Box 99, Folder 12.

30. Blackburne to Baker, 23 June 1906, RU 74, Box 99, Folder 12.

31. Salmon to Baker, Autopsy Report, March 26, 1903; Baker to Salmon, April 3, 1903, RU 74, Box 98, Folder 3.

32. Salmon to Baker, Autopsy Report, May 22, 1903, RU 74, Box 98, Folder 3.

33. Harrington to Baker, Autopsy Report, October 21, 1903, RU 74, Box 98, Folder 3.

34. Harrington to Baker, Autopsy Report, February 15, 1904, RU 74, Box 98, Folder 4.

35. Autopsy Reports: Harrington to Baker, April 27, 1904 & July 30, 1904; Salmon to Baker, July 13, 1904, RU 74, Box 98, Folder 4.

36. Autopsy Reports: Harrington to Baker, August 6 & 24, 1904, October 20 & 26, 1904, November 17, 1904; Salmon to Baker, October 13, 1904, December 15, 1904, RU 74, Box 98, Folder 4.

37. Autopsy Reports: Harrington to Baker, January 14 & 30, 1905; Salmon to Baker, March 24, 1905, June 13, 1905, July 28, 1905; Melvin to Baker, December 5, 1905, RU 74, Box 98, Folder 5.

38. Autopsy Reports: Melvin to Baker, January 8 & 10 & 17, 1906, March 2 & 7 & 23, 1906, April 18 & 19, 1906, May 14, 1906, June 16 & 26, 1906, August 8 & 28, 1906, December 11, 1906; Snowler to Baker, October 24, 1906, November 23, 1906, RU 74, Box 98, Folder 6.

39. Hamlet, Folder 2, 1907, 2. Autopsy Reports: Snowler to Baker, January 30, 1907, March 11 & 13, 1907, April 8, 1907, May 16, 1907, June 6, 1907, November 2 & 4, 1907, December 21, 1907; Washburn to Baker, April 30, 1907, May 9, 1907, August 23, 1907, RU 74, Box 98, Folder 7.

40. Daniel, *Captain of Death*, 101.

41. Snowler to Baker, 23 March 1907, RU 74, Box 98, Folder 7.

42. The data in the graph comes from two sources: autopsy reports for 1903–1907 (Baker received between two and four every month) and official reports titled "Report of the Superintendent of the National Zoological Park," written by Baker at the end of every fiscal year, for 1908–1916. The first collection is noted above. The second can be found in "Letters to the Smithsonian Institution, Director's Outgoing Correspondence to Smithsonian Institution Administrators, 1900–1931," RU 74, Box 29, Volumes 8, 9, 10, 11.

43. Chris Otter, *Diet for a Large Planet: Industrial Britain, Food Systems, and World Ecology* (Chicago: University of Chicago Press, 2020), 114.

44. True to Baker, 26 September 1912, RU 74, Box 99, Folder 12.

45. True to Baker, 26 September 1912.

46. Baker to True, 30 September 1912, RU 74, Box 99, Folder 12.

47. Otter, *Diet for a Large Planet*, 115; Keir Waddington, *The Bovine Scourge: Meat, Tuberculosis and Public Health, 1850–1914* (Woodbridge, UK: Boydell, 2006), 43.

48. Baker to True, 30 September 1912.

49. Baker to True, 30 September 1912.

50. Hamlet, Folder 3, 1915, 4.

51. Hamlet, Folder 2, 1905, 3–5.

52. Turner to Rathbun, 16 September 1912, RU 74, Box 99, Folder 12.

53. Turner to Rathbun, 16 September 1912.

54. Baker to Mitchell, Baker to Cadwalader, 19 October 1912; Baker to Hornaday, Baker to Stephan, 21 October 1912, RU 74, Box 99, Folder 12.

55. Stephan to Baker, 25 October 1912, RU 74, Box 99, Folder 12.

56. Hornaday to Baker, 25 October 1912, RU 74, Box 99, Folder 12; Blair to Hornaday, 25 January 1914, WCSA, Collection 1012, Box 27, 4–5.

57. *Nineteenth Annual Report of the New York Zoological Society, 1914*, NYZS (January 1915): 39, 76; William Bridges, *Gathering of Animals: An Unconventional History of the New York Zoological Society* (New York: Harper & Row, 1974), 323–324; Hornaday to Madison Grant, 17 January 1914; W. Reid Blair to Hornaday, 15 January 1914; Hornaday to W. Reid Blair, 25 February 1914, WCSLA, Bronx Zoo, Collection 1012, Box 27. On rats: opportunistic rodents used zoos to their benefit and posed a challenge for zoo directors and keepers. When mysterious diseases plagued zoo animals, rats were framed as the culprits. Hornaday believed so strongly that rats killed his famous primate collection that he urged zoo employees to shoot them so they could be tested for tuberculosis, and he even took time to express frustration in writing when their shots missed. Man up and hit your target!

58. Cadwalader to Baker, 7 November 1912, RU 74, Box 99, Folder 12.

59. "Primates" chart in Herbert Fox, *Report of the Penrose Research Laboratory: Formerly Laboratory and Museum of Comparative Pathology of the Zoological Society of Philadelphia*, 1939, 40, REF SF 996 P45, TUSCRC.

60. Mitchell served as secretary of the Zoological Society of London from 1903 to 1935, and he established the Whipsnade Wild Animal Park, a safari park located in Bedfordshire, thirty miles north of London, in 1926. Interestingly, Mitchell's inspiration for this safari park began with a visit to Hornaday's zoo, which framed itself as a conservation center in the heart of the Bronx. Mitchell was also a prolific zoological author, publishing *Thomas Henry Huxley: A Sketch of His Life and Work* (1900), *The Nature of Man* (1904), *The Childhood of Animals* (1912), and *Materialism and Vitalism in Biology* (1930). See D. P. Crook, "Peter Chalmers Mitchell and Antiwar Evolutionism in Britain during the Great War," *Journal of the History of Biology* 49, no. 2 (June 1989): 325–356.

61. Mitchell to Brown, October 26, November 25, 1908, Acc# 1150, Box RES A1, Items #22, #27; Mitchell to Brown, February 3, 1909, Acc# 631, Box RES A1, Item #34, PZA.

62. Mitchell to Baker, 29 October 1912, RU 74, Box 99, Folder 12.

63. Mitchell to Baker, 29 October 1912.

64. Peter Atkins, *A History of Uncertainty: Bovine Tuberculosis in Britain, 1850 to the Present* (London: Winchester University Press, 2016).

65. See Lise Wilkinson, *Animals and Disease: An Introduction to the History of Comparative Medicine* (New York: Cambridge University Press, 1992); and Anne Hardy, "Animals, Disease, and Man: Making Connections," *Perspectives in Biology and Medicine* 46, no. 2 (Spring 2003): 200–215. On rabies, or hydrophobia, see Harriet Ritvo, *The Animal Estate: The English and Other Creatures in the Victorian Age* (Cambridge, MA: Harvard University Press, 1987); Bill Wasik and Monica Murphy, *Rabid: A Cultural History of the World's Most Diabolical Virus* (New York: Viking, 2012); and Neil Pemberton and Michael Worboys, *Mad*

Dogs and Englishmen: Rabies in Britain, 1830–2000 (London: Palgrave Macmillan, 2013). On anthrax, see Susan D. Jones, *Death in a Small Package: A Short History of Anthrax* (Baltimore, MD: Johns Hopkins University Press, 2010).

66. One cited figure was that 6.44 percent of tuberculosis deaths in England and Wales was caused by bovine tuberculosis. Frederick Hobday, "Certain Diseases of the Cow and Their Interest to the Physician," *British Medical Journal* (February 1923): 314.

67. Otter, *Diet for a Large Planet*, 115.

68. Worboys, *Spreading Germs*, 221–225. On bovine tuberculosis, see Angela Cassidy, *Vermin, Victims and Disease: British Debates over Bovine Tuberculosis and Badgers* (London: Palgrave Macmillan, 2019); Keir Waddington, "Mad and Coughing Cows: Bovine Tuberculosis, BSE and Health in Twentieth Century Britain," in *Meat, Medicine and Human Health in the Twentieth Century*, ed. David Cantor, Christian Bonah, and Matthias Dörries (London: Routledge, 2015), 159–177; Alan L. Olmstead and Paul W. Rhode, "An Impossible Undertaking: The Eradication of Bovine Tuberculosis in the United States," *Journal of Economic History* 64, no. 3 (September 2004): 734–772; Alan L. Olmstead and Paul W. Rhode, "The 'Tuberculosis Cattle Trust': Disease Contagion in an Era of Regulatory Uncertainty," *Journal of Economic History* 64, no. 4 (December 2004): 929–963; Alan L. Olmstead and Paul W. Rhode, "Not on My Farm! Resistance to Bovine Tuberculosis Eradication in the United States," *Journal of Economic History* 67, no. 3 (September 2007): 768–809; and Barbara Orland, "Cow's Milk and Human Disease: Bovine Tuberculosis and the Difficulties Involved in Combating Animal Diseases," *Food & History* 1, no. 1 (January 2003): 179–202.

69. P. J. Atkins, "White Poison? The Social Consequences of Milk Consumption, 1850–1930," *Social History of Medicine* 5, no. 2 (August 1992): 207–227.

70. Anonymous letter, 3–4, RU 74, Box 99, Folder 12.

71. Anonymous letter, 5–6, 9–10, and unnumbered page, RU 74, Box 99, Folder 12.

72. Daniel, *Captain of Death*, 115–117.

73. Schmid to unknown zoo employee, 8 May 1914, RU 74, Box 99, Folder 12.

74. "Illness at the Zoo," *New Haven Evening Register*, July 16, 1898, 6.

75. "Tuberculosis in Monkeys," *Idaho Statesman*, May 28, 1893, 3.

76. "Princess Agwan Dead," *Philadelphia Inquirer*, November 6, 1899, 14.

77. "He Is Doctor to Wild Beasts," *Daily Express*, September 22, 1901, 17.

78. "Zoological Parks Barbarous," *Kansas City Star*, November 10, 1903, 12.

79. "Would Abolish Zoos," *Evening Herald*, November 11, 1903, 3.

80. "Menelik's Gifts to President Die," *Duluth News Tribune*, January 7, 1905, 2.

81. Hutson to Baker, 21 January 1913, RU 74, Box 99, Folder 12.

82. Baker to Hutson, 10 February 1913, RU 74, Box 99, Folder 12.

83. Hutson to Baker, 15 February 1913, RU 74, Box 99, Folder 12.

84. Garrison to Baker, 6 August 1914, RU 74, Box 99, Folder 12.

85. National Tuberculosis Association, *A Tuberculosis Directory: Containing a*

List of Institutions, Associations and Other Agencies Dealing with Tuberculosis in the United States and Canada, compiled by Philip Peter Jacobs (New York: National Association for the Study and Prevention of Tuberculosis, 1911): 47, 107, 136.

86. B. H. Waters, "Destruction and Repair in Pulmonary Tuberculosis," *Journal of the American Medical Association* 75, no. 18 (1920): 1187–1190.

87. Harrington to Baker, 1 July 1914, RU 74, Box 99, Folder 12. Despite their omnipresence on operation tables, animals have been largely ignored within the historiography of surgery. As an exception, and a call-to-action, see Thomas Schlich, Eric Mykhalovskiy, and Melanie Rock, "Animals in Surgery—Surgery in Animals: Nature and Culture in Animal-Human Relationship and Modern Surgery," *History and Philosophy of the Life Sciences* 31, no. 3/4 (2009): 321–354.

88. Harrington to Baker, 1 July 1914.

89. On Lister and his students, see Anne M. Crowther and Marguerite W. Dupree, *Medical Lives in the Age of Surgical Revolution* (Cambridge: Cambridge University Press, 2007).

90. For an illuminating and skeptical critique of this method, see, for example, "The Nascent Iodine Treatment of Tuberculosis," *Hospital* 56, no. 1464 (July 1914): 405.

91. On the history of the Lincoln Park Zoo, see Mark Rosenthal, Carol Tauber, and Edward Uhlir, *The Ark in the Park: The Story of the Lincoln Park Zoo* (Springfield: University of Illinois Press, 2003). Researchers should see I. J. Bryan's sixty-five-page manuscript "History of Lincoln Park," found at LPDA.

92. Burton J. Hendrick, "Oxygenizing a City: An Attempt to Make Two and a Quarter Million People Work and Play Under Sanitary Conditions," *McClure's Magazine* 35, no. 4 (August 1910): 373–387.

93. Sean Kheraj, "The Great Epizootic of 1872–73: Networks of Animal Disease in North American Urban Environments," *Environmental History* 23, no. 3 (July 2018): 498. Also, on Melosi, see his now-classic "Humans, Cities, and Nature: How Do Cities Fit in the Material World?" *Journal of Urban History* 36, no. 1 (January 2010): 3–21.

94. "Consumption Kills Snake After Two Years' Fast" (*Chicago American*, August 14, 1901), CPNC, Vol. 1, CPD Box 6A: 39, CPLSC.

95. "Consumption Killed Ostrich" (*Chicago Daily News*, December 5, 1901); "Museum Buys an Ostrich" (*Chicago Eagle*, December 5, 1901), CPNC, Vol. 1, CPD Box 6A: 120, CPLSC.

96. "Mourning in Lincoln Park" (*Chicago Daily News*, January 20, 1902), CPNC, Vol. 1, CPD Box 6A: 142, CPLSC.

97. "Leo Africanus, Victim of Consumption, is Slain" (*Chicago Chronicle*, April 26, 1904); "Lincoln Park Lion Chloroformed" (*Milwaukee Journal*, April 26, 1904), CPNC, Vol. 3, CPD Box 7C: 27, 29, CPLSC.

98. On women reformers in Chicago, see Maureen A. Flanagan, *Seeing with Their Hearts: Chicago Women and the Vision of the Good City, 1871–1933* (Princeton, NJ: Princeton University Press, 2002).

99. "Says Zoos Are Barbarous" (*Chicago Chronicle*, November 10, 1903), CPNC, Vol. 2, CPD Box 8C: 221, CPLSC.

100. Hendrick, "Oxygenizing a City," 373.

101. Virginia Gardner, "A Boy from Mississippi Led Chicago to Health," "'The Issue of The Mississippi Doctor Dedicated to Dr. W. A. Evans" (*Mississippi Doctor*, October 1934): 10–11, Evans, W. A., 1865–1946 [Collected Papers on Medical Subjects], UCL.

102. Gardner, "Boy from Mississippi," 11; Albert Nelson Marquis, *Who's Who in Chicago: The Book of Chicagoans* (Chicago: A.N. Marquis & Co., 1926), 281.

103. Gardner, "Boy from Mississippi," 10; "Doctors Decide Miss Dooley Died from Acute Digestion" (*Chicago Inter Ocean*, December 24, 1905), CPNC, Vol. 4, CPD Box 6E: 28, CPLSC.

104. Hendrick, "Oxygenizing a City," 373.

105. "Wins More Air for Monkeys" (*Chicago Tribune*, October 15, 1904) CPNC, Vol. 3, CPD Box 7C: 86, CPLSC.

106. Kelly L. Kraus, Vivien C. Pellis, and Sergio M. Pellis, "Targets, Tactics, and Cooperation in the Play Fighting of Two Genera of Old World Monkeys (*Mandrillus* and *Papio*): Accounting for Similarities and Differences," *International Journal of Comparative Psychology* 32 (January 2019): 1–31.

107. Hendrick, "Oxygenizing a City," 373–374.

108. Hendrick, "Oxygenizing a City," 374.

109. "Snow and Ice Have No Terrors for Monkeys That Are Taking the Open Air Cure," January 16, 1912, *Cincinnati Times Star*, Scrapooks, Box #108464628, CZA.

110. Baker to Evans, 3 August 1910; Baker to DeVry, 25 August 1910, RU 74, Box 19, Folder 4.

111. Daniel, *Captain of Death*, 92–93.

112. R. L. Riley, C. C. Mills, W. Nyka, N. Weinstock, P. B. Storey, L. U. Sultan, M. C. Riley, and W. F. Wells, "Aerial Dissemination of Pulmonary Tuberculosis: A Two-Year Study of Contagion in a Tuberculosis Ward," *American Journal of Epidemiology* 70, no. 2 (September 1959): 185–196.

113. Hendrick, "Oxygenizing a City," 374. While the "open air" movement of Lincoln Park was scientifically justified, pathogenically framed, and widely publicized, moving animals outdoors to prevent "miasmic" diseases has precedence across zoo history, as Andy Flack shows for the Bristol Zoo in 1902 and as Abigail Woods shows for the London Zoo in the mid-nineteenth century. See Andrew Flack, *The Wild Within: Histories of a Landmark British Zoo* (Charlottesville: University of Virginia Press, 2018), 100–101; Woods, "Doctors in the Zoo," 36.

114. Hendrick, "Oxygenizing a City," 374. On captive breeding and bison in the Bronx, see Gregory Dehler, *The Most Defiant Devil: William Temple Hornaday and His Controversial Crusade to Save American Wildlife* (Charlottesville: University of Virginia Press, 2013), 74–85; William Bridges, *A Gathering of Animals: An Unconventional History of the New York Zoological Society* (New York:

Harper and Row, 1974), 71–74, 265–266; and Mark V. Barrow Jr., "Teetering on the Brink of Extinction: The Passenger Pigeon, the Bison, and American Zoo Culture in the Late Nineteenth and Early Twentieth Centuries," in *The Ark and Beyond: The Evolution of Zoo and Aquarium Conservation*, ed. Ben A. Minteer, Jane Maienschein, James P. Collins (Chicago: University of Chicago Press, 2018), 57–62.

115. Contract between commissioners of Lincoln Park and Edwin H. Clark (architect), "Steam Heating & Ventilation Specifications: Monkey House," Box titled "Lincoln Specifications," LPDA.

116. "Heating Specification for the Lion House for the Commissioners of Lincoln Park," "Lion House, 1912," Box titled "Lincoln Specifications," LPDA.

117. W. A. Evans, *Report of the Department of Health of the City of Chicago for the Years 1907, 1908, 1909, 1910* (Chicago: Cameron Amberg's Co., 1911): 41, CPLMRC.

118. "Proceedings of the City Council," "Regular Meeting, Monday, January 27, 1908," *Proceedings of the City Council of the City of Chicago for the Council Year 1907–1908*, Vol. II (Chicago: John F. Higgins Print, 1908): 3775, CPLMRC.

119. W. A. Evans, "Bulletin, Chicago School of Sanitary Instruction Devoted to the Dissemination of Advice and Information of the Department of Health," series 8, vol. 12, February 20, 1909, Chicago Department of Health, Bulletins, Volume 3, CPL, MRC.

120. "Proceedings of the City Council," "November 23, 1908," *Proceedings of the City Council of the City of Chicago for the Council Year 1908–1909* (Chicago: John F. Higgins Print, 1909): 1841, CPLMRC.

121. Harold Platt, *Shock Cities: The Environmental Transformation and Reform of Manchester and Chicago* (Chicago: University of Chicago Press, 2005), 442–491.

122. W. A. Evans, "Hygienic Principles of Ventilation" (New York: William Wood & Company, 1910), 10–12, Evans, W. A., 1865–1946 [Collected Papers on Medical Subjects], UCL.

123. W. A. Evans, "Ventilation of Cars" (New York: William Wood & Company, 1910): 9–10, Evans, W. A., 1865–1946 [Collected Papers on Medical Subjects], UCL.

124. Hendrick, "Oxygenizing a City," 380–387. For a fascinating exploration of "open air," see Tim Steffen Altenhof, "Breathing Space: The Architecture of Pneumatic Beings," PhD diss., Yale University, 2018.

125. W. A. Evans, "Spanish Influenza, Pneumonia and Tuberculosis" (Chicago: Pamphlet reprinted from *The Lancet*, 1928): 10, Evans, W. A., 1865–1946 [Collected Papers on Medical Subjects], UCL.

126. For a survey of these historiographical debates, see Keir Waddington, *An Introduction to the Social History of Medicine* (London: Palgrave Macmillan, 2011), 254–268.

127. In fact, the field's name changed twice—from "hygiene and sanitary science" to simply "sanitary science" in 1913, which changed to "public health" in

1921. Typewriter Bio card, Folder titled "Evans, William Augustus," NUA. For more on the new definition of "public health" and Evans's role in its creation, see W. A. Evans, "Disease and Health" (Philadelphia: American Baptist Publication Society, 1917), Evans, 1865–1946 [Collected Papers on Medical Subjects], UCL.

128. "Medical School Announcement" (Evanston: Northwestern University, September 1917): 34, in the bounded *N.W.U.M.S. Catalogues, 1916–1925*, NUA.

129. William Evans, "Public Health Notes," 15–16, folder titled "Evans, William Augustus," NUA.

130. For more, see Kathryn Lougheed, *Catching Breath: The Making and Unmaking of Tuberculosis* (New York: Bloomsbury, 2017); and Christian W. McMillen, *Discovering Tuberculosis: A Global History, 1900 to the Present* (New Haven, CT: Yale University Press, 2015).

131. "Evans Speaks at Opening of College Here" (*Dispatch*, Columbus, MS, September 14, 1936), folder titled "Evans, William Augustus," NUA. Evans had the status of emeritus professor until his death in 1953.

132. "The Issue of The Mississippi Doctor Dedicated to Dr. W. A. Evans" (*Mississippi Doctor*, October 1934), Evans, W. A., 1865–1946 [Collected Papers on Medical Subjects], UCL.

133. Abigail Woods, Michael Bresalier, Angela Cassidy, and Rachel Mason Dentinger acknowledge this as well in the introduction to their powerful volume. *Animals and the Shaping of Modern Medicine* marks one of the first attempts in the history of medicine to interrogate animals as living, breathing, and historically contingent subjects. This volume argues that "modern medicine, as developed in the West over the last two centuries, was a more-than-human endeavor, whose boundaries with veterinary medicine and biology were porous and in a constant state of flux" (13). These authors follow Robert G. W. Kirk and Michael Worboys's (timely yet largely ignored) call for a history of medicine focused on interspecies interaction. See "Medicine and Species: One Medicine, One History?" in *The Oxford Handbook of the History of Medicine*, ed. Mark Jackson (Oxford: Oxford University Press, 2011), 561–577.

Conclusion. The National Zoo Movement

1. Baker to DeVry, 10, 16, & 31 February 1910, RU 74, Box 19, Folder 3; Baker to DeVry, 7 January 1909, RU 74, Box 18, Folder 4; Baker to Wm. Bartels Co., 8 May 1909, RU 74, Box 19, Folder 1; Baker to DeVry, 11 & 24 January 1907, RU 74, Box 17, Folder 3; Baker to DeVry, 2 January 1906, RU 74, Box 16, Folder 4; Baker to DeVry, 2 May 1902, RU 74, Box 13, Folder 2.

2. Parker to Baker, 12 June 1901, RU 74, Box 100, Folder 1.

3. Baker to Parker, 15 July 1901, RU 74, Box 100, Folder 1.

4. "List of Zoological Gardens," RU 74, Box 100, Folder 1, 1–9.

5. Dyer to Schmid [forwarded to Baker], 28 September 1901, RU 74, Box 100, Folder 1.

6. Canley to Baker, 29 October 1901, RU 74, Box 100, Folder 1.

7. Baker to "The Honorable Mayor," Newark, NJ, 30 October 1901, RU 74,

Box 100, Folder 1; Elizabeth Hanson, *Animal Attractions: Nature on Display in American Zoos* (Princeton, NJ: Princeton University Press, 2002): 197. This is the standard wording for all the letters sent.

8. "List of Zoological collection now at City Park, Denver, Colorado," RU 74, Box 100, Folder 1.

9. Baker to Wright, 21 November 1901, RU 74, Box 100, Folder 1.

10. Trimble to Baker, 27 November 1901, RU 74, Box 100, Folder 2.

11. Hicks to Baker, 6 November 1901, RU 74, Box 100, Folder 1.

12. Williams to Baker, 4 December 1901, RU 74, Box 100, Folder 2.

13. Clarke to Baker, 24 December 1901, RU 74, Box 100, Folder 2.

14. Doyle to Baker, 16 November 1901; Baker to Doyle, 23 November 1901, RU 74, Box 100, Folder 3.

15. For more on the relations between the Silver Lake Park and the National Zoo see: RU 74, Box 57, Folders 3–12. Letters to, from, and about Lodge fill these folders, mostly concerning black bear trading and breeding. Baker even directed the attention of famous zoologist C. Hart Merriam to the Silver Lake Park regarding the birth of black bears in captivity.

16. Green to Baker, 15 January 1902, RU 74, Box 100, Folder 3.

17. Morres to Baker, 15 November 1901; Baker to Cornish, 25 November 1901, RU 74, Box 100, Folder 3.

18. Perry to Baker, 4 November 1901, RU 74, Box 100, Folder 5.

19. Reiser to Baker, 20 November 1901, RU 74, Box 100, Folder 5.

20. Flood to Baker, 28 & 29 November 1901; Crowley to Baker, 24 September 1903, RU 74, Box 100, Folder 5.

21. Power to Baker, 7 March 1902, RU 74, Box 100, Folder 3.

22. Hagen to Baker, 28 March 1902, RU 74, Box 100, Folder 3.

23. Hamann to Baker, 8 November 1901; Barstow to Baker, 9 November 1901, RU 74, Box 100, Folder 1; Mayor to Baker, 23 December 1901; Byrne to Baker, 30 November 1901, RU 74, Box 100, Folder 2.

24. Fitzgerald to Baker, 23 November 1901; Baker to "The Honorable Mayor," Schenectady, New York [scribbled response in margins], 26 November 1901; Artz to Baker, 23 November 1901; Palmer to Baker, 10 December 1901, RU 74, Box 100, Folder 2.

25. Eager to Baker, 25 March 1902, RU 74, Box 100, Folder 3.

26. Halmer to Baker, 28 March 1902, RU 74, Box 100, Folder 4.

27. Steed to Baker, 28 June 1902, RU 74, Box 100, Folder 5.

28. Choir to Baker, 19 November 1901; Haas to Choir [Forwarded to Baker], 21 November 1901, RU 74, Box 100, Folder 5.

29. Choir to Baker, 11 March 1902, RU 74, Box 100, Folder 5.

30. Baker to Walsh, 17 March 1902, RU 74, Box 100, Folder 5.

31. Smith, "Binghamton, N.Y.," 2 November 1903, RU 74, Box 100, Folder 6.

32. Hillard to Baker, 15 July 1902, RU 74, Box 100, Folder 6.

33. Ludd to Baker, 31 October 1903, RU 74, Box 100, Folder 6.

34. H. Lowitz, "Portland, Oregon," 12 November 1903, RU 74, Box 100, Folder 6.

35. Smith, "Portland, Maine," 4 November 1903, RU 74, Box 100, Folder 6.

36. Starin, "Glen Island, N. Y.," 9 November 1903, RU 74, Box 100, Folder 6.

37. Many zoo historians have critiqued this simplistic view. See, for example, Jeffrey Hyson, "Jungles of Eden: The Design of American Zoos," in *Environmentalism in Landscape Architecture*, ed. Michel Conan (Washington, DC: Dumbarton Oaks, 2000), 23–44.

38. Jesse Donahue and Erik Trump, *Political Animals: Public Art in American Zoos and Aquariums* (Lanham, MD: Lexington Books, 2007): 32.

39. "Zoological Collections of U.S.," compiled by Baker (Smithsonian Institution, National Zoological Park, 1903), "Preface," RU 74, Box 100, Folder 5. Only the preface of this report exists.

40. "Statistics of Zoological Collections," RU 74, Box 100, Folder 6.

41. Hanson, *Animal Attractions*, 3.

42. Baker to Morgan, 24 May 1905, RU 74, Box 100, Folder 7.

43. The names here represent the leading animal dealers in the United States at the turn of the twentieth century, and they deserve a study of their own. Baker's list goes as follows: Frank B. Armstrong (Brownsville, Texas), A. H. Nicholson (Orlando, Florida), Mrs. C. F. Latham, Grant (Brevard County, Florida), Chester A. Lamb (Kansas City, Missouri), Charles Payne (Wichita, Kansas), C. E. Mallory (Buffalo Center, Iowa), C. P. Forge (Carman, Manitoba), Howard Eaton (Wolf, Wyoming), William Bartels (New York City), Louis Ruhe (New York City), and Cecil French (Washington, DC). Baker to Morgan, 27 May 1905, RU 74, Box 100, Folder 7.

44. Clarke to Arthur Baker, 26 November 1906, RU 74, Box 100, Folder 7.

45. Underwood to Frank Baker, 2 January 1906, RU 74, Box 100, Folder 7.

46. Baker to Underwood, 25 January 1906, RU 74, Box 100, Folder 7.

47. Widmann to Baker, 12 September 1910, RU 74, Box 100, Folder 9.

48. Baker to Widmann, 15 November 1910, RU 74, Box 100, Folder 9.

49. Stinchcomb to Baker, 9 April 1906, RU 74, Box 100, Folder 7.

50. Questionnaire answers, RU 74, Box 100, Folder 7.

51. Baker to "The Honorable Mayor," Cleveland, Ohio, 30 October 1901, RU 74, Box 100, Folder 5.

52. Leslie to Baker, 26 September 1903, RU 74, Box 100, Folder 7.

53. Jason M. Colby, *Orca: How We Came to Know and Love the Ocean's Greatest Predator* (New York: Oxford University Press, 2018): 5, 213.

54. Nigel Rothfels, "Prison, Spectacle, Refuge: Modern Zoos Are Proud of Their Contribution to Animal Conservation But Will Always Be Haunted by Their Histories," *Aeon* (November 2019), https://aeon.co/essays/have-modern-zoos-really-put-their-history-behind-them.

55. Matthew Chrulew, "Saving the Golden Lion Tamarin," in *Extinction Studies: Stories of Time, Death, and Generations*, ed. Deborah Bird Rose, Thom

van Dooren, and Matthew Chrulew (New York: Columbia University Press, 2017), 73.

56. Jack Halberstam, *Wild Things: The Disorder of Desire* (Durham, NC: Duke University Press, 2020).

57. Thom van Dooren, *The Wake of Crows: Living and Dying in Shared Worlds* (New York: Columbia University Press, 2019), 57.

58. Juno Salazar Parreñas, *Decolonizing Extinction: The Work of Care in Orangutan Rehabilitation* (Durham, NC: Duke University Press, 2018), 27–28.

59. Rothfels, "Prison, Spectacle, Refuge."

60. Amphibians represent some of the most threatened species on the planet. And not only have zoos seen far more success with amphibian breeding and reintroduction than the breeding and reintroduction of other species, but a frog or salamander habitat is more possible to replicate in the confined space of the zoo than the habitats of virtually all mammals and birds. Nonetheless, since amphibians do not capture public attention, zoos devote far too little attention to them. See, for example, Joseph R. Mendelson III, "Frogs in Glass Boxes: Responses of Zoos to Global Amphibian Extinction," in *The Ark and Beyond: The Evolution of Zoo and Aquarium Conservation*, ed. Ben A. Minteer, Jane Maienschein, and James P. Collins (Chicago: University of Chicago Press, 2018), 298–310.

61. These topics become increasingly important in the twentieth century, after the timeline of this book. See Parreñas, *Decolonizing Extinction*, 83–103; Irus Braverman, *Zooland: The Institution of Captivity* (Stanford, CA: Stanford University Press, 2013): 159–184; and Carrie Friese, *Cloning Wild Life: Zoos, Captivity, and the Future of Endangered Animals* (New York: New York University Press, 2013).

62. See Jesse Donahue, "Zoo Animal Citizens," in *Increasing Legal Rights for Zoo Animals: Justice on the Ark*, ed. Jesse Donahue (Lanham, MD: Lexington Books, 2017), 1–24.

63. Zoos staged human exhibits until the middle of the twentieth century. Shouldn't the zoo be responsible for teaching, working through, and learning from this history through and with its public? See, for example, Pascal Blanchard, Nicolas Bancel, Gilles Boëtsch, and Nanette Jacomijn Snoep, eds., *Human Zoos: Science and Spectacle in the Age of Empire* (Liverpool: Liverpool University Press, 2008). See also Pamela Newkirk, *Spectacle: The Astonishing Life of Ota Benga* (New York: Harper Collins, 2015).

Bibliography

Acampora, Ralph, ed. *Metamorphoses of the Zoo: Animal Encounter after Noah.* Lanham, MD: Rowman & Littlefield, 2010.

———. "Zoos and Eyes: Contesting Captivity and Seeking Successor Practices." *Society and Animals* 13, no. 1 (2005): 69–88.

Adams, Carol. *The Sexual Politics of Meat: A Feminist-Vegetarian Critical Theory*, 20th anniversary ed. New York: Continuum, 2010.

Adichie, Chimamanda Ngozi. "The Danger of a Single Story." *TED Talk*, July 2009, https://www.ted.com/talks/chimamanda_ngozi_adichie_the_danger_of_a_single_story?language=en.

Agamben, Giorgio. *The Open: Man and Animal.* Trans. Kevin Attell. Stanford, CA: Stanford University Press, 2003.

Ahamed, A. M. Riyas. "Activity Time Budget of the Asian Elephant (Elephas Maximus Linn.) in the Wild." *Trends in Biosciences* 8, no. 12 (January 2015): 3024–3028.

Ahonen, Heidi, Robert G. Harcourt, Adam J. Stow, and Isabelle Charrier. "Geographic Vocal Variation and Perceptual Discrimination Abilities in Male Australian Sea Lions." *Animal Cognition* 21, no. 2 (March 2018): 235–243.

Ahuja, Neel. *Bioinsecurities: Disease Interventions, Empire, and the Government of Species.* Durham, NC: Duke University Press, 2016.

Allen, Garland E. "'Culling the Herd': Eugenics and the Conservation Movement in the United States, 1900–1940." *Journal of the History of Biology* 46, no. 1 (2013): 31–72.

Allen, J. A. "Historical and Nomenclatorial Notes on North American Sheep." In *Bulletin of the American Museum of Natural History*, ed. J. A. Allen, 1–29, vol. 31. New York: Trustees of the AMNH, 1912.

Aloi, Giovanni. *Speculative Taxidermy: Natural History, Animal Surfaces, and Art in the Anthropocene.* New York: Columbia University Press, 2018.

Altenhof, Tim Steffen. "Breathing Space: The Architecture of Pneumatic Beings." PhD diss., Yale University, 2018.

Alves, Abel. *The Animals of Spain: An Introduction to Imperial Perceptions and Human Interaction with Other Animals, 1492–1826.* Leiden: Brill, 2011.

Ames, Eric. *Carl Hagenbeck's Empire of Entertainments.* Seattle: University of Washington Press, 2009.

Anderson, John D. *A History of Aerodynamics and Its Impacts on Flying Machines.* Cambridge: Cambridge University Press, 1998.

———. *Inventing Flight: The Wright Brothers and Their Predecessors*. Baltimore, MD: Johns Hopkins University Press, 2004.

Anderson, Virginia DeJohn. *Creatures of Empire: How Domestic Animals Transformed Early America*. New York: Oxford University Press, 2004.

Andrei, Mary Anne. *Nature's Mirror: How Taxidermists Shaped America's Natural History Museums and Saved Endangered Species*. Chicago: University of Chicago Press, 2020.

Arendt, Hannah. *Men in Dark Times*. San Diego: Harcourt Brace, 1968.

Arsić, Branka. *Bird Relics: Grief and Vitalism in Thoreau*. Cambridge, MA: Harvard University Press, 2016.

Atkins, Peter. *A History of Uncertainty: Bovine Tuberculosis in Britain, 1850 to the Present*. London: Winchester University Press, 2016.

———. "White Poison? The Social Consequences of Milk Consumption, 1850–1930." *Social History of Medicine* 5, no. 2 (August 1992): 207–227.

Atterton, Peter, and Tamra Wright, eds. *Face to Face with Animals: Levinas and the Animal Question*. Albany: SUNY Press, 2019.

Baldwin, Peter. *Contagion and the State in Europe, 1830–1930*. Cambridge: Cambridge University Press, 1999.

Barad, Karen. *Meeting the Universe Halfway: Quantum Physics and the Entanglement of Matter and Meaning*. Durham, NC: Duke University Press, 2007.

Baratay, Eric, and Elisabeth Hardouin-Fugier. *Zoo: A History of Zoological Gardens in the West*. London: Reaktion Books, 2002.

Barnes, David S. *The Making of a Social Disease: Tuberculosis in Nineteenth-Century France*. Berkeley: University of California Press, 1995.

Barrow, Mark V., Jr. *Nature's Ghosts: Confronting Extinction from the Age of Jefferson to the Age of Ecology*. Chicago: University of Chicago Press, 2009.

———. *A Passion for Birds: American Ornithology after Audubon*. Princeton, NJ: Princeton University Press, 1998.

———. "The Specimen Dealer: Entrepreneurial Natural History in America's Gilded Age," *Journal of the History of Biology* 33, no. 3 (December 2000): 493–534.

———. "Teetering on the Brink of Extinction: The Passenger Pigeon, the Bison, and American Zoo Culture in the Late Nineteenth and Early Twentieth Centuries." In *The Ark and Beyond: The Evolution of Zoo and Aquarium Conservation*, eds. Ben A. Minteer, Jane Maienschein, and James P. Collins, 57–62. Chicago: University of Chicago Press, 2018.

Bates, Barbara. *Bargaining for Life: A Social History of Tuberculosis, 1876–1938*. Philadelphia: University of Pennsylvania Press, 1992.

Bechtel, Stefan. *Mr. Hornaday's War: How A Peculiar Victorian Zookeeper Waged a Lonely Crusade for Wildlife That Changed the World*. New York: Beacon Press, 2012.

Beck, Benjamin. "Reintroduction, Zoos, Conservation, and Animal Welfare." In *Ethics on the Ark: Zoos, Animal Welfare, and Wildlife Conservation*, eds. Bryan G. Norton, Michael Hutchins, Elizabeth F. Stevens, and Terry L. Maple, 155–163. Washington, DC: Smithsonian Institution Press, 1995.

Beers, Diane. *For the Prevention of Cruelty: The History and Legacy of Animal Rights Activism in the United States*. Athens: Ohio University Press, 2006.

Bender, Daniel E. *The Animal Game: Searching for Wildness at the American Zoo*. Cambridge, MA: Harvard University Press, 2016.

Bennett, Joshua. *Being Property Once Myself: Blackness and the End of Man*. Cambridge, MA: Harvard University Press, 2020.

Bensaude-Vincent, Bernadette. "A Genealogy of the Increasing Gap between Science and the Public." *Public Understanding of Science* 10, no. 1 (January 2001): 99–113.

Benson, Etienne. "Animal Writes: Historiography, Disciplinarity, and the Animal Trace." In *Making Animal Meaning*, ed. Linda Kalof and Georgina M. Montgomery, 3–16. East Lansing: Michigan State University Press, 2011.

———. "The Urbanization of the Eastern Gray Squirrel in the United States." *Journal of American History* 100, no. 3 (December 2013): 691–710.

———. *Wired Wilderness: Technologies of Tracking and the Making of Modern Wildlife*. Baltimore, MD: Johns Hopkins University Press, 2010.

Berger, John. "Why Look at Animals?" In *About Looking*. New York: Pantheon Books, 1980.

Berman, Marshall. *All That Is Solid Melts into Air: The Experience of Modernity*. New York: Penguin, 1988.

Bertram, Brian C. R. *The Ostrich Communal Nesting System*. Princeton, NJ: Princeton University Press, 1992.

Biehler, Dawn Day. *Pests in the City: Flies, Bedbugs, Cockroaches, and Rats*. Seattle: University of Washington Press, 2013.

Blanchard, Pascal, Nicolas Bancel, Gilles Boëtsch, and Nanette Jacomijn Snoep, eds. *Human Zoos: Science and Spectacle in the Age of Empire*. Liverpool: Liverpool University Press, 2008.

Blight, David W. *Race and Reunion: The Civil War in American Memory*. Cambridge, MA: Harvard University Press, 2001.

Blunt, Wilfrid. *The Ark in the Park: The Zoo in the Nineteenth Century*. London: Hamish Hamilton, 1976.

Boisseron, Bénédicte. *Afro-Dog: Blackness and the Animal Question*. New York: Columbia University Press, 2018.

Bostock, Stephen St. C. *Zoos and Animal Rights: The Ethics of Keeping Animals*. New York: Routledge, 1993.

Bowler, Peter J. *The Eclipse of Darwinism: Anti-Darwinian Evolutionary Theories in the Decades Around 1900*. Baltimore, MD: Johns Hopkins University Press, 1983.

———. "Revisiting the Eclipse of Darwinism." *Journal of the History of Biology*, 38 no. 1 (Spring 2005): 19–32.

Bowler, Peter J., and Iwan Rhys Morus. *Making Modern Science: A Historical Survey*. Chicago: University of Chicago Press, 2005.

Brantz, Dorothee, ed. *Beastly Natures: Animals, Humans, and the Study of History*. Charlottesville: University of Virginia Press, 2010.

Braverman, Irus. "Saving Species, One Individual at a Time: Zoo Veterinarians Between Welfare and Conservation." *Humanimalia* 9, no. 1 (September 2018): 1–27.

———. *Zooland: The Institution of Captivity*. Stanford, CA: Stanford University Press, 2013.

Bridges, William. *A Gathering of Animals: An Unconventional History of the New York Zoological Society*. New York: Harper & Row, 1974.

Brinkley, Douglas. *The Wilderness Warrior: Theodore Roosevelt and the Crusade for America*. New York: Harper Collins, 2009.

Broglio, Ron. *Surface Encounters: Thinking with Animals and Art*. Minneapolis: University of Minnesota Press, 2011.

Brown, Frederick L. *The City Is More Than Human: An Animal History of Seattle*. Seattle: University of Washington Press, 2016.

Bruce, Gary. *Through the Lion Gate: A History of the Berlin Zoo*. New York: Oxford University Press, 2017.

Bryant, Gregory A., and Athena C. Aktipis. "The Animal Nature of Spontaneous Human Laughter." *Evolution and Human Behavior* 35, no. 4 (July 2014): 327–335.

Bryder, Linda. *Below the Magic Mountain: A Social History of Tuberculosis in Twentieth-Century Britain*. New York: Oxford University Press, 1988.

Bulliet, Richard W. *The Camel and the Wheel*. New York: Columbia University Press, 1990.

———. *Hunters, Herders, and Hamburgers: The Past and Future of Human-Animal Relationships*. New York: Columbia University Press, 2005.

Burleigh, Nina. *The Stranger and the Statesman: James Smithson, John Quincy Adams, and the Making of America's Greatest Museum; The Smithsonian*. New York: Harper Collins, 2003.

Bushong, William. *Rock Creek Park, District of Columbia: Historic Resource Study*. Washington, DC: Department of the Interior, National Park Service, 1990.

Calarco, Matthew. *Zoographies: The Question of the Animal from Heidegger to Derrida*. New York: Columbia University Press, 2008.

Cardona, Pere-Joan, Martí Català, and Clara Prats. "Origin of Tuberculosis in the Paleolithic Predicts Unprecedented Population Growth and Female Resistance." *Scientific Reports* 10, no. 42 (January 2020).

Carson, Gerald. *Men, Beasts, and Gods: A History of Cruelty and Kindness to Animals*. New York: Charles Scribner's Sons, 1972.

Cart, Theodore Whaley. "The Lacey Act: America's First Nationwide Wildlife Statute." *Forest History* 17, no. 3 (October 1973): 4–13.

Cassidy, Angela. *Vermin, Victims and Disease: British Debates over Bovine Tuberculosis and Badgers*. London: Palgrave Macmillan, 2019.

Castleton, Kenneth B. *Petroglyphs and Pictographs of Utah, Vol. 1 and Vol. 2*. Salt Lake City: Utah Museum of Natural History, 2002.

Chen, Mel Y. *Animacies: Biopolitics, Racial Mattering, and Queer Affect*. Durham, NC: Duke University Press, 2012.

Chrulew, Matthew. "Saving the Golden Lion Tamarin." In *Extinction Studies: Stories of Time, Death, and Generations*, ed. Deborah Bird Rose, Thom van Dooren, and Matthew Chrulew, 49–88. New York: Columbia University Press, 2017.

Cincinnati, Noah. "Too Sullen for Survival: Historicizing Gorilla Extinction, 1900–1930." In *The Historical Animal*, ed. Susan Nance, 166–183. Syracuse, NY: Syracuse University Press, 2015.

Clark, J. F. M. *Bugs and the Victorians*. New Haven, CT: Yale University Press, 2009.

Clifford, Hatley. "A More-Than-Human Green: National Parks and Animality in Marianne Moore's 'An Octopus.'" In *Modernism in the Green: Public Greens in Modern Literature and Culture*, ed. Julia E. Daniel and Margaret Konkol. New York: Routledge, 2020.

Cockram, Sarah, and Andrew Wells, eds. *Interspecies Interactions: Animals and Humans from the Middle Ages to Modernity*. London: Routledge, 2018.

Coen, Deborah R. *The Earthquake Observers: Disaster Science from Lisbon to Richter*. Chicago: Chicago University Press, 2012.

Cohen, Jon. *Almost Chimpanzee: Searching for What Makes Us Human, in Rainforests, Labs, Sanctuaries, and Zoos*. New York: Henry Holt, 2010.

Colby, Jason M. *Orca: How We Came to Know and Love the Ocean's Greatest Predator*. New York: Oxford University Press, 2018.

Coleman, Jon. *Vicious: Wolves and Men in America*. New Haven, CT: Yale University Press, 2004.

Coleman, William. *Biology in the Nineteenth Century: Problems of Form, Function, and Transformation*. New York: John Wiley & Sons, 1971.

Conn, Steve. *Museums and American Intellectual Life, 1876–1926*. Chicago: University of Chicago Press, 1998.

Cooter, Roger. *The Cultural Meaning of Popular Science: Phrenology and the Organization of Consent in Nineteenth-Century Britain*. Cambridge: Cambridge University Press, 1984.

Cooter, Roger, and Stephen Pumfrey. "Separate Spheres and Public Places: Reflections on the History of Science Popularization and Science in Popular Culture." *History of Science* 32 (September 1994): 237–267.

Cowan, Ian McTaggart. "Distribution and Variation in the Native Sheep of North America." *American Midland Naturalist* 24, no. 3 (November 1940): 505–580.

Cowie, Helen. *Exhibiting Animals in Nineteenth-Century Britain: Empathy, Education, Entertainment*. London: Palgrave Macmillan, 2014.

———. *Llama*. London: Reaktion, 2017.

Cowperthwaite, Gabriela, director. *Blackfish*. Manny O. Productions, 2013.

Cranz, Galen. *The Politics of Park Design: A History of Urban Parks in America*. Cambridge, MA: MIT Press, 1982.

Crook, D. P. "Peter Chalmers Mitchell and Antiwar Evolutionism in Britain during the Great War." *Journal of the History of Biology* 49, no. 2 (June 1989): 325–356.

Crouch, Tom. *The Bishop's Boys: A Life of Wilbur and Orville Wright*. New York: Norton, 2003.

———. *A Dream of Wings: Americans and the Airplane*. New York: W. W. Norton, 2002.

Crowther, M. Anne. "Lister at Home and Abroad: A Continuing Legacy." *Notes and Records of the Royal Society of London* 67, no. 3 (September 2013): 281–294.

Crowther, M. Anne, and Marguerite W. Dupree. *Medical Lives in the Age of Surgical Revolution*. Cambridge: Cambridge University Press, 2007.

cummings, e. e. "The Secret of the Zoo Exposed." In *E. E. Cummings: A Miscellany*, ed. George J. Firmage. New York: Argophile Press, 1958.

Cunfer, Geoff, and Bill Waiser, eds. *Bison and People on the North American Great Plains: A Deep Environmental History*. College Station: Texas A&M University Press, 2016.

Dale, Joshua Paul, Joyce Goggin, Julia Leyda, Anthony P. Mcintyre, and Diane Negra, eds. *The Aesthetics and Affects of Cuteness*. New York: Routledge, 2017.

Daniel, Thomas M. *Captain of Death: The Story of Tuberculosis*. Rochester, NY: University of Rochester Press, 1997.

Davis, Janet. "Cockfight Nationalism: Blood Sport and the Moral Politics of American Empire and Nation Building." *American Quarterly* 65, no. 3 (September 2013): 549–574.

———. *The Gospel of Kindness: Animal Welfare and the Making of Modern America*. Oxford: Oxford University Press, 2016.

Davis, Susan G. *Spectacular Nature: Corporate Culture and the Sea World Experience*. Berkeley: University of California Press, 1997.

Daum, Andreas. "Varieties of Popular Science and the Transformations of Public Knowledge: Some Historical Reflections," *Isis* 100, no. 2 (June 2009): 319–332.

Dayan, Colin. *The Law Is a White Dog: How Legal Rituals Make and Unmake Persons*. Princeton, NJ: Princeton University Press, 2011.

———. *With Dogs at the Edge of Life*. New York: Columbia University Press, 2016.

Dehler, Gregory. *The Most Defiant Devil: William Temple Hornaday and His Controversial Crusade to Save American Wildlife*. Charlottesville: University of Virginia Press, 2013.

Deleuze, Gilles, and Felix Guattari. *A Thousand Plateaus: Capitalism and Schizophrenia*, trans. Brian Massumi. Minneapolis: University of Minnesota Press, 1987, 2003.

Derrida, Jacques. *The Animal That Therefore I Am*, ed. Marie-Louise Mallet, trans. David Wills. New York: Fordham University Press, 2008.

———. *Of Grammatology*. Trans. Gayatri Chakravorty Spivak. Baltimore, MD: Johns Hopkins University Press, 1997.

de Waal, Frans. *Are We Smart Enough to Know How Smart Animals Are?* New York: Norton, 2016.

Dolph, James A. "Bringing Wildlife to the Millions: William Temple Hornaday, the Early Years, 1854–1896." PhD diss., University of Massachusetts, 1975.

Dolph, James A., and C. Ivar Dolph. "The American Bison: His Annihilation

and Preservation." *Montana: The Magazine of Western History* 25, no. 3 (Summer 1975): 14–25.
Domańska, Ewa. "Posthumanist History." In *Debating New Approaches to History*, ed. Marek Tamm and Peter Burke, 327–352. London: Bloomsbury, 2019.
Donahue, Jesse, ed. *Increasing Legal Rights for Zoo Animals: Justice on the Ark*. Lanham, MD: Lexington Books, 2017.
———. "Zoo Animal Citizens." In *Increasing Legal Rights for Zoo Animals: Justice on the Ark*, ed. Jesse Donahue, 1–24. Lanham, MD: Lexington Books, 2017.
Donahue, Jesse, and Erik Trump. *American Zoos During the Depression: A New Deal for Animals*. Jefferson, NC: McFarland Press, 2010.
———. *Political Animals: Public Art in American Zoos and Aquariums*. Lanham, MD: Lexington Books, 2007.
Donovan, Victoria M., Samantha P. H. Dwinnell, Jeffrey L. Beck, Caleb P. Roberts, Justin G. Clapp, Greg S. Hiatt, Kevin L. Monteith, and Dirac Tidwell. "Fire-Driven Landscape Heterogeneity Shapes Habitat Selection of Bighorn Sheep." *Journal of Mammalogy* 102, no. 3 (June 2021): 757–771.
Dormandy, Thomas. *The White Death: A History of Tuberculosis*. New York: New York University Press, 1999.
Duarte, F. J. *Fundamentals of Quantum Entanglement*. Bristol: Institute of Physics Publishing, 2019.
Dugatkin, Lee Alan. *Mr. Jefferson and the Giant Moose: Natural History in Early America*. Chicago: University of Chicago Press, 2009.
Dunlap, Thomas R. "The Realistic Animal Story: Ernest Thompson Seton, Charles Roberts, and Darwinism." *Forest & Conservation History* 36, no. 2 (April 1992): 56–62.
Dunway, Finis. "Hunting with the Camera: Nature Photography, Manliness, and Modern Memory, 1890–1930." *Journal of American Studies* 34, no. 2 (August 2000): 207–230.
Edgerton, David. *Shock of the Old: Technology and Global History since 1900*. London: Profile Books, 2011.
Egan, Timothy. *The Big Burn: Teddy Roosevelt and the Fire That Saved America*. New York: Mariner Books, 2010.
Eicher, John H., and David J. Eicher. *Civil War High Commands*. Stanford, CA: Stanford University Press, 2001.
Esposito, Roberto. "Politics and Human Nature." *Angelaki* 16, no. 3 (September 2011): 77–84.
Ewing, Heather. *The Lost World of James Smithson: Science, Revolution, and the Birth of the Smithsonian*. New York: Bloomsbury, 2007.
Fagan, Brian. *The Intimate Bond: How Animals Shaped Human History*. New York: Bloomsbury, 2015.
Falls, J. B. "Playback: A Historical Perspective." In *Playback and Studies of Animal Communication*, ed. P. K. McGregor, 11–33. New York: Plenum Press, 1992.
Faust, Drew Gilpin. *This Republic of Suffering: Death and the American Civil War*. New York: Vintage Books, 2008.

Ferguson, Moira. *Animal Advocacy and Englishwomen, 1780–1900: Patriots, Nation, and Empire*. Ann Arbor: University of Michigan Press, 2001.

Ferrando, Francesca. *Philosophical Posthumanism*. London: Bloomsbury, 2019.

Few, Martha, and Zeb Tortorici, eds. *Centering Animals in Latin American History*. Durham, NC: Duke University Press, 2013.

Flack, Andrew. "Dark Trails: Animal Histories Beyond the Light of Day." *Environmental History* 27, no. 2 (April 2022).

———. *The Wild Within: Histories of a Landmark British Zoo*. Charlottesville: University of Virginia Press, 2018.

Flanagan, Maureen A. *Seeing with Their Hearts: Chicago Women and the Vision of the Good City, 1871–1933*. Princeton, NJ: Princeton University Press, 2002.

Flinterud, Guro. "The Rise and Fall of Polar Bear Knut." In *Zoo Studies: A New Humanities*, ed. Tracy McDonald and Daniel Vandersommers, 191–210. Montreal: McGill-Queens University Press, 2019.

Flores, Dan. *Coyote America: A Natural and Supernatural History*. New York: Basic Books, 2016.

Foucault, Michel. *The Order of Things: An Archeology of the Human Sciences*. New York: Vintage Books, 1994.

Fraser, Nancy. "Rethinking the Public Sphere: A Contribution to the Critique of Actually Existing Democracy." In *Habermas and the Public Sphere*, ed. Craig Calhoun, 109–142. Cambridge, MA: MIT Press, 1992.

Freeberg, Ernest. *A Traitor to His Species: Henry Bergh and the Birth of the Animal Rights Movement*. New York: Basic Books, 2020.

Friese, Carrie. *Cloning Wild Life: Zoos, Captivity, and the Future of Endangered Animals*. New York: New York University Press, 2013.

Fudge, Erica. *Brutal Reasoning: Animals, Rationality and Humanity in Early Modern England*. Ithaca, NY: Cornell University Press, 2006.

———. "A Left-Handed Blow: Writing the History of Animals." In *Representing Animals*, ed. Nigel Rothfels. Bloomington: Indiana University Press, 2002.

———. *Perceiving Animals: Humans and Beasts in Early Modern English Culture*. Chicago: University of Illinois Press, 2002.

Fuller, Luca H., and Seth W. Donahue. "Material Properties of Bighorn Sheep (Ovis Canadensis) Horncore Bone with Implications for Energy Absorption Impacts." *Journal of the Mechanical Behavior of Biomedical Materials* 114 (February 2021).

Fyfe, Aileen, and Bernard Lightman, eds. *Science in the Marketplace: Nineteenth-Century Sites and Experiences*. Chicago: University of Chicago Press, 2007.

Gaarder, Emily. *Women and the Animal Rights Movement*. New Brunswick: Rutgers University Press, 2011.

Gamito-Marques, Daniel. "A Space of One's Own: Barbosa du Bocage, the Foundation of the National Museum of Lisbon, and the Construction of a Career in Zoology (1851–1907)." *Journal of the History of Biology* 51, no. 2 (June 2018): 223–257.

Geertz, Clifford. *The Interpretations of Cultures: Selected Essays*. New York: Basic Books, 1973.

Gibson, Abe. "Dog Days at the Zoo, Part I." Smithsonian Institution Archives (September 2012). Accessed May 15, 2020. http://siarchives.si.edu/blog/dog-days-zoo-part-i.

———. "Dog Days at the Zoo, Part II." Smithsonian Institution Archives (September 2012). Accessed May 25, 2020. http://www.siarchives.si.edu/blog/dog-days-zoo-part-ii.

———. *Feral Animals in the American South: An Evolutionary History*. Cambridge: Cambridge University Press, 2016.

Gieryn, Thomas. *Cultural Boundaries of Science: Credibility on the Line*. Chicago: University of Chicago Press, 1999.

———. *Truth Spots: How Places Make People Believe*. Chicago: University of Chicago Press, 2018.

Goldstein, Daniel. "'Yours for Science': The Smithsonian Institution's Correspondents and the Shape of Scientific Community in Nineteenth-Century America." *Isis* 85, no. 4 (December 1994): 573–599.

Gould, Stephen Jay. *Ontogeny and Phylogeny*. Cambridge, MA: Harvard University Press, 1985.

Govindrajan, Radhika. *Animal Intimacies: Interspecies Relatedness in India's Central Himalayas*. Chicago: University of Chicago Press, 2018.

Graetz, Rick, and Susie Graetz. *Crow Country: Montana's Crow Tribe of Indians*. Billings, MT: Northern Rockies Publishing Company, 2000.

Greenblatt, Stephen. *Marvelous Possessions: The Wonder of the New World*. Chicago: University of Chicago Press, 1991.

Grigson, Caroline. *Menagerie: The History of Exotic Animals in England, 1100–1837*. Oxford: Oxford University Press, 2016.

Gruen, Lori. *Entangled Empathy: An Alternative Ethic for Our Relationships with Animals*. Brooklyn: Lantern Books, 2015.

Gruffudd, Pyrs. "'A Crusade Against Consumption': Environment, Health and Social Reform in Wales, 1900–1939." *Journal of Historical Geography* 21, no. 1 (January 1995): 39–54.

Guenther, Lisa. *Solitary Confinement: Social Death and Its Afterlives*. Minneapolis: University of Minnesota Press, 2013.

Guerrini, Anita. *The Courtiers' Anatomists: Animals and Humans in Louis XIV's Paris*. Chicago: University of Chicago Press, 2015.

Gulliford, Andrew. *The Woolly West: Colorado's Hidden History of Sheepscapes*. College Station: Texas A&M University Press, 2018.

Halberstam, Jack. *Wild Things: The Disorder of Desire*. Durham, NC: Duke University Press, 2020.

Hallions, Richard P. *Taking Flight: Inventing the Aerial Age, from Antiquity through the First World War*. New York: Oxford University Press, 2003.

Hamera, Judith. *Parlor Ponds: The Cultural Work of the American Home Aquarium, 1850–1970*. Ann Arbor: University of Michigan Press, 2012.

Hanson, Elizabeth. *Animal Attractions: Nature on Display in American Zoos*. Princeton, NJ: Princeton University Press, 2002.

Haraway, Donna J. *Primate Visions: Gender, Race, and Nature in the World of Modern Science*. New York: Routledge, 1989.

———. "Primatology Is Politics by Other Means." *PSA: Proceedings of the Biennial Meeting of the Philosophy of Science Association* 1984, no. 2 (Chicago: University of Chicago Press, 1984): 489–524.

———. *Staying with the Trouble: Making Kin in the Chthulucene*. Durham, NC: Duke University Press, 2016.

———. *When Species Meet*. Minneapolis: University of Minnesota Press, 2007.

Harding, Sandra. *Sciences from Below: Feminisms, Postcolonialities, and Modernities*. Durham, NC: Duke University Press, 2008.

Hardy, Anne. "Animals, Disease, and Man: Making Connections." *Perspectives in Biology and Medicine* 46, no. 2 (Spring 2003): 200–215.

Harrison, Peter. "Descartes on Animals." *Philosophical Quarterly* 42, no. 167 (April 1992): 219–227.

Harvey, David. *Paris, Capital of Modernity*. London: Taylor & Francis, 2003.

Heholt, Ruth, and Melissa Edmundson, eds. *Gothic Animals: Uncanny Otherness and the Animal With-Out*. Cham, Switzerland: Palgrave, 2020.

Heintzman, Alix. "E Is for Elephant: Jungle Animals in Late Nineteenth-Century Picture Books." *Environmental History* 19, no. 3 (July 2014): 553–563.

Heise, Ursula K. *Imagining Extinction: The Cultural Meanings of Endangered Species*. Chicago: University of Chicago Press, 2016.

Hilgartner, Stephen. "The Dominant View of Science Popularization: Conceptual Problems, Political Uses." *Social Studies of Science* 20, no. 3 (August 1990): 519–539.

Hochadel, Oliver. "Science at the Zoo: An Introduction." *Centaurus: Journal of the European Society for the History of Science* 64, no. 3 (2022): 561–590.

Hochschild, Adam. *King Leopold's Ghost: A Story of Greed, Terror, and Heroism in Colonial Africa*. New York: Houghton Mifflin, 1998.

Hofmeyr, Isabel. "J is for Jackal (and Dingo)." In *Animalia: An Anti-Imperial Bestiary for Our Times*, ed. Antoinette Burton and Renisa Mawani, 96–100. Durham, NC: Duke University Press, 2020.

Horowitz, Helen L. "The National Zoological Park: 'City of Refuge' or Zoo?" *Records of the Columbia Historical Society, Washington, D.C.* 49 (1973/1974): 405–429.

Hyson, Jeffrey. "Jungles of Eden: The Design of American Zoos." In *Environmentalism in Landscape Architecture*, ed. Michel Conan, 23–44. Washington, DC: Dumbarton Oaks, 2000.

———. "Urban Jungles: Zoos and American Society." PhD diss., Cornell University, 1999.

Iacobbo, Karen, and Michael Iacobbo. *Vegetarian America: A History*. Westport, CT: Praeger, 2004.

Isenberg, Andrew C. *The Destruction of the Bison: An Environmental History, 1750–1920*. New York: Cambridge University Press, 2000.

———. "The Returns of the Bison: Nostalgia, Profit, and Preservation." *Environmental History* 2, no. 2 (April 1997): 179–196.
Israel, Paul. *Edison: A Life of Invention*. New York: John Wiley, 1998.
Ito, Takashi. *London Zoo and the Victorians, 1828–1859*. Cambridge: Cambridge University Press, 2014.
Jackson, Zakiyyah Iman. *Becoming Human: Matter and Meaning in an Antiblack World*. New York: New York University Press, 2020.
Jacobs, Nancy J. *Birders of Africa: History of a Network*. New Haven, CT: Yale University Press, 2016.
Jacoby, Karl. *Crimes against Nature: Squatters, Poachers, Thieves, and the Hidden History of American Conservation*. Berkeley: University of California Press, 2001.
Jakab, Peter L. *Visions of a Flying Machine*. Washington, DC: Smithsonian Institution, 1990.
James, William. *The Principles of Psychology*. In *The Works of William James*, ed. Frederick H. Burkhardt, Fredson Bowers, and Ignas K. Skrupskelis, vol. 8–10. Cambridge, MA: Harvard University Press, 1981.
Jamieson, Dale. "Against Zoos." In *In Defense of Animals*, ed. Peter Singer, 132–143. New York: Blackwell, 1984.
Janney, Caroline E. *Remembering the Civil War: Reunion and the Limits of Reconciliation*. Chapel Hill: University of North Carolina Press, 2013.
Jansen, Brian D., Paul R. Krausman, Kirby D. Bristow, James R. Heffelfinger, and James C. deVos. "Surface Mining and Ecology of Desert Bighorn Sheep." *Southwestern Naturalist* 54, no. 4 (December 2009): 430–438.
Jardine, Nicholas, James A. Secord, and E. C. Spary, eds. *Cultures of Natural History*. Cambridge: Cambridge University Press, 1996.
Jarvis, Gary. "The Road Not Taken: Humanitarian Reform and the Origins of Animal Rights in Britain and the United States, 1883–1919." PhD diss., University of Iowa, 2009.
Johnston, David W. *The History of Ornithology in Virginia*. Charlottesville: University of Virginia Press, 2003.
Jones, Susan D. *Death in a Small Package: A Short History of Anthrax*. Baltimore, MD: Johns Hopkins University Press, 2010.
———. *Valuing Animals: Veterinarians and Their Patients in Modern America*. Baltimore, MD: Johns Hopkins University Press, 2003.
Jørgensen, Dolly. *Recovering Lost Species in the Modern Age: Histories of Longing and Belonging*. Cambridge, MA: MIT Press, 2019.
———. "Tracking Animals in a Pandemic." In "Reflections: Environmental History in the Era of COVID-19." *Environmental History* 25, no. 4 (October 2020): 626–631.
Judd, Richard W. *Common Lands, Common People: The Origins of Conservation in New England*. Cambridge, MA: Harvard University Press, 1997.
Kasson, John F. *Amusing the Million: Coney Island at the Turn of the Century*. New York: Hill and Wang, 1978.

Kean, Hilda. "Challenges for Historians Writing Animal-Human History: What Is Really Enough?" *Anthrozoös* 25, no. 3 (August 2012): 57–72.

Kean, Hilda, and Philip Howell, eds. *The Routledge Companion to Animal-Human History*. New York: Routledge, 2018.

Kendrick, Keith M. "Social Cognition in Sheep." *Animal Sentience* 25, no. 40 (2019).

Kerber, Linda. "Separate Spheres, Female Worlds, Woman's Place: The Rhetoric of Women's History." *Journal of American History* 75, no. 1 (June 1988): 9–39.

Kete, Kathleen. *The Beast in the Boudoir: Petkeeping in Nineteenth-Century Paris*. Berkeley: University of California Press, 1994.

———. "Introduction: Animals and Human Empire." In *Cultural History of Animals in the Age of Empire*, ed. Kathleen Kete, 1–24. New York: Berg, 2007.

Kete, Mary Louise. *Sentimental Collaborations: Mourning and Middle-Class Identity in Nineteenth-Century America*. Durham, NC: Duke University Press, 2000.

Khan, Amina. *Adapt: How Humans Are Tapping into Nature's Secrets to Design and Build a Better Future*. New York: St. Martin's Press, 2017.

Kheraj, Sean. "The Great Epizootic of 1872–73: Networks of Animal Disease in North American Urban Environments." *Environmental History* 23, no. 3 (July 2018): 495–521.

Kim, Claire Jean. *Dangerous Crossings: Race, Species, and Nature in a Multicultural Age*. New York: Cambridge University Press, 2015.

Kimmerer, Robin Wall. *Braiding Sweetgrass: Indigenous Wisdom, Scientific Knowledge and the Teachings of Plants*. Minneapolis, MN: Milkweed Editions, 2013.

King, Amy M. *Bloom: The Botanical Vernacular in the English Novel*. Oxford: Oxford University Press, 2003.

Kirk, Jay. *Kingdom Under Glass: A Tale of Obsession, Adventure, and One Man's Quest to Preserve the World's Greatest Animals*. New York: Picador, 2010.

Kirk, Robert G. W., and Michael Worboys. "Medicine and Species: One Medicine, One History?" In *The Oxford Handbook of the History of Medicine*, ed. Mark Jackson, 561–577. Oxford: Oxford University Press, 2011.

Kohler, Robert E. *Inside Science: Stories from the Field in Human and Animal Science*. Chicago: University of Chicago Press, 2019.

———. *Lords of the Fly: Drosophila Genetics and the Experimental Life*. Chicago: University of Chicago Press, 1994.

Kohlstedt, Sally Gregory. "History in a Natural History Museum: George Brown Goode and the Smithsonian Institution." *Public Historian* 10, no. 2 (Spring 1988): 7–26.

Kraus, Kelly L., Vivien C. Pellis, and Sergio M. Pellis. "Targets, Tactics, and Cooperation in the Play Fighting of Two Genera of Old World Monkeys (*Mandrillus* and *Papio*): Accounting for Similarities and Differences." *International Journal of Comparative Psychology* 32 (January 2019): 1–31.

Krebber, André, and Mieke Roscher, eds. *Animal Biography: Re-framing Animal Lives*. Cham, Switzerland: Palgrave Macmillan, 2018.

Kroeger, Brooke. *Nellie Bly: Daredevil, Reporter, Feminist.* Pittsburgh, PA: Three Rivers Press, 1995.

Kuhn, Thomas. *The Structure of Scientific Revolutions.* Chicago: University of Chicago Press, 1962.

Kuriyama, Shigehisa. *The Expressiveness of the Body and the Divergence of Greek and Chinese Medicine.* New York: Zone Books, 2002.

Lankford, John. *American Astronomy: Community, Careers, and Power, 1859–1940.* Chicago: University of Chicago Press, 1997.

Largent, Mark A. "The So-Called Eclipse of Darwinism." *Transactions of the American Philosophical Society* 99, no. 1 (2009): 3–21.

Lasch, Christopher. *The New Radicalism in America, 1889–1963: The Intellectual as a Social Type.* New York: Norton, 1997.

Latour, Bruno. *Reassembling the Social: An Introduction to Actor-Network Theory.* Oxford: Oxford University Press, 2005.

———. *Science in Action: How to Follow Scientists and Engineers Through Society.* Cambridge, MA: Harvard University Press, 1988.

———. *We Have Never Been Modern,* trans. Catherine Porter. Cambridge, MA: Harvard University Press, 1993.

Latour, Bruno, and Dipesh Chakrabarty. "Conflicts of Planetary Proportions—A Conversation." *Journal of the Philosophy of History* 14, no. 3 (November 2020): 419–454.

Lerner, Steve. *Sacrifice Zones: The Front Lines of Toxic Chemical Exposure in the United States.* Cambridge, MA: MIT Press, 2010.

Leroi, Armand Marie. *The Lagoon: How Aristotle Invented Science.* New York: Bloomsbury, 2014.

Levy, Sharon. "The Evolution of Tuberculosis: Genetic Analysis Offers New Insight on the Spread of an Ancient Disease." *BioScience* 62, no. 7 (July 2012): 625–629.

Li, Chien-hui. *Mobilizing Traditions in the First Wave of the British Animal Defense Movement.* London: Palgrave Macmillan, 2019.

Lightman, Bernard. *Victorian Popularizers of Science: Designing Nature for New Audiences.* Chicago: University of Chicago Press, 2007.

Longshore, Kathleen, Chris Lowrey, and Daniel B. Thompson. "Detecting Short-Term Responses to Weekend Recreation Activity: Desert Bighorn Sheep Avoidance of Hiking Trails." *Wildlife Society Bulletin* 37, no. 4 (December 2013): 698–706.

Lorimer, Jamie. *Wildlife in the Anthropocene: Conservation after Nature.* Minneapolis: University of Minnesota Press, 2015.

Lougheed, Kathryn. *Catching Breath: The Making and Unmaking of Tuberculosis.* New York: Bloomsbury, 2017.

Lundblad, Michael. *The Birth of a Jungle: Animality in Progressive-Era U.S. Literature and Culture.* New York: Oxford University Press, 2013.

Lutes, Jean Marie. "Into the Madhouse with Nellie Bly: Girl Stunt Reporting in Late Nineteenth-Century America." *American Quarterly* 54, no. 2 (June 2002): 217–253.

Lynteris, Christos. "Plague Masks: The Visual Emergence of Anti-Epidemic Personal Protection Equipment." *Medical Anthropology* 37, no. 6 (August 2018): 442–457.

Macdonald, Sharon, ed. *The Politics of Display: Museums, Science, Culture.* New York: Routledge, 1998.

Malamud, Randy. "Al Gore, *Blackfish*, and Me: Eco-activist Progress and Prospects for the Future." In *Zoo Studies: A New Humanities*, ed. Tracy McDonald and Daniel Vandersommers, 262–275. Montreal: McGill-Queens University Press, 2019.

———. *Reading Zoos: Representations of Animals and Captivity.* New York: New York University Press, 1998.

Margulis, Lynn. *Symbiotic Planet: A New Look at Evolution.* New York: Basic Books, 1998.

Marks, Robert. *The Origins of the Modern World: A Global and Ecological Narrative from the Fifteenth to the Twenty-First Century*, 2nd ed. New York: Rowman & Littlefield, 2007.

Marx, Leo. *The Machine in the Garden: Technology and the Pastoral Ideal in America.* New York: Oxford University Press, 1964.

Mason, Jim. "Animals: From Souls and the Sacred in Prehistoric Times to Symbols and Slaves in Antiquity." In *A Cultural History of Animals in Antiquity*, ed. Linda Kalof, 17–45. Oxford: Berg, 2007.

Mbembe, Achille. *On the Postcolony*, trans. Steven Rendell. Berkeley: University of California Press, 2001.

McCormick, Michael. "Rats, Communications, and Plague: Toward an Ecological History." *Journal of Interdisciplinary History* 34, no. 1 (Summer 2003): 1–25.

McCuaig, Katherine. *The Weariness, the Fever and the Fret: The Campaign Against Tuberculosis in Canada, 1900–1950.* Montreal: McGill-Queens University Press, 1999.

McDonald, Tracy. "Sculpting Dinah with the Blunt Tools of the Historian." In *Zoo Studies: A New Humanities*, ed. Tracy McDonald and Daniel Vandersommers, 93–118. Montreal: McGill-Queens University Press, 2019.

McDonald, Tracy, and Daniel Vandersommers, eds. *Zoo Studies: A New Humanities.* Montreal: McGill-Queens University Press, 2019.

McGerr, Michael. *A Fierce Discontent: The Rise and Fall of the Progressive Movement in America, 1870–1920.* New York: Oxford University Press, 2003.

McHugh, Susan. *Animal Stories: Narrating across Species Lines.* Minneapolis: University of Minnesota Press, 2011.

McMillen, Christian W. *Discovering Tuberculosis: A Global History, 1900 to the Present.* New Haven, CT: Yale University Press, 2015.

McNeill, John. "Bats, Battiness, and the COVID-19 Pandemic." In "Reflections: Environmental History in the Era of COVID-19." *Environmental History* 25, no. 4 (October 2020): 635–639.

McNeur, Catherine. *Taming Manhattan: Environmental Battles in the Antebellum City.* Cambridge, MA: Harvard University Press, 2014.

McShane, Clay, and Joel Tarr. *The Horse in the City: Living Machines in the Nineteenth Century*. Baltimore, MD: Johns Hopkins University Press, 2007.

Melosi, Martin. "Humans, Cities, and Nature: How Do Cities Fit in the Material World?" *Journal of Urban History* 36, no. 1 (January 2010): 3–21.

Meloy, Ellen. *Eating Stone: Imagination and the Loss of the Wild*. New York: Vintage Books, 2005.

Mendelson, Joseph R., III. "Frogs in Glass Boxes: Responses of Zoos to Global Amphibian Extinction." In *The Ark and Beyond: The Evolution of Zoo and Aquarium Conservation*, ed. Ben A. Minteer, Jane Maienschein, and James P. Collins, 298–310. Chicago: University of Chicago Press, 2018.

Mengozzi, Chiara, ed. *Outside the Anthropological Machine: Crossing the Human-Animal Divide and Other Exit Strategies*. New York: Routledge, 2020.

Meretoja, Hanna. *The Ethics of Storytelling: Narrative Hermeneutics, History, and the Possible*. New York: Oxford University Press, 2018.

Midgley, Mary. *Animals and Why They Matter*. Athens: University Press of Georgia, 1983.

Mighetto, Lisa. *Wild Animals and American Environmental Ethics*. Tucson: University of Arizona Press, 1991.

Mikhail, Alan. *The Animal in Ottoman Egypt*. New York: Oxford University Press, 2014.

Milgrom, Melissa. *Still Life: Adventures in Taxidermy*. New York: Houghton Mifflin Harcourt, 2010.

Miller, Ian Jared. *The Nature of the Beasts: Empire and Exhibition at the Tokyo Imperial Zoo*. Berkeley: University of California Press, 2013.

Minteer, Ben A., Jane Maienschein, and James P. Collins, eds. *The Ark and Beyond: The Evolution of Zoo and Aquarium Conservation*. Chicago: University of Chicago Press, 2018.

Mionczynski, John. *The Pack Goat*. Boulder: Pruett Publishing, 1992.

Mitman, Gregg. *Reel Nature: America's Romance with Wildlife on Film*. Cambridge, MA: Harvard University Press, 1999.

———. "When Nature *Is* the Zoo: Vision and Power in the Art and Science of Natural History," *Osiris* 11 (1996): 117–143.

Mizelle, Brett. *Pig*. London: Reaktion, 2011.

Mizelle, Brett, Mieke Roscher, and André Krebber, eds. *Handbook of Historical Animal Studies*. Berlin: De Gruyter, 2021.

Montgomery, Georgina M. "Place, Practice and Primatology: Clarence Ray Carpenter, Primate Communication and the Development of Field Methodology, 1931–1945." *Journal of the History of Biology* 38, no. 3 (Autumn 2005): 495–533.

Moore, Donald E., III. "Reintroductions and Animal Welfare: Laws and Regulations." In *Increasing Legal Rights for Zoo Animals*, ed. Jesse Donahue. Lanham, MD: Lexington Books, 2017.

Moore, Madeleine. "Red Blood and Red Tape: Building the Collections of the National Zoological Park, 1887–1908." Honor's thesis, Georgetown University, 2017.

Morris, James McGrath. *Pulitzer: A Life in Politics, Print, and Power*. New York: Harper Collins, 2010.

Morton, David. *Off the Record: The Technology and Culture of Sound Recording in America*. New Brunswick: Rutgers University Press, 2000.

Morton, Timothy. *Being Ecological*. Cambridge, MA: MIT Press, 2018.

———. *Dark Ecology: For a Logic of Coexistence*. New York: Columbia University Press, 2016.

———. *Ecology Without Nature: Rethinking Environmental Aesthetics*. Cambridge, MA: Harvard University Press, 2007.

———. *Hyperobjects: Philosophy and Ecology After the End of the World*. Minneapolis: University of Minnesota Press, 2013.

———. *Shelley and the Revolution in Taste: The Body and the Natural World*. New York: Cambridge University Press, 1994.

Moss, Andrew, Eric Jensen, and Markus Gusset. *A Global Evaluation of Biodiversity Literacy in Zoo and Aquarium Visitors*. Gland, CH: World Association of Zoos and Aquariums Executive Office, 2014.

Mullan, Bob, and Garry Marvin. *Zoo Culture: The Book about Watching People Watch Animals*. 2nd edition. Urbana: University of Illinois Press, 1998.

Myers, Greg. "Discourse Studies of Scientific Popularization: Questioning the Boundaries." *Discourse Studies* 5, no. 2 (May 2003): 265–279.

Nagel, Thomas. "What Is It Like to Be a Bat?" *Philosophical Review* 83, no. 4 (October 1974): 435–50.

Nance, Susan. "Animal History: The Final Frontier?" *American Historian* (November 2015): 29–34.

———. *Animal Modernity: Jumbo the Elephant and the Human Dilemma*. London: Palgrave Macmillan, 2015.

———. *Entertaining Elephants: Animal Agency and the Business of the American Circus*. Baltimore, MD: Johns Hopkins University Press, 2013.

——, ed. *The Historical Animal*. Syracuse: Syracuse University Press, 2015.

Nash, Roderick Frazier. *The Rights of Nature: A History of Environmental Ethics*. Madison: University of Wisconsin Press, 1989.

Neff, Nichole. "The Belly of the Beast: The Uncanny Shark." *Gothic Studies* 18, no. 2 (November 2016): 52–61.

Newkirk, Pamela. *Spectacle: The Astonishing Life of Ota Benga*. New York: Harper Collins, 2015.

Nichols, Catherine A. *Exchanging Objects: Nineteenth-Century Museum Anthropology at the Smithsonian Institution*. New York: Berghahn, 2021.

Nietzsche, Friedrich. *Beyond Good and Evil: Prelude to a Philosophy of the Future*. Trans. Walter Kaufmann. New York: Vintage, 1989.

———. *The Gay Science*. Trans. Walter Kaufmann. New York: Penguin, 1974.

Norton, Bryan G., Michael Hutchins, Elizabeth F. Stevens, and Terry L. Maple, eds. *Ethics on the Ark: Zoos, Animal Welfare, and Wildlife Conservation*. Washington, DC: Smithsonian Institution Press, 1995.

Novick, Peter. *That Noble Dream: The 'Objectivity Question' and the American Historical Profession*. Cambridge: Cambridge University Press, 1988.

Nyhart, Lynn. "Civic and Economic Zoology in Nineteenth-Century Germany: The 'Living Communities' of Karl Mobius." *Isis* 89, no. 4 (December 1998): 605–630.

———. *Modern Nature: The Rise of the Biological Perspective in Germany*. Chicago: University of Chicago Press, 2009.

Obenchain, Theodore G. *The Victorian Vivisection Debate: Frances Power Cobbe, Experimental Science and the "Claim of Brutes."* New York: McFarland, 2012.

O'Gorman, Emily, and Andrea Gaynor. "More-Than-Human Histories." *Environmental History* 25, no. 4 (August 2020): 711–735.

Ohrem, Dominik. "The Ends of Man: The Zooanthropological Imaginary and the Animal Geographies of Westward Expansion in Antebellum America." In *American Beasts: Perspectives on Animals, Animality and U.S. Culture, 1776–1920*, ed. Dominik Ohrem, 245–278. Berlin: Neofelis Verlag, 2017.

Okasha, Samir. "Biological Altruism." In *The Stanford Encyclopedia of Philosophy*, ed. Edward N. Zalta. Summer 2020 edition. https://plato.stanford.edu/archives/sum2020/entries/altruism-biological/.

Olmstead, Alan L., and Paul W. Rhode. "An Impossible Undertaking: The Eradication of Bovine Tuberculosis in the United States." *Journal of Economic History* 64, no. 3 (September 2004): 734–772.

———. "Not on My Farm! Resistance to Bovine Tuberculosis Eradication in the United States." *Journal of Economic History* 67, no. 3 (September 2007): 768–809.

———. "The 'Tuberculosis Cattle Trust': Disease Contagion in an Era of Regulatory Uncertainty." *Journal of Economic History* 64, no. 4 (December 2004): 929–963.

Orland, Barbara. "Cow's Milk and Human Disease: Bovine Tuberculosis and the Difficulties Involved in Combating Animal Diseases." *Food & History* 1, no. 1 (January 2003): 179–202.

Orr, Oliver H., Jr. *Saving American Birds: T. Gilbert Pearson and the Founding of the Audubon Movement*. Gainesville: University Press of Florida, 1992.

Ott, Katherine. *Fevered Lives: Tuberculosis in American Culture since 1870*. Cambridge, MA: Harvard University Press, 1996.

Otter, Chris. *Diet for a Large Planet: Industrial Britain, Food Systems, and World Ecology*. Chicago: University of Chicago Press, 2020.

———. *The Victorian Eye: A Political History of Light and Vision in Britain, 1800–1910*. Chicago: University of Chicago Press, 2008.

Painter, Nell Irvin. *Standing at Armageddon: A Grassroots History of the Progressive Era*. New York: Norton, 2008.

Parreñas, Juno Salazar. *Decolonizing Extinction: The Work of Care in Orangutan Rehabilitation*. Durham, NC: Duke University Press, 2018.

Parrish, Susan Scott. *American Curiosity: Cultures of Natural History in the*

Colonial British Atlantic World. Chapel Hill: University of North Carolina Press, 2006.
Paulson, Tom. "Epidemiology: A Mortal Foe." *Nature* 502 (October 2013).
Pauly, Philip J. *Biologists and the Promise of American Life*. Princeton, NJ: Princeton University Press, 2000.
Pearson, Susan. *The Rights of the Defenseless: Protecting Animals and Children in Gilded Age America*. Chicago: University of Chicago Press, 2011.
Pemberton, Neil, and Michael Worboys. *Mad Dogs and Englishmen: Rabies in Britain, 1830—2000*. London: Palgrave Macmillan, 2013.
Pickering, Andrew. *Constructing Quarks: A Sociological History of Particle Physics*. Chicago: University of Chicago Press, 1984.
Platt, Harold. *Shock Cities: The Environmental Transformation and Reform of Manchester and Chicago*. Chicago: University of Chicago Press, 2005.
Poliquin, Rachel. *Beaver*. London: Reaktion Books, 2015.
Pouillard, Violette. "Gust (ca. 1952–1988), or A History from Below of the Changing Zoo." In *Zoo Studies: A New Humanities*, ed. Tracy McDonald and Daniel Vandersommers, 167–190. Montreal: McGill-Queens University Press, 2019.
Powell, Miles. *Vanishing America: Species Extinction, Racial Peril, and the Origins of Conservation*. Cambridge, MA: Harvard University Press, 2016.
Price, Jennifer. *Flight Maps: Adventures with Nature in Modern America*. New York: Basic Books, 1999.
Price, Mark R. Stanley, and John E. Fa. "Reintroduction from Zoos: A Conservation Guiding Light or a Shooting Star?" In *Zoos in the 21st Century: Catalysts for Conservation*, ed. Alexandra Zimmermann, Matthew Hatchwell, Leslie Dickie, and Chris West, 155–177. Cambridge, MA: Cambridge University Press, 2007.
"Project CETI." Cetacean Translation Initiative. 2021. https://www.projectceti.org/.
Purdy, Jedediah. *After Nature: A Politics for the Anthropocene*. Cambridge, MA: Harvard University Press, 2015.
Qvarnström, Martin, Ahlberg, Per E., Niedźwiedzki, Grzegorz. "Tyrannosaurid-like Osteophagy by a Triassic Archosaur." *Scientific Reports* 9, no. 925 (January 2019).
Raber, Karen. *Animal Bodies, Renaissance Culture*. Philadelphia: University of Pennsylvania Press, 2013.
Rader, Karen. *Making Mice: Standardizing Animals for American Biomedical Research, 1900—1955*. Princeton, NJ: Princeton University Press, 2004.
Rader, Karen, and Victoria E. M. Cain, eds. *Life on Display: Revolutionizing U.S. Museums of Science and Natural History in the Twentieth Century*. Chicago: University of Chicago Press, 2014.
Radick, Gregory. "Morgan's Canon, Garner's Phonograph, and the Evolutionary Origins of Language and Reason." *British Journal for the History of Science* 33, no. 1 (March 2000): 3–23.

———. "Primate Language and the Playback Experiment, in 1890 and 1980." *Journal of the History of Biology* 38, no. 3 (Autumn, 2005): 461–493.

———. *The Simian Tongue: The Long Debate about Animal Language*. Chicago: University of Chicago Press, 2008.

Raj, Kapil. "Beyond Postcolonialism . . . and Postpositivism: Circulation and the Global History of Science." *Isis* 104, no. 2 (June 2013): 337–347.

Ramey, Rob Roy, II. "Evolutionary Genetics and Systematics of North American Mountain Sheep: Implications for Conservation." PhD diss., Cornell University, 1993.

Rees, Amanda. "Reflections on the Field: Primatology, Popular Science and the Politics of Personhood." *Social Studies of Science* 37, no. 6 (December 2007): 881–907.

Reiger, John F. *American Sportsmen and the Origins of Conservation*, 3rd ed. Corvallis: Oregon State University Press, 2000.

Regan, Tom. *The Case for Animal Rights*. Berkeley: University of California Press, 1983.

Rich, Jeremy. *Missing Links: The African and American Worlds of R.L. Garner, Primate Collector*. Athens: University of Georgia Press, 2012.

Richards, John F. *The World Hunt: An Environmental History of the Commodification of Animals*. Berkeley: University of California Press, 2014.

Richards, Robert J. *Darwin and the Emergence of Evolutionary Theories of Mind and Behavior*. Chicago: University of Chicago Press, 1987.

Ringmar, Erik. "Audience for a Giraffe: European Expansionism and the Quest for the Exotic." *Journal of World History* 17, no. 4 (December 2006): 375–397.

Ritvo, Harriet. *The Animal Estate: The English and Other Creatures in the Victorian Age*. Cambridge, MA: Harvard University Press, 1987.

———. *The Platypus and the Mermaid, and Other Figments of the Classifying Imagination*. Cambridge, MA: Harvard University Press, 1997.

Robbins, Louise E. *Elephant Slaves and Pampered Parrots: Exotic Animals in Eighteenth-Century Paris: Animals, History, Culture*. Baltimore, MD: Johns Hopkins University Press, 2002.

Robichaud, Andrew. *Animal City: The Domestication of America*. Cambridge, MA: Harvard University Press, 2019.

Robin, Libby, and Will Steffen. "History for the Anthropocene." *History Compass* 5, no. 5 (July 2007): 1694–1719.

Rodgers, Daniel T. *Atlantic Crossings: Social Politics in a Progressive Age*. Cambridge, MA: Harvard University Press, 1998.

Rominger, Eric M., Heather A. Whitlaw, Darrel L. Weybright, William C. Dunn, and Warren B. Ballard. "The Influence of Mountain Lion Predation on Bighorn Sheep Translocations." *Journal of Wildlife Management* 68, no. 4 (October 2004): 993–999.

Roscher, Mieke. "Animals as Signifiers: Re-Reading Michel Foucault's *The Order of Things* as a Genealogical Working Tool for Historical Human-Animal

Studies." In *Beyond the Human-Animal Divide: Creaturely Lives in Literature and Culture*, ed. Dominik Ohrem and Roman Bartosch, 189–214. New York: Palgrave Macmillan, 2017.

Rose, Deborah Bird. *Wild Dog Dreaming: Love and Extinction*. Charlottesville: University of Virginia Press, 2012.

Rose, Deborah Bird, Thom van Dooren, and Matthew Chrulew, eds. *Extinction Studies: Stories of Time, Death, and Generations*. New York: Columbia University Press, 2017.

Rosenthal, Mark, Carol Tauber, and Edward Uhlir. *The Ark in the Park: The Story of the Lincoln Park Zoo*. Springfield: University of Illinois Press, 2003.

Ross, Stephan R., and Katie L. Gillespie. "Influences on Visitor Behavior at a Modern Immersive Zoo Exhibit." *Zoo Biology* 28, no. 5 (September 2009): 462–472.

Rothfels, Nigel. "The Antelope Collectors." In *Zoo Studies: A New Humanities*, ed. Tracy McDonald and Daniel Vandersommers, 45–64. Montreal: McGill-Queens University Press, 2019.

———. "Catching Animals." In *Animals in Human Histories: The Mirror of Nature and Culture*, ed. Mary J. Henninger-Voss, 182–228. Rochester, NY: University of Rochester Press, 2002.

———. *Elephant Trails: A History of Animals and Cultures*. Baltimore, MD: Johns Hopkins University Press, 2021.

———. "Prison, Spectacle, Refuge: Modern Zoos Are Proud of Their Contribution to Animal Conservation but Will Always Be Haunted by Their Histories." *Aeon* (November 2019). https://aeon.co/essays/have-modern-zoos-really-put-their-history-behind-them.

———. *Savages and Beasts: The Birth of the Modern Zoo*. Baltimore, MD: Johns Hopkins University Press, 2002.

Rothman, Sheila M. *Living in the Shadow of Death: Tuberculosis and the Social Experience of Illness in American History*. New York: Basic Books, 1994.

Russell, Edmund. *Evolutionary History: Uniting History and Biology to Understand Life on Earth*. Cambridge: Cambridge University Press, 2011.

Sachs, Aaron. *The Humboldt Current: Nineteenth-Century Exploration and the Roots of American Environmentalism*. New York: Viking, 2006.

Saha, Jonathan. "Colonizing Elephants: Animal Agency, Undead Capital and Imperial Science in British Burma." *BJHS Themes* 2 (April 2017): 169–189.

Sausman, Karen A. "Survival of Captive-Born *Ovis canadensis* in North American Zoos." *Zoo Biology* 3, no. 2 (1984): 111–121.

Sax, Boria. *The Mythical Zoo: An Encyclopedia of Animals in World Myth, Legend, and Literature*. Santa Barbara: ABC-CLIO, 2001.

Schickore, Jutta. *About Method: Experimenters, Snake Venom, and the History of Writing Scientifically*. Chicago: University of Chicago Press, 2017.

Schlich, Thomas, Eric Mykhalovskiy, and Melanie Rock. "Animals in Surgery—Surgery in Animals: Nature and Culture in Animal-Human Relationship and

Modern Surgery." *History and Philosophy of the Life Sciences* 31, no. 3/4 (2009): 321–354.

Schultz, P. Wesley, and Florian G. Kaiser. "Promoting Pro-Environmental Behavior." In *The Oxford Handbook of Environmental and Conservation Psychology*, ed. Susan Clayton, 556–580. New York: Oxford University Press, 2012.

Schuyler, David. *The New Urban Landscape: The Redefinition of City Form in Nineteenth-Century America*. Baltimore, MD: Johns Hopkins University Press, 1986.

Secord, Anne. "Science in the Pub: Artisan Botanists in Early Nineteenth-Century Lancashire." *History of Science* 32 (September 1994): 269–315.

Secord, James. "Knowledge in Transit." *Isis* 95, 4 (December 2004): 654–672.

———. *Victorian Sensation: The Extraordinary Publication, Reception, and Secret Authorship of Vestiges of the Natural History of Creation*. Chicago: University of Chicago Press, 2000.

Senior, Matthew. "Psychotic Humans, Psychotic Animals: The Zoo and the Mental Hospital, 1656–1664." In *Zoo Studies: A New Humanities*, ed. Tracy McDonald and Daniel Vandersommers, 19–44 (Montreal: McGill-Queens University Press, 2019).

———. "The Souls of Men and Beasts, 1630–1764." In *A Cultural History of Animals in the Age of Enlightenment*, ed. Matthew Senior, 23–45. Oxford: Berg Publishers, 2007.

Sepkoski, David. *Catastrophic Thinking: Extinction and the Value of Diversity from Darwin to the Anthropocene*. Chicago: University of Chicago Press, 2020.

Seshadri, Kalpana Rahita. *HumAnimal: Race, Law, Language*. Minneapolis: University of Minnesota Press, 2012.

Shaviro, Steven. *The Universe of Things: On Speculative Realism*. Minneapolis: University of Minnesota Press, 2014.

Shorter, J. R., and O. Rueppell. "A Review on Self-destructive Defense Behaviors in Social Insects." *Insectes Sociaux* 59, no. 1 (December 2011): 1–10.

Singer, Peter. *Animal Liberation: A New Ethics for Our Treatment of Animals*. New York: HarperCollins, 1975.

Sivasundaram, Sujit. "Imperial Transgressions: The Animal and Human in the Idea of Race." *Comparative Studies of South Asia, Africa and the Middle East* 35, no. 1 (May 2015): 156–172.

———. "Trading Knowledge: The East India Company's Elephants in India and Britain." *Historical Journal* 48, no. 1 (March 2005): 27–63.

Sloterdijk, Peter. *Foams*, trans. Wieland Hoban. Cambridge, MA: MIT Press, 2016.

Smith, Diane. *Yellowstone and the Smithsonian: Centers of Wildlife Conservation*. Lawrence: University Press of Kansas, 2017.

Smith, Doug. "Should Humans Rule the World? Using Wolves to Connect with Nature." TED Talk at Big Sky, MO, January 26, 2019. https://www.youtube.com/watch?v=kCu34aAyBFc.

Smith, Kimberly K. *The Conservation Constitution: The Conservation Movement and Constitutional Change, 1870–1930*. Lawrence: University Press of Kansas, 2019.

Snyder, Noel F. R., Scott R. Derrickson, Steven R. Beissinger, James W. Wiley, Thomas B. Smith, William D. Toone, and Brian Miller. "Limitations of Captive Breeding in Endangered Species Recovery." *Conservation Biology* 10, no. 2 (April 1996): 338–348.

Sorenson, John. *Ape*. London: Reaktion, 2009.

Stallwood, Kim. "Topsy: The Elephant We Must Never Forget." In *Animal Biography: Re-framing Animal Lives*, ed. André Krebber and Mieke Roscher, 227–242. Cham, Switzerland: Palgrave Macmillan, 2018.

Steinmo, Sven. "Historical Institutionalism." In *Approaches and Methodologies in the Social Sciences: A Pluralist Perspective*, ed. Donatella Della Porta and Michael Keating, 118–138. Cambridge: Cambridge University Press, 2008.

Stempniewicz, Lech. "Polar Bears Observed Climbing Steep Slopes to Graze on Scurvy Grass in Svalbard." *Polar Research* 36, no. 1 (June 2017).

Stengers, Isabelle, and Vinciane Despret. *Women Who Make a Fuss: The Unfaithful Daughters of Virginia Woolf*, trans. April Knutson. Minneapolis: University of Minnesota Press, 2014.

Stern, Alexandra Minna. *Eugenic Nation: Faults and Frontiers of Better Breeding in Modern America*. Oakland: University of California Press, 2015.

Stiegler, Bernard. *Technics and Time, 1: The Fault of Epimetheus*. Trans. Richard Beardsworth and George Collins. Stanford, CA: Stanford University Press, 1998.

Suzuki, Yuka. *The Nature of Whiteness: Race, Animals, and Nation in Zimbabwe*. Seattle: University of Washington Press, 2017.

Swanson, Drew. "Mountain Meeting Ground: History at an Intersection of Species." In *The Historical Animal*, ed. Susan Nance, 240–258. Syracuse, NY: Syracuse University Press, 2015.

Swart, Sandra. "African Studies." In *Handbook of Historical Animal Studies*, ed. Mieke Roscher, André Krebber, and Brett Mizelle, 85–99. Berlin: De Gruyter, 2021.

Sweet, Timothy. *Extinction and the Human: Four American Encounters*. Philadelphia: University of Pennsylvania Press, 2021.

Szczygielska, Marianna. "Elephant Empire: Zoos and Colonial Encounters in Eastern Europe." *Cultural Studies* 34, no. 5 (September 2020): 789–810.

———. "Pandas and the Reproduction of Race and Heterosexuality in the Zoo." In *Zoo Studies: A New Humanities*, ed. Tracy McDonald and Daniel Vandersommers, 211–236. Montreal: McGill-Queens University Press, 2019.

Tamm, Marek, and Zoltán Boldizsár Simon. "More-Than-Human History: Philosophy of History at the Time of the Anthropocene." In *Philosophy of History Twenty-First-Century Perspectives*, ed. Jouni-Matti Kuukkanen, 198–215. London: Bloomsbury, 2021.

Taylor, Dorceta E. *The Rise of the American Conservation Movement: Power, Privilege, and Environmental Protection*. Durham, NC: Duke University Press, 2016.

Thompsell, Angela. *Hunting Africa: British Sport, African Knowledge and the Nature of Empire*. New York: Palgrave Macmillan, 2015.

Thorsen, Liv Emma. "Animal Matter in Museums: Exemplifying Materiality." In *The Routledge Companion to Animal-Human History*, ed. Hilda Kean and Philip Howell, 171–193. London: Routledge, 2019.

Tober, James A. *Who Owns the Wildlife?: The Political Economy of Conservation in Nineteenth-Century America*. Westport, CT: Greenwood Press, 1981.

Tobin, James. *To Conquer the Air: The Wright Brothers and the Great Race for Flight*. New York: Free Press, 2004.

Tom, Erica. "Humanizing Animals: Talking about Second Chances, Horses, and Prisoners." In *Racial Ecologies*, ed. Leilani Nishime and Kim D. Hester Williams, 123–138. Seattle: University of Washington Press, 2018.

Topham, Jon. "Rethinking the History of Science Popularization/Popular Science." In *Popularizing Science and Technology in the European Periphery, 1800–2000*, ed. Faidra Papanelopoulou, Augustí Nieto-Galan, and Enrique Perdiguero, 1–20. Burlington, VT: Ashgate, 2009.

Tortorici, Zeb. "Animal Archive Stories: Species Anxieties in the Mexican National Archive." In *The Historical Animal*, ed. Susan Nance, 75–98. Syracuse: Syracuse University Press, 2015.

Traisnel, Antoine. *Capture: American Pursuits and the Making of a New Animal Condition*. Minneapolis: University of Minnesota Press, 2020.

Tsing, Anna. *The Mushroom at the End of the World*. Princeton, NJ: Princeton University Press, 2015.

Tsing, Anna, Heather Swanson, Elaine Gan, and Nils Bubandt, eds. *Arts of Living on a Damaged Planet*. Minneapolis: University of Minnesota Press, 2017.

Tsing, Anna L., Jennifer Deger, Alder Keleman Saxena, and Feifei Zhou. "What Is the Anthropocene?" In *Feral Atlas: The More-Than-Human Anthropocene*, ed. Tsing, Deger, Saxena, Zhou. Published online. Stanford, CA: Stanford University Press, 2021. https://feralatlas.supdigital.org/?cd=true&bdtext=feral-atlas-and-the-more-than-human-anthropocene.

Turner, James. *Reckoning with the Beast: Animals, Pain, and Humanity in the Victorian Mind*. Baltimore, MD: Johns Hopkins University Press, 2000.

Tweed, Katherine. "Rock-Climbing Bears." *Science World* 67, no. 3 (October 2010).

Tyler, Tom. *CIFERAE: A Bestiary in Five Fingers*. Minneapolis: University of Minnesota Press, 2012.

Uddin, Lisa. *Zoo Renewal: White Flight and the Animal Ghetto*. Minneapolis: University of Minnesota Press, 2015.

Uekoetter, Frank. *The Green and the Brown: A History of Conservation in Nazi Germany*. Cambridge: Cambridge University Press, 2006.

Ugarković, Damir, Nera Fabijanić, Kristijan Tomljanović, Davor Krmpotić, and Nikolina Kelava Ugarković. "Microhabitat Characteristics of Brown Bear Den Areas." *Baltic Forestry* 26, no. 2 (2020): 176–184.

Vaeth, J. Gordon. *Langley: Man of Science and Flight*. New York: Ronald Press Company, 1966.

van Dooren, Thom. *Flight Ways: Life and Loss on the Edge of Extinction*. New York: Columbia University Press, 2014.

———. *The Wake of Crows: Living and Dying in Shared Worlds*. New York: Columbia University Press, 2019.

Van Horn, Gavin. *The Way of Coyote: Shared Journeys in Urban Wilds*. Chicago: University of Chicago Press, 2018.

Vandersommers, Daniel. "Animal Activism and the Zoo-Networked Nation." *Humanimalia* 6, no. 2 (March 2015): 111–165.

———. "Failed Zoo Experiments: Primatology, Aeronautics, and the Animality of 'Modern' Science, 1891–1903." In *Zoo Studies: A New Humanities*, ed. Tracy McDonald and Daniel Vandersommers, 65–92. Montreal: McGill-Queens University Press, 2019.

———. "Narrating Animal History from the Crags: A Turn-of-the-Century Tale about Mountain Sheep, Resistance, and a Nation." *Journal of American Studies* 51, no. 3 (August 2017): 751–777.

———. "The Sectionalism of the National Zoo: Animals, Language, Politics, and Laughter." In *Animal Histories of the Civil War Era*, ed. Earl Hess, 228–239. Baton Rouge: Louisiana State University Press, 2022.

Varga, Donna. "Babes in the Woods: Wilderness Aesthetics in Children's Stories and Toys, 1830–1915." *Society and Animals* 17 (January 2009): 187–205.

———. "Winnie: Troubling the Idealization of the Bear as Childhood Innocent." In *Positioning Pooh: Edward Bear after 100 Years*, ed. J. Harrison. Jackson: University Press of Mississippi, 2021.

Veltre, Thomas. "Menageries, Metaphors and Meanings." In *New Worlds, New Animals: From Menagerie to Zoological Park in the Nineteenth Century*, eds. R. J. Hoage and William A. Deiss, 19–29. Baltimore, MD: Johns Hopkins University Press, 1996.

Vetter, Jeremy. *Field Life: Science in the American West during the Railroad Era*. Pittsburgh: University of Pittsburgh Press, 2016.

Von Drehle, David. *Triangle: The Fire That Changed America*. New York: Grove Press, 2003.

Waddington, Keir. *The Bovine Scourge: Meat, Tuberculosis and Public Health, 1850–1914*. Woodbridge, UK: Boydell, 2006.

———. *An Introduction to the Social History of Medicine*. London: Palgrave Macmillan, 2011.

———. "Mad and Coughing Cows: Bovine Tuberculosis, BSE and Health in Twentieth Century Britain." In *Meat, Medicine and Human Health in the Twentieth Century*, ed. David Cantor, Christian Bonah, and Matthias Dörries, 159–177. London: Routledge, 2015.

Walker, Brett. *The Lost Wolves of Japan*. Seattle: University of Washington Press, 2005.

Wang, Jessica. *Mad Dogs and Other New Yorkers: Rabies, Medicine, and Society in an American Metropolis, 1840–1920*. Baltimore, MD: Johns Hopkins University Press, 2019.

Warren, Louis S. *The Hunter's Game: Poachers and Conservationists in Twentieth-Century America*. New Haven, CT: Yale University Press, 1999.

Wasik, Bill, and Monica Murphy. *Rabid: A Cultural History of the World's Most Diabolical Virus*. New York: Viking, 2012.

Webster, John. "Sentient Animals Do Not Just Live in the Present." *Animal Sentience* 25, no. 10 (2019).

Wehausen, John D., and Rob Roy Ramey II. "Cranial Morphometric and Evolutionary Relationships in the Northern Range of *Ovis Canadensis*." *Journal of Mammalogy* 81, no. 1 (February 2000): 145–161.

Wehausen, John D., Vernon C. Bleich, and Rob R. Ramey II. "Correct Nomenclature for Sierra Nevada Bighorn Sheep." *California Fish and Game* 91, no. 3 (June 2005): 216–218.

Weir, Kathryn, and Ted Gott. *Gorilla*. London: Reaktion Books, 2012.

Wemmer, Christen, and Catherine Christen, eds. *Elephants and Ethics: Toward a Morality of Coexistence*. Baltimore, MD: Johns Hopkins University Press, 2008.

Whitley, David S. *The Art of the Shaman: Rock Art of California*. Salt Lake City: University of Utah Press, 2000.

Wilkinson, Lise. *Animals and Disease: An Introduction to the History of Comparative Medicine*. New York: Cambridge University Press, 1992.

Williams, Caroline Randall. "1619 and Beyond: Explorations in Atlantic Slavery and Its American Legacy." Lecture at Center for Historical Research, The Ohio State University, November 5, 2021.

Windsor, Mary P. *Reading the Shape of Nature: Comparative Zoology at the Agassiz Museum*. Chicago: University of Chicago Press, 1990.

Winter, Alison. *Mesmerized: Powers of Mind in Victorian Britain*. Chicago: University of Chicago Press, 1998.

Wischermann, Clemens, Aline Steinbrecher, and Philip Howell, eds. *Animal History in the Modern City: Exploring Liminality*. London: Bloomsbury, 2018.

Wolfe, Cary. *Zoontologies: The Question of the Animal*. Minneapolis: University of Minnesota Press, 2003.

Woods, Abigail. "Doctors in the Zoo: Connecting Human and Animal Health in British Zoological Gardens, 1828–1890." In *Animals and the Shaping of Modern Medicine: One Health and Its Histories*, ed. Abigail Woods, Michael Bresalier, Angela Cassidy, and Rachel Mason Dentinger, 27–69. London: Palgrave Macmillan, 2018.

Woods, Abigail, Michael Bresalier, Angela Cassidy, and Rachel Mason Dentinger, eds. *Animals and the Shaping of Modern Medicine: One Health and Its Histories*. London: Palgrave Macmillan, 2018.

Worboys, Michael. “Joseph Lister and the Performance of Antiseptic Surgery.” *Notes and Records of the Royal Society of London* 67, no. 3 (September 2013): 199–209.

———. *Spreading Germs: Disease Theories and Medical Practice in Britain, 1865–1900*. New York: Cambridge University Press, 2000.

Worster, Donald. *A River Running West: The Life of John Wesley Powell*. New York: Oxford University Press, 2002.

You, Chengcheng. “The Necessity of an Anthropomorphic Approach to Children’s Literature.” *Children’s Literature in Education* 52, no. 2 (June 2021): 183–199.

———. “Representing Zoo Animals: The Other-than-Anthropocentric in Anthony Brown’s Picture Books.” *Lion and the Unicorn* 43, no. 1 (January 2019): 22–41.

Žižek, Slavoj. *The Parallax View*. Cambridge, MA: MIT Press, 2006.

———. *Sex and the Failed Absolute*. London: Bloomsbury, 2019.

———. *The Sublime Object of Ideology*. London: Verso, 1989.

Zupančič, Alenka. *What Is Sex?* Cambridge, MA: MIT Press, 2017.

Index